MATHEMATICAL MODELLING IN GEOGRAPHICAL INFORMATION SYSTEM, GLOBAL POSITIONING SYSTEM AND DIGITAL CARTOGRAPHY

LIST OF THE CONTRIBUTORS

Prof. H.S. Sharma, Professor of Geography (Retd.), University of Rajasthan, Jaipur.

Dr. S.K. Pathan, Head, Land Use Planning & Photogrammetry Division, SAC, Ahemdabad.

Dr. J.R. Sharma, Director, RRSSC/ISRO, Jodhpur.

Smt. Rajshree V. Bothale, Scientist, RRSSC/ISRO, Jodhpur

Dr. A.K. Bera, Scientist, RRSSC/ISRO, Jodhpur.

Dr. M.P. Punia, Head, Remote Sensing Division, Birla Institute of Science and Technology, Jaipur.

Dr. Ami Rawal, Department of Geography, M.D.S. University, Baroda.

Sh. S. Pathak, Scientist, RRSSC/ISRO, Jodhpur.

Dr. D. Dutta, Scientist, RRSSC/ISRO, Jodhpur.

Sh. Vinod M. Bothale, Scientist, RRSSC/ISRO, Jodhpur.

Dr. B.K. Bhadra, Scientist, RRSSC/ISRO, Jodhpur.

Dr. [illegible]kesh Paliwal, Scientist, RRSSC/ISRO, Jodhpur.

Dr. Milap Punia, Scientist, IIRS, Dehradun.

Dr. Rajeev Jain, Scientist, State Remote Sensing Application Centre, Jodhpur.

Dr. R.K. Gupta, Director, Rajasthan Cancer Foundation, Jaipur.

Sh. S.S. Solanki, Jaipur Institute of Information & Technology, Jaipur.

Sh. V.S.P. Sinha, Scientist, Birla Institute of Science and Technology, Jaipur.

Dr. Sandeep Goyal, Scientist, State Remote Sensing Application Centre, Bhopal.

Mathematical Modelling in Geographical Information System, Global Positioning System and Digital Cartography

Edited By
Prof. H.S. Sharma
Dr. D.R. Ram
Dr. Rama Prasad
Dr. P.R. Binda

CONCEPT PUBLISHING COMPANY, NEW DELHI-110 059

Cataloging in Publication Data—DK
Courtesy: D.K. Agencies (P) Ltd. <docinfo@dkagencies.com>

Training Programme on Mathematical Modelling in GIS/ GPS and Digital Cartography (2005 : Jaipur, India)
Mathematical modelling in geographical information system, global positioning system and digital cartography : Training Programme on Mathematical Modelling in GID/ GPS and Digital Cartography, 1st Feb. to 2nd March, 2005, Department of Geography, University of Rajasthan, Jaipur / edited by H.S. Sharma ... [et al.] ; sponsored by Department of Science and Technology, Government of India.
p. cm.
Includes bibliographical references.
Includes index.
ISBN 8180693422

1. Geographic information systems—Mathematical models—Congresses. 2. Global Positioning System—Mathematical models—Congresses. 3. Digital mapping—Mathematical models—Congresses. I. Sharma, H. S. (Hari Shanker), 1945- II. India. Dept. of Science and Technology. III. Title.

DDC 910.285 22

ISBN 81-8069-342-2

First Published 2006

Published and Printed by
Ashok Kumar Mittal
Concept Publishing Company
A/15-16, Commercial Block, Mohan Garden
New Delhi-110059 (INDIA)

Phones : 25351460, 25351794
Fax : 091-11-25357103
Email : publishing@conceptpub.com

PREFACE

Geographical Information System, GPS and Digital Cartography have become a slogan in the 21st Scientific literature. These have attained a status of multi-discipline. The thrust area basically meant for geographers and geo-scientists has been taken over by scientists of cognate disciplines. As a matter of fact, the thrust area has developed very fast. The computer multinationals, electronic industries and programmers have accepted this idea as a challenge.

With the possibility of using vector and raster information for GIS, there has been an attempt to use the Remote Sensing data at least one of the inputs. GIS does provide a right opportunity for much desired integration of Remote Sensing data with collateral data (P. Nag and M. Kundrat, 1998). GIS is widely used in many industries, application centres and universities. It has become more integrated and less modular. With the fast changing scenario in private sector, it is becoming possible to bring different applications in one integrated system. The integration of GIS, image processing and GPS data is ready example.

With these above considerations in mind a proposal was submitted to Department of Science and Technology, Government of India, New Delhi for funding a month training programme on Mathametical Modelling in GIS/GPS and Digital Cartography to provide basic training to young students of geography, geology and cognate discipline.

The manual consists of 34 chapters on various aspects of mathematical modelling and related problems on digital cartography, GIS and GPS delivered by resource persons during training programme from 1st February to 2nd March, 2005. In introduction H.S. Sharma defines geographical information system and GPS. S.K. Pathan discusses the need of computer knowledge in geography, use of Remote Sensing in Mathematical Modelling, data preparation for spatial analysis and mathematical modelling, application and automation development to further exploit existing spacialised software from chapter second to fifth. In chapter sixth J.R. Sharma introduces the Indian Space Programme towards National Development. Smt. Rajshree V. Bothale in chapter seventh describes Spatial Modelling in Raster and Vector Data. Cartographic data exchange in different file format has been discussed by A.K. Bera in chapter eighth. M.P. Punia from chapter ninth to eleven has described Digital Cartography, Hardware and Software in GIS and Geometrical Rectification of Digital Cartographic Data. Spatial modelling through GIS has been discussed by Ami Rawal in chapter twelfth. In chapter thirteenth Digital Cartography as the Basic Requirement for Geographers has been given by S. Pathak. D. Dutta discusses Mathematical Modelling in grid analysis in chapter fourteenth. Data Management Model through GIS and Project Organisation have been discussed by Vinod M. Bolhale and B.K. Bhadra in chapter fifteen and sixteen respectively. Model of Data Management in GIS has been discussed by Rakesh Paliwal. From chapter eighteen to twenty Milap Punia discusses Model of three-dimension view in GIS, Mathematical Model in Slope analysis and Generation of Digital Terrain Models. Coordination system transformations of the cartographic data, ARC/INFO—a typical GIS system and attachment of attribute data to existing cartographic information have been discussed by M.P. Punia from chapter from chapter twenty one to twenty three. In chapter twenty four and twenty five Rajeev Jain has discussed GIS as a tool for geographical study and mathematical model from statistical processing of GIS. Visual interpretation of remote sensing data and cartographic visualization have been analysed by Milap Punia in chapter twenty sixth and twenty seven. Use of GIS to understand cancer distribution has been given by R.K. Gupta in chapter twenty eight. Topology building of the Cartographic information and Acquisition and evaluation of map and cartographic data from data sheet have been examined by Milap Punia in chapter twenty nine and thirty. Basic concepts of GPS have been introduced by S.S. Solanki in chapter thirty one. V.S.P. Sinha in chapter thirty two analyses model generation, management and transformation of Raster data for landuse study. In chapter thirty three and thirty four Sandeep Goyal examines the GIS modeling for landuse study and acquisition and evaluation of map and cartographic data from data sheet.

We take this opportunity to express our gratitude to Prof. K.L. Sharma, Vice chancellor, University of Rajasthan, Jaipur for his continual guidance and support through out the training programme. We extend our sincere thanks to all contributors and resource persons who despite of their busy schedule, have extended help as per our lecture programme. We would like to thank Dr. M.K. Khandelwal, Head of the Department of Geography and other colleagues of the department for their support. The financial support by Department of Science and Technology, Government of India, New Delhi, for conduct of a training course that lead to the genesis of this manual is gratefully acknowledged.

Last but not least, we thank all the authors and publishers whose published works have been discussed and used in this manual.

Editors
Prof. H.S. Sharma
Dr. D.R. Ram
Dr. Rama Prasad
Dr. P.R. Binda

CONTENTS

1

MODELLING THROUGH GEOGRAPHICAL INFORMATION SYSTEM AND GLOBAL SYSTEMS

H.S. Sharma

INTRODUCTION

There have been so many attempts to define GIS that it is difficult to select one definitive definition. Maguire (1991) offers a list of 11 different definitions. This variety can be explained, as Pickles (1995) suggests, by the fact that any definition of GIS will depend on who is giving it, and their background and viewpoint. Pickles also considers that definitions of GIS are likely to change quickly as technology and applications develop further. Some of the shorter definitions give an idea of what a GIS is, albeit in a superficial way. For example Rhind (1989: 28) proposes that GIS is 'a computer system that can hold and use data describing places on the Earth's surface'. Fuller definitions give more idea of what GIS can do, as well as what they are. Those provided by Burrough (1986: 6): 'a set of tools for collecting, storing, retrieving at will, transforming, and display spatial data from the real world for a particular set of purpose', and the Department of the Environment (1987: 132): 'a system for capturing, storing, checking, integrating, manipulating, analysing and displaying data which are spatially referenced to the Earth', fall into this category.

In general, the definitions GIS cover three main components. They reveal that GIS is a *computer system*. This implies more than just a series of computer boxes sitting on a desk, but includes hardware (the physical parts of the computer itself and associated peripherals—plotters and printers), software (the computer programs that run on the computer) and appropriate procedures (or techniques and orders for task implementation). They also tell us that GIS uses *spatially referenced* or *management and analysis tasks* on these data, including their input and output. The Department of the Environment (1987) lists the capabilities that a 'well-designed GIS' should be able to provide:

1. Quick and easy access to large volumes of data.
2. The ability to:
   - select detail by area or theme;
   - link or merge one data set with another;

- analyse spatial characteristics of data;
- search for particular characteristics or features in an area;
- update data quickly and cheaply; and
- model data and assess alternatives.

3. Output capabilities (maps, graphs address lists and summary statistics) tailored to meet particular needs.

In short, GIS can be used to add value to spatial data. By allowing data to be organized and viewed efficiently, by integrating them with other data, by analysis and by the creation of new data that can be operated on in turn, GIS creates useful information to help decision making. A GIS can, as was alluded to in the house hunting case study, be described as a form of spatial decision support system.

Some authors consider that there are important elements of a GIS in addition to those common to the definitions above. For example, Burrough (1986) suggests that GIS have three main elements: 'computer hardware, application software modules, and a proper organizational context'. Ohers, such as Maguire (1989), stress that data are the most important part of GIS. In practice, none of the main elements (the computer system, data or processing tools) will function as a GIS in isolation, so all might be considered of equal importance. However, it is perhaps the nature of the data used, and the attention given to the processing and interpretation of these data, that should lie at the centre of any definition of GIS.

GIS draws on concepts and ideas from many different disciplines. The term Geographic Information Science has been adopted to refer to the science behind the systems. Geographic Information Science draws on disciplines as diverse as cartography, cognitive science, computer science engineering, environmental sciences, geodesy, landscape architecture, law, photogrammetry, public policy, remote sensing, statistics and surveying. Geographic Information Science involves the study of the fundamental issues arising from the creation, handling, storage and use of geographic information (Longley *et al.*, 2001), but it also examines the impacts of GIS on individuals and society and the influences of society on GIS (Table) (Goodchild, 1997).

Goodchild (1997) offers a useful summary of key concepts that help with the definition of GIS:

- Geographical information is information about place on the Earth's surface.
- Geographical information technologies include global positioning system (GPS), remote sensing and geographic information system.
- Geographical information systems are both computer systems and software.
- Geographical information Science is the science behind GIS technology.

There is a almost as much debate over the components of a GIS as there is about its definition. At the simplest level, a GIS can be viewed as a software package, the components being the various tools used to enter, manipulate, analyse and output data. At the other extreme, the components of a GIS include: the *computer system* (hardware and operating system), the *software, spatial data, data management and analysis procedures* and the people to operate the GIS. In addition, a GIS cannot operate in isolation from an application area, which has its own tradition of ideas and procedures. It is this more comprehensive perspective that is adopted here (Ian Heywood, Sarah Cornelius and Steve Carves, 2003).

Table 1.1: Application areas for GIS

Activity	*Application*
Socio-economic/government	Health Local government Transport planning Service planning Urban management
Defence agencies	Target site identification Tactical support planning Mobile command modelling Intelligence data integration
Commerce and business	Market share analysis Insurance Fleet management Direct marketing Target marketing Retail site location
Utilities	Network management Service provision Telecommunications Emergency repairs
Environmental management	Landfill site selection and mineral mapping potential Pollution monitoring Natural hazard assessment Resource management Environmental impact assessment

COMPUTER SYSTEMS AND SOFTWARE

GIS run on the whole spectrum of computer systems ranging from portable personal computers (PCs) to multi-user supercomputers, and are programmed in a wide variety of software languages. Systems are available that use dedicated and expensive workstations, with monitors and digitizing tables built in; others will run on bottom-of-the-range PCs. In all cases, there are a number of elements that are essential for effective GIS operation. These include (after Burrough, 1986):

- the presence of a processor with sufficient power to run the software;
- sufficient memory for the storage of large volumes of data;
- a good quality, high-resolution colour graphics screen; and
- data input and output devices (for example, digitizers, scanners, keyboard, printers and plotters).

Likewise, there are a number of essential software elements that must allow the user to input, store, manage, transform, analyse and output data. Although GIS generally fit all these requirements, there on-screen appearance (user interface) may be very different. Some systems require instruction to be typed at a command line, while others have 'point and click' menus operated using a mouse. The type of interface individual users find easier to operate is largely a matter of personal preference and experience.

Unlike the issue of software functionality there is limited discussion of hardware and interface technology in this paper. This is because these technologies to be changing so rapidly that any discussion would soon be out of date. If the reader is interested in the latest technical advances in GIS hardware and interfaces then World Wide Web hardware and interfaces the World Wide Web (WWW) addresses are provided for a number of leading GIS developers and product directories on the website.

SPATIAL DATA

All GIS software has been designed to handle spatial data (also referred to as geographical data). Spatial data are characterized by information about position, connections with other features and details of non-spatial characteristics (Burrough, 1986; Department of the Environment, 1987). For example, spatial data about one of weather stations may include:

- latitude and longitude as a geographical reference. This reference can be used to deduce relationships with nearby features of interest. If the latitude and longitude of a weather station are known, the relative position of other weather stations can be deduced, along with proximity to ski slopes and avalanche areas;
- connection details such as which service roads, lifts and ski trails would allow the meteorologist access to the weather station;
- non-spatial (or attribute) data, for instance details of the amount of snowfall, temperature, wind speed and direction. In a similar way spatial data about a ski piste may include:
- a series of spatial references to describe position;
- details of other runs that cross or join the ski piste;
- attribute data such as the number of skiers using the piste and its standard of difficulty.

The spatial referencing of spatial data is important and should be considered at the outset of any GIS project. If an inappropriate referencing system is used, this may restrict future use of the GIS (Openshaw, 1990).

The traditional method of representing the geographic space occupied by spatial data is as a series of thematic layers. Consider, for example, traditional cartographic maps that may be available for an individual area. There may be a map for geology, one for soils and a topographic map showing cultural and environmental features on the surface. Computer models of space frequently use a similar approach. for example, the house hunting GIS example discussed earlier contained layers of data including insurance, transport, schools and urban-rural land use. This was the first method of modelling space to be developed. This method, known as the layer-based approach, is still used by most GIS.

An alternative method of representing reality in a computer is to consider that space is populated by discrete 'objects' (Goodchild, 1995). Goodchild (1995) presents the example of a utility company which needs to map and manage a vast array of telegraph poles, connection boxes and cables. Each of these may be regarded as a discrete object, and there is empty space between the objects. This method is known as the object-oriented approach.

Spatial data, represented as either layer or objects, must be simplified before they can be stored in the computer. A common way of doing this is to break down all geographic features into three basic entity types (an entity is a component or building block used to help data organization). These are points, lines and areas. Lines can be used to represent features such as roads, rivers and ski lifts. Area features are used to represent geographical ones, which may be observable in the real world or may be artificial constructs (such as administrative areas). Points, lines and areas can all be used to represent surfaces. For example, spot heights or contour lines can be used to create a surface mode of the landscape. In addition, points (representing junctions) and lines (representing roads) can be used to create a network model. These representation of real-world phenomena are normally held in a GIS according to one of two models-raster (sometimes referred to as grid or tesseral) or vector (Dale and McLaughlin, 1988; Peuquet,1990). These two different approaches are compared in Figure 1. The raster model is particularly applicable where remotely sensed images are used (since the data are collected in this format) and is considered the most appropriate choice for modelling continuous geographic phenomena such as snow depth. The vector model is more appropriate for mapping discrete geographic entities such as road and river networks and administrative boundaries.

DATA MANAGEMENT AND ANALYSIS PROCEDURES

The functions that a GIS should be able to perform include data input, storage, management, transformation, analysis and output. *Data Input* is the process of converting data from its existing form to one that can be used by the GIS (Aronoff, 1989). It is the procedure of encoding data into a computer-readable form and writing the data to the GIS database. This process should include verification procedures to check that the data are correct and transformation procedures to allow data from different sources to be used. GIS need to handle two types of data—graphical data and non-spatial attribute data. The graphical data describe the spatial characteristics of the real-world feature being modelled. In some cases, particularly when area and line features are used to model real-world features, the graphical data may include information about the linkages between them. For example, if the boundary of an area feature such as a car park is also a snow fence that prevents skiers from overshooting the nursery slopes this information may be stored with the graphical data. Non-spatial attribute data describe what the features represent (for instance, a set of points may represent hotels). In addition, further non-spatial attribute data may be stored which provide extra information about the hotels (standard, number of rooms and restaurant facilities).

Data input and updating are frequently the most expensive and time-consuming part of any GIS project and their importance and complexity should never be underestimated. Approximately 80 per cent of the duration of many large-scale GIS projects is concerned with data input and management. Aronoff (1989) estimates that the construction of a large database could cost five to ten times more than the GIS software and hardware.

The *data management* functions necessary in any GIS facilitate the storage, organization and retrieval of data using a database management system (DBMS). A DBMS is a set of computer programs for organizing information, at the core of which will be a database. Database applications that have no GIS component include management of payrolls, bibliographies, and airline and travel agency booking systems. In the same way that DBMS organize these different types of data they can be used to handle both the graphical and non-graphical elements of spatial data. An ideal GIS DBMS should provide support for multiple users and multiple databases, allow efficient updating, minimized repeated (or redundant) information and allow data independence, security and integrity (Smith *et al.*, 1987).

It is the ability of GIS to *transform* spatial data, for example from one entity type (points, lines and areas) to another, and to perform *spatial analysis*, that distinguishes GIS from other types of information systems.

Transformation is the process of changing the representation of a single entity, or a whole set of data. In GIS, transformation may involve changing the projection of a map layer or the correction of systematic errors resulting from digitizing. In addition, it may be necessary to convert data held as rasters to vectors or *vice versa.*

Aronoff (1989) classifies GIS analysis procedures into three types:

1. Those used for storage and retrieval. For example, presentation capabilities may allow the display of a solid map of the area of interest.
2. Constrained queries that allow the user to look at patterns in their data. Using queries, only sandy soil could be selected for viewing or further analysis.
3. Modelling procedures, or functions, for the prediction of what data might be at a different time and place. Predictions could be made about which soils would be highly vulnerable to erosion in high winds of during flooding or the type of soil present in an unmapped area.

Transformation and analysis procedures can also be classified based on the amount of data analysed. Data in GIS are normally held in a series of layers. For instance, a 1: 50,000 topographic map might be digitized to create a series of layer—one layer for road data, one for buildings, one for recreational interest (parking, picnic sites and youth hostels) and additional layers for soils and population data. Data layers normally contain data of only one entity type, that is point, or line, or area data. Analysis can be carried out either on one layer at a time, or on two or more layers in combination.

The form of *data output* used will depend on cost constraints, the audience to whom the results are directed and the output facilities available. A local authority may produce simple tables, graphs and maps for the communication of important points to councillors, whilst professional map makers may produce detailed plots for publication. In other cases, data may be output in digital form for transfer to another software package for statistical analysis, desktop publishing or further analysis. However, most GIS output is in the form of maps. These may be displayed on screen for immediate communication to individuals or small groups, photographed, stored digitally or plotted to produce permanent hard copy. Most GIS provide the ability to design screen formats and forms for plotting and these help to ensure that all maps have titles, keys, north arrows and scales, just as with traditional cartographic output.

The facilities available for map design can be very extensive, incorporating a myriad of different colours, symbols and line styles, and it is often possible to design additional symbols for your own use. This can make map design very time-consuming, but effective. The audience for any GIS product is an important consideration when designing output, but generally it is best to keep products clear and simple.

PEOPLE AND GIS

Most definitions of GIS focus on the hardware, software, data and analysis components. However, no GIS exists in isolation from the organizational context, and there must always be people to plan, implement and operate the system as well as make decisions based on the output. GIS projects range from small research applications where one user is responsible for design and implementation and output, to international corporate distributed systems, where teams of staff interact with the GIS in many different ways. In most organizations the introduction of GIS is an important event, a major change bringing with it the need for internal restructuring, retraining of staff and improved information flows. Research has been undertaken to highlight the factors that promote successful GIS and it has been suggested that in certain business sectors, innovative flexible organizations with adequate resource and straightforward applications are more likely to succeed (Campbell and Masser, 1995). However, not all GIS are successful. There is evidence that many systems fail, and more are under-used (Cornelius and Medyckyj Scott, 1991). So the issues surrounding how to choose a system and how to implement it successfully require examination.

GIS technology is now well-established and, has been in use since the 1960s. The growth in application areas and products through the later years of the twentieth century has helped GIS to become an accepted tool for the management and analysis of spatial data. This trend is set to continue as computer technology continues to improve with faster and more powerful machines and as more data become available in digital formats directly compatible with GIS. In addition, the striking advances in related technologies such as surveying and field data collection, visualization and database management technology are likely to influence this growth.

There have been some notable failures in GIS. Sometimes data difficulties or other technical problems have set back system developments and application; however, there are also human and organizational problems at the root of GIS failures. Before we can begin to appreciate these fully, to ensure that our GIS applications are successful, it is important to have a good understanding of what a GIS can do and the data it is working with.

SPATIAL DATA

All geographical Information systems are computer representation of some aspect of the real world. GIS present a simplified view of the world as it would be impossible to represent reality in its entirety in a computer. The simplified view of the real world adopted by GIS is often termed a model. A model is 'a synthesis of data' (Haggett and Chorley, 1967) which is used as a 'mean of "getting to grips" with systems whose spatial scale or complexity might otherwise put them beyond our mental grasp' (Hardisty *et al.*, 1993). GIS is used to help build models where it would be impossible to synthesize the data by any other means. Models also contain our ideas about how or why elements of the real world interact in a particular way. Therefore, a GIS populated with data and ideas about how these data interact is a spatial model. Haggett and Chorley (1967) point out that a spatial model places emphasis on reasoning about the real world by means of translation in space. This is exactly the reason why GIS is used to solve geographical problems.

Before looking at how spatial models are constructed using a GIS it is necessary to consider the material. In the context of our Martian and the Car analogy, this is where we explain to the Martian about the fuel a car requires to make it run. First, however, it is necessary to review our understanding of the term 'data' and take a closer look at the distinction between data and information.

Data are observations we make from monitoring the real world. Data are collected as facts evidence that may be processed to give the meaning and turn them into information. There a clear distinction between data and information although the two terms are often used interchangeable. To help appreciate the distinction it is perhaps easiest to think of data as raw numbers such as those you might see listed in a table. All you see are numbers that have no particular meaning. To make the numbers useful you need to add context. For somebody else to interpret your tables or lists of figures they would need to know to what the data refer and which scale or unit of measurement has been used for recording the data. With these details the data become information. Therefore, information is data with meaning and context added (Hanold, 1972).

There are a wide variety of data sources, though all data fall into one of two categories: primary or secondary. Counts of skiers using a particular ski run are an example of primary data collected through first-hand observation. Secondary data will have been collected by another individual or organization, for example consumer surveys of customers buying ski equipment. Many secondary data sources are published and include maps, population census details and meteorological data (Griffith and Amerhein, 1991).

All primary and secondary data have three modes or dimensions: temporal, thematic and spatial. For all data it should be possible to identify each of these three modes:

- Temporal.
- Thematic.
- Spatial.

The temporal dimension provides a record of when the data were collected and the thematic dimension describes the character of the real-world feature to which the data refer. Additional thematic data for the avalanche accident might relate to the size and type of the avalanche. In GIS the thematic data are often referred to as non-spatial or attribute data.

The spatial dimension of data can be regarded as the values, character strings or symbols that convey to the user information about the location of the feature being observed. In this case, the spatial reference used is a textual description that would only be of use to those who are familiar with the area. However, because GIS have no 'local knowledge' all spatial data used in GIS must be given a mathematical spatial reference. One of the most common is a map co-ordinate. Here, a co-ordinate pair (x, y) is used to locate the position of a feature on a uniform grid placed on a map.

It is common to find the term temporal data used to describe data organized and analysed according to time, thematic data used for data organized and analysed by theme, and spatial data for data organised and analysed by location. However, even though one dimension may be used to organize data, the other dimensions will still be present.

GIS place great emphasis on the use of the spatial dimension for turning data into information, which, in turn, assists our understanding of geographic phenomena. Therefore, we consider next the characteristics of spatial data in detail and examine how the map metaphor has shaped these characteristics. This is followed by a review of the thematic dimension of spatial data and a review of a range of sources of spatial data, including surveys, aerial photographs, satellite images and field data sources.

MAPS AND THEIR INFLUENCE ON THE CHARACTER OF SPATIAL DATA

The traditional method for storing, analysing and presenting spatial data is the map. The map is of fundamental importance in GIS as a source of data, a structure for storing data and a device for analysis and display. Perhaps more importantly, maps have shaped the way most of us think about space in two dimensions. Therefore, understanding maps and how they are produced is an essential starting point for exploring the characteristics of spatial data.

Maps take many different forms and come at a range of different scales. Example range from simple sketch maps, such as those used to show colleagues and friends how to get to a party, to the more complex topographic and thematic maps that can be found in national atlases.

It is common to make a distinction between thematic and topographic maps. Thematic maps show data relating to a particular theme or topic, such as soil, geology, geomorphology, land use, population or transport. Topographic maps contain a diverse set of data on different themes. Thus, land use, relief and cultural features may all appear on the same topographic map. Unwin (1981) argues that the topographic map is simply a 'composite of many different kinds of maps'.

Even though there are many different types of maps the mapping process is of a general nature. During this process the cartographer must (after Robinson *et al.*, 1995):

- establish the *purpose* the map is to serve;
- define the *scale* at which the map is to be produced;
- select the *feature* (spatial entities) from the real world which must be portrayed on the map;
- choose a method for the *representation* of these features (points, line and areas);
- *generalize* these features for representation in two dimensions;
- adopt a *map projection* for placing these features onto a flat piece of paper;
- apply a *spatial referencing system* to locate these features relative to each other; and
- *annotate* the map with keys, legends and text to facilitate use of the map.

When a cartographer produces a map, an underlying geometric structure is created which allows the user to describe the relationships between features. It will be clear, for instance, that islands lie within lakes, that fields are adjacent and that if you follow a particular road you will reach a certain destination. This structure, known as *topology,* is based on the geometric relationships of objects. Following the map-making process outlined above helps the cartographer shape the character of the final map. The key characteristics which result from this process are described in more detail below.

PURPOSE

All maps, and other sources of spatial data, are generated with a purpose in mind. In most cases that purpose is to turn data into information which will be communicated to a third party. Naturally, such a map can have a strong influence over the user.

In some cases maps may have a single purpose. Other maps, such as the topographic maps produced by national mapping agencies, aim to meet the needs of a wide range of users, ranging from utility companies to outdoor enthusiasts. In this case, the maps will be more geographically accurate than the propaganda maps described above, but they will still contain generalized data to enable them to be of wide generic use. These generalizations will limit their use for certain applications. A utility company is unlikely to use national maps, on their own, to plan the detailed installation of a new set of electricity cables, because these maps do not contain details about the location of the existing cable network.

Whilst not strictly a spatial characteristics in its own right, the purpose for which a spatial data set has been created will influence the quality and spatial detail provided by the data set. An appreciation of the purpose behind the prerequisite for judging whether or not the data are appropriate for use in a particular situation. Linked closely to purpose is the idea of scale.

SCALE

Virtually all sources of spatial data, including maps, are smaller than the reality they represent (Monmonier, 1991; Keates, 1982). Scale gives an indication of how much smaller than reality a map is. Scale can be defined as the ratio of a distance of the map to the corresponding distance on the ground (Martin,1996). An alternative definition is offered by Laurini and Thompson (1992) as the order of magnitude or level of generalization at which phenomena exist or are perceived or observed. Scale can be expressed in one of three ways: as a ratio scale, a verbal scale or a graphical scale.

Ratio	1 : 5,000	1:1,000,000
Verbal (nominal)	1 cm represents 50m	1 cm represents 10 km
Graphical	0 100 200 km	0 10 20 30 40 km

Examples of ratio scales are 1:5,000 and 1:5,000,000. At a scale of 1:5,000 a 1 mm line on the map represents a 5,000 mm line on the ground. In the same fashion a line of 1 m on the map represents a line of 5,000 m on the ground; the units do not matter as long as they are the same. A verbal scale would express the scale in words, for example '1 cm represents 50 m'. Finally, a graphic scale (or scale bar) is usually drawn on the map to illiterate the distance represented visually. Graphic scales are frequently used on computer maps. They are useful where changes to the scale are implemented quickly and interactively by the user. In such cases, recalculating scale could be time-consuming, and the ratios produced (which may not be whole numbers) may be difficult to interpret. Redrawing a graphic scale in proportion to the map is relatively straightforward and simple to understand. However, it is often possible in GIS to specify the scale at which you require your maps using a ratio representation.

Standard topographic maps contain examples of verbal, ratio and graphical scales. It should be remembered that small-scale maps (for example, 1: 250,000 or 1:1,000,000) are those which cover large areas. Conversly, large-scale maps (for example, 1:10,000 or 1:25,000) cover small areas and contain large amounts of details. With some data used in GIS, such as aerial photographs or satellites imager, the scale is not immediately obvious and may have to be calculated by the user. Scale is also important when using spatial entities (points, line and areas) to represent generalized two-dimensional version of real-world feature.

SPATIAL ENTITIES

Traditionally, maps have used symbols to represent real-world features. Examination of a map will reveal three basic symbol types: points, lines and areas (Monmonier, 1991). Each is a simple two-dimensional model that can be used to represent a feature in the real-world. These simple models have been developed by cartographers to allow them to portray three-dimensional features in two dimensions on a piece of paper (Laurini and Thompson, 1992; Martin, 1996).

The representation of real-world features using the point, line and area entity types appears relatively straightforward. However, the method chosen to represent a spatial feature will depend on the scale used. Consider the way cities are represented on maps of different scales. On a world map a point would be the most appropriate method of representation, given the number of cities to be included. However, at national and regional scales a point could provide an oversimplified view of the extent of the geographical area covered by a city. A point used here would tell us nothing about the relative size of cities, so it is more likely that the cartographer would choose to represent the cities using areas. At the local scale even the area spatial entity may be considered too simplistic and the cartographer may choose to build up a representation of the city using a mixture of point, line and area entities. Points may be used for the representation of the features such as telephone boxes, areas for residential blocks and parks, and lines for road networks. Choosing the appropriate entity to represent real-world features is often difficult.

GENERALIZATION

All spatial data are a generalization or simplification of real-world features. In some instances generalization is needed because data are required at a particular scale. In other cases generalization is introduced by the limitations of the technical procedures used to produce data. The grain size of photographic film, or the resolution of a remote-sensing device, will determine the level of detail discernible in the resulting air photo or satellite image. Generalization may also be introduced directly by human intervention in order to improve the clarity of an image or to enhance its major theme.

All the data sources used in GIS-aerial photographs, satellite images, census data and particularly map-contain inherent generalization. This simplification of detail is necessary in order to maintain clarity. If a cartographer wishes to depict the course of a river on a map, decisions need to be taken regarding the amount of detail to be included. These decisions are largely governed by the scale of the map—the level of detail shown is proportional to scale. At very small scales (for example, 1:20,000,000), a large river like the Mississippi in the USA may appear as a single blue line. This line will show the approximate course of the river from its source in the northern state of Minnesota to its exit into the Gulf of Mexico. All bends and turns in the river will have been smoothed to create a simple, easy to understand map. At larger scales (for example, 1:5,000,000) it is possible to show something of the meandering nature of this great river and its network of tributaries: the red, Arkansas, Missouri and Ohio. At even larger scales (for example, 1:250,000) it becomes possible to indicate width, river banks, small bends and meander cut-offs. At larger scales still (for example, 1:50,000) it may be possible to indicate depth and positions of sandbanks and shoals that might be important for navigation purposes. As the scale increases, the cartographer has greater scope for including more detail is referred to as *scale-related generalization* and is illustrated in figure.

Decisions regarding what features to include on the final map and which to leave out also need to be made by the cartographer. If the cartographer were to include every single tributary of the Mississippi river network on the 1:20,000,000 scale map, the map would be covered by dense blue linework and impossible to read. For the sake of clarity, the cartographer has to be selective about drawing map features.

Another problem facing the cartographer is how to depict features in proportion to their size on the ground. If a river is drawn as a line 0.5 mm thick on a 1:1,000,000 scale map, this would imply that the river is 500 m wide. In the width of line features drawn on the map were determined rigidly by the map scale, then most features on small-scales map could not be seen with the naked eye. For example, a 50m wide river on the 1:1,000,000 scale map would have to be drawn using a line only 0.05 mm wide. Similarly, if a road running along the bank of the river were to be depicted accurately in this fashion, it would need to be drawn on top of the river on the map. In order to make the road distinguishable from the river, the cartographer has to displace the road to leave a gap between it and the river. To cope with these and other problems relating to the necessary generalization of map features, cartographers have adopted a broad code of practice relating to selection, simplification, displacement and smoothing.

For the GIS analyst to make use of simple spatial entities (points, lines and areas) it is necessary to locate them in two dimensions. The analyst, like the cartographer, must treat the world as a flat surface to achieve this. While this is a gross generalization, for most purpose it works well. Moreover, current technology provides us with no other realistic choice (though future development in three-dimensional modelling and virtual reality may change this). The method by which the 'world is laid flat' is to use a map projection.

Map Projections

Map Projections transfer the spherical Earth on to a two dimensional surface. In doing so they approximate the true shape of the Earth. This process introduces errors into spatial data, the character of which will vary depending on the projection method chosen. Some projection will cause distance between spatial entities to be preserved while direction is distorted. In other cases, shape may be preserved at the expense of accurate area estimates. One way to visualize the problem of representing a spherical world in two dimension is to imagine a plastic beach ball overprinted with a map of the world showing lines of latitude and longitude. The inflated ball is a globe, with the countries in their correct locations and shown as area entities with correct relative shapes and size. Imagine that you have to deflate the ball and lay it flat on a table while still displaying all the countries. The only way to do this is to cut the beach ball into pieces. In doing this you would find that the distance between countries will be altered and their shape distorted. The principle is the same with map projections.

If you imagine the beach ball has a hole at the 'north pole' large enough for a light bulb to be inserted, it is transformed into a light fitting .When the light is switched on, an image of the surface is projected onto the walls of the room. Careful examination of the images on the walls reveals that the centre of the image reflects the globe most accurately. It is on this simple concept that the whole range of map projections is based.

Today's there are a wide range of map projections in use, and there were even more used in the past. Different map projections are used in different parts of the world for mapping different sized areas and for different applications. Think again of the globe as a light fitting. The picture of the Earth from our 'globe light' will vary depending upon the shape of the room in which the light is placed.

In a circular room, assuming our globe is hanging from the North Pole, there will be a continuous picture of the Earth. Countries nearest the equator will appear in their true relative geographical positions. The equator is the line of latitude nearest the wall and so represents the line of true scale, along which distance (and consequently the map scale) are not distorted. However, at the top or bottom of the wall the location of the countries is distorted, with the distance between countries increased. Our view of the poles will be very distorted, or missing altogether. In fact, if the poles are included on our projected map the points representing the north and south poles become so distorted as to be projected as a straight line equal in length to that of the equator.

In a square room with flat walls, only a part of the Earth's surface will be visible on any one wall. This will depend on the position of the light. The view will be of half the globe or less. Distortion will be similar to that in the circular room except that, since the single wall is straight and not curved, the image of the world will be distorted at all four edges of the wall and not just the top and bottom.

If the room is shaped like a tepee (with circular walls tapering towards a point at the apex of the structure) the line of true scale is no longer the equator, as in the circular room, but some line of latitude nearer the north pole that lies at a tangent to the sloping walls of the tepee. The exact line of latitude will depend on the angle of the tepee walls: the steeper the walls, the lower the latitude (nearer the equator); the shallower the walls; the higher the latitude (nearer the pole). Not all projections of this kind have the north pole uppermost; projections can have the south pole uppermost or may have the apex of the tepee centred over a different point altogether.

The circular room is equivalent to the family of cylindrical projections (which include the Mercator projection) where the surface of the Earth is projected onto a cylinder, which encompasses the globe. This projection is very suitable for making maps of an area which have only a small extent in longitude. It has been chosen as the basic projection for use by the Ordnace Survey to map the U.K. The transverse Mercator projection has the advantage of maintaining scale, shape areas, and bearing for small areas. This explains why it has become a popular projection for mapping small areas of the globe. The single wall illustrates the *azimuthal* family of projections and the tepee is equivalent to the conic family.

Many of the map-based spatial data sources used in GIS have a projection associated with them. To undertake meaningful analysis it is necessary to know something about the projections being used. The result of analyses will be affected in different ways by different map projections. If a GIS application requires the accurate calculation of areas, then using a projection which distorts area is obviously not suitable. When using data at large scales (covering small areas) the effects may be slight, but at small scales (covering large areas) the effects can be suitable. Finally, since one of the functions of a GIS is to allow the integration of data from different sources, the ability to alter projections is a fundamental ability of many GIS. There are hundreds of different map projections and some GIS seem to offer the capability to re-project data for most of these. Only the most common projections have been considered above.

SPATIAL REFERENCING

A referencing system is used to locate a feature on the Earth's surface or a two-dimensional representation of this surface such as a map. There are a number of characteristics that a referencing system should have. These include stability, the ability to show points, lines and areas, and the ability to measure length, size (area) and shape (Dale and McLaughlin, 1998). Several methods of spatial referencing exist, all of which can be grouped into three categories:

- Geographic co-ordinate systems;
- Rectangular co-ordinate systems; and
- Mon co-ordinate systems.

The only true *geographic co-ordinates* are latitude and longitude. The locations of a point on the Earth's surface can be defined by a reference using latitude and longitude. Lines of longitude (also known as meridians) start at one pole and radiate outwards until they converge at the opposite pole. Conceptually they can be thought of as semicircles. If you slice a globe along two opposing lines of longitude you will always cut the globe in half. The arbitrary choice for a central line of lingitude is that which runs through the Royal Observatory in Greenwich in England, and is hence known as the Greenwich meridian or the prime meridian. Lines of longitude are widest apart at the eqator and closest together at the poles. The relative distance between lines of longitude where they intersect lines of latitude (or parallels) is always equal. However, the real distance will vary depending on the line of latitude which is intersected. For example, the distance between the lines of longitude intersecting the same parallel will increase towards the equator, with the maximum distance existing at the equator itself.

Lines of longitude lie at right angles to lines of longitude and run parallel to one another. Each line of longitude represents a circle running round the globe. Each circle will have a different circumference and area depending on where it lies relative to the two poles. The circle with the greatest circumference is known as the equator (or central parallel) and lies equidistant from the two poles. At the two poles te lines of latitude are represented by a single point—the pole.

Using lines of longitude and latitude any point on the Earth's surface can be located by a reference given in degrees and minutes. For example the city of Moscow represented as a point can be given a geographical co-ordinate references using latitude and longitude of 55 degree 45 minutes north and 36 degree 0 minutes east (55° 45′N 36°0′E). The first set of number, 55° 45′N, represents latitude. The N informs us that Moscow can be found north of the equator. The second set of numbers, 36° 0′E, tells us that Moscow lies to the east of the prime meridian. Therefore, the N and E together give the quarter of the globe in which Moscow is located. The line of latitude on which Moscow lies is given by the degrees and minutes of this latitude away from the equator. Finally, the line of longitude on which Moscow lies must be identified. The shortest distance between two points on the Earth's surface is known as the great circle distance.

The latitude and longitude referencing system assumes that the earth is a perfect sphere. Unfortunately this is not correct. The earth is actually an oblate spheroid somewhat like an orange with flatter poles and outward bulges in equatorial regions. To complicate matters further the surface of the Earth is far from smooth and regular, as you will appreciate if you have visited any mountainous areas. At small scales these minor blemishes and imperfections in shape can be ignored. However, when dealing with large-scale maps of a small portion of the earth's surface it is essential to make local corrections for these factors.

At present, most of the spatial data available for use in GIS exist in two-dimensional form. In order to make use of these data a referencing system which uses *rectangular co-ordinates* is required. To obtain these a map graticule, or grid, is placed on top of the map. This graticule is obtained by projecting the lines of latitude or longitude from our representation of the world as a globe onto a flat surface using a map projection. The lines of latitude and longitude become the grid lines on a flat map. As already pointed out, the problem is that when you project from a sphere onto a flat surface the image becomes distored. When small areas are being studied there will be only minor distortion in the layout of the grid. However, when large areas of the globe are projected on to a flat surface, the grid will tear and stretch, therefore, all rectangular co-ordinate systems are designed to allow the mapping of specific geographical regions.

Non co-ordinate systems provide spatial references using a descriptive code rather than a co-ordinate. Postal codes, widely used throughout the world, are an example. Some postal codes are fully numeric such as the American ZIP codes, while others are alphanumeric, as in the case of the UK postcode. All have the same basic purpose: to increase the efficiency of mail sorting and delivery rather than to be an effective spatial referencing system for GIS users. There are several advantages to such systems: they are important to the postal services and are, therefore, maintained and updated, and they offer coverage of all area, where people reside and work. Providing that individual codes do not refer to single addresses, they also provide a degree as a referencing system.

In the western United State another non-co-ordinate referencing system is often used. This is known as the Public Land Survey System (PLSS). Here, there has been a recursive subdivision of the land into quarter sections. By knowing which section you are in, you can reference yourself to the earth's surface (DeMers, 1997). Other non-co-ordinate referencing systems in use are based on administrative areas, for example the units used for aggregation and presentation of population census data in different countries. For referencing within smaller areas, unique feature reference may be used, for instance the property reference numbers used by a local authority, or the pipeline references used by a utility company.

All spatial referencing systems have problems associated with them. some are specific to the referencing system, such as the updating problems with postcodes or the difficulties caused by geographical co-ordinates with respect to map projections. However, some of the problems stem from the nature of the spatial entities which require referencing:

- Spatial entities may be mobile. Animals, cars and people move, therefore, any spatial reference they are tagged with only to represent their known location at a particular time.
- Spatial entities may change. Rivers meander, roads can be relocated and policy areas redefined.
- The same object may be referenced in different ways. A house may be represented and referenced as both a point and an area on maps of different scales.

All additional problem for the GIS user is the large number of different spatial referencing systems in use. Choosing an appropriate referencing system can be difficult, and it will frequently be necessary to integrate data collected using different referencing systems. This can be problematic. For some referencing system, however, links have been developed to allow the integration and conversion of data.

TOPOLOGY

In GIS *topology* is the term used to describe the geometric characteristics of objects which do not change under transformations such as stretching or bending and are independent of any co-ordinate system (Bernhardsen, 1992). The topological characteristics of an object are also independent of scale of measurement (Hrisman,1997). Topology, as it relates to spatial data, consists of three elements: adjacency, containment and connectivity (Burrough, 1986).

Adjacency and containment describe the geometric relationships which exist between area features. Area can be described as being 'adjacent' when they share a common boundary. Containment is an extension of the adjacency theme and describes area features which may be wholly contained within another area feature such as an island within a lake. Connectivity is a geometric property used to are usually connected together to form a road network through which traffic can flow.

An understanding of the geometric relationships between spatial entities is important for analysis and integration in GIS. Without knowledge of how entities are geometrically related to each other, it is impossible to answer questions such as 'What is the shortest route from A to B ?' or 'How many ski slopes lie within or are next to zones of high avalanche risk?'

THEMATIC CHARACTERISTICS OF SPATIAL DATA

- Attributes are the non-spatial data associated with point, line and area entities. Attributes are the characteristic of an entity (Laurini and Thompson, 1992).

Each spatial entity may have more than one attribute associated with it. For example, a point representing the hotel may have a number of other attributes: the number of rooms; the standard of accommodation; the name and address of the owner. Attributes give additional information about the character of the entities. They also allow certain GIS operations to be performed where it is the characteristics of the entities which are under scrutiny. Questions such as 'Where are all the hotels with fewer than 20 bedrooms of luxury standard?' figure the analysis of attribute data associated with the point entities used to represent the location of hotels.

On an *interval* scale the difference between numbers is meaningful but the scale dose not have a real origin. Temperatures, in degrees Celsius, are a good example of data which are collected using an interval scale. On a temperature scale it is possible to say that there is a 10-degree difference between a thermometer which records a value of 10 degrees and one which records a value of 20 degrees. Thus, differences can be calculated. However it would be incorrect to say that 20 degrees on the Celsius scale is not a true zero. There is still a temperature when the thermometer reads zero! Negative numbers are also possible on an interval scale.

On a *ratio* scale measurements can have an absolute or real zero, and the difference between the numbers is significant. Snow depth is an example. It is impossible to have a negative value for snow depth. However, something is known about relationships between data, for example a snow pack which is 3 m deep is twice as deep as one which is 1.5 m deep.

One of the problems with the scales of measurement used for the collection of attribute data is that the distribution between the various scales is not always obvious. Many data used in GIS are nominal or ordinal. It is important to take care when using these data in an analytical context. If the scale of measurement is not known, or the GIS user is aware of what scale has been used or what operations can be carried out on data from that scale, the GIS is unlikely to indicate when impossible or meaningless operations have been carried out, on a computer numbers are all the same and will be treated in the same ways. So, ranked scores for city sizes may be added. Two different soil types might will have a numerical code to tag them to the appropriate area in the GIS. If clay soils have the value 2 and sandy soils 3 on a nominal scale, multiplying them together to give soil class 6 would be a meaningless operation. On the other hand, population and area (both on a ratio scale) can be divided to give populations density, or elevation at one point may be subtracted from elevation at another point to give difference in elevation. In the second case the data have been collected using the interval scale.

OTHER SOURCES OF SPATIAL DATA

So far in this chapter we have considered the characteristics of spatial data and their thematic dimension. To do this we have drawn heavily on the map metaphor. However, there area number of other sources of spatial data, including census and survey data, aerial photographs, satellite images and global positioning systems, which have additional special characteristics. These are reviewed below.

Census and Survey Data

Census and survey data are collections of related information. They may be spatial in character if each item in the collection has a spatial reference which allows its location on the surface of the Earth to be identified. Examples are population census, employment data agricultural census data or marketing data.

Population census data normally have some element of spatial referencing. Knowing how many people are in a given country may be useful in its own right, but details of, where the population are, will be of additional interest. The UK census of population is one such survey which provides a 'snapshot' of the people of Great Britain and Northern Ireland on a particular day once every 10 years (Openshaw, 1995). This is described in Box 2.6 together with examples from the USA. Most population censuses use a hierachical series of spatial units to publish data. In the UK the smallest unit is the enumeration district (ED). This is the area over which one census enumerator delivers and collects census forms. Since normally this covers at least 150 households, this relatively small unit will guarantee privacy for individual respondents. Enumeration districts aggregate into wards and wards into local government districts (Martin,1996). In the UK, census data are not usually released in map form, but in tables for the spatial areas you request. However, since a spatial reference is attached linking the data to the areal units of collection, the data are immediately useful for spatial studies. Census data are only one example of the myriad of spatial data which are collected through the use of survey techniques.

Arial Photographs

Arial photography was the first method of remote sensing. It is the capturing of images from a position above the Earth's surface, or without contact with the object of interest (Curran, 1989; Mather, 1991). Unlike a map, which is a model of the Earth's surface and contains only a selection of data, an aerial photograph is a 'snapshot' of the earth at a particular instant in time. As such it contains a mass of data and it is necessary to carry out some form of interpretation to make effective use of the information portrayed. Aerial photographs may be used in GIS as a background for other data, to give those data spatial context and to aid interpretation. Alternatively, the user may abstract information on land use, vegetation type, moisture or heat levels or other aspects of the landscape from the photograph. Aerial photographs are particularly useful for monitoring change, since repeated photographs of the same area are relatively inexpensive, For example, Gun *et al.* (1994) have monitored changes in land use, particularly peat extraction, in Country Fermanagh, Northern Ireland, form a time series of photograph. Interpretation of a sequence of photographs may allow the dating of events such as major floods which cause changes to the landscape.

Curran (1989) identifies six characteristics of aerial photographs which make them of immense value as a data source for GIS:

- Wide availability;
- Low cost (compared with other remotely sensed images);
- Wide area views;
- Time-freezing ability;
- High spectral and spatial resolution; and
- Three-dimensional perspective.

Additionally, aerial photographs can be used to obtain data not available from other secondary sources, such as the location and extent of new housing estates, or the extent of forest fires. One characteristic of aerial photograph that constitutes possible disadvantage is the fact that they do not provide spatially referenced data. Spatial referencing has to be added to features on the image by reference to other sources such as paper maps. Several different types of aerial photographs are available, from simple black and white, which may be used for a wide variety of purpose, to colour and thermal infrared for heat identification.

The angle at which the photograph was taken is important. A photograph is referred to as *vertical* if taken directly below the aeroplane, and *oblique* if taken at an angle. Oblique photographs generally cover larger area and are cheaper than vertical photographs. Vertical photographs are, however, the most widely used for GIS applications.

Before any aerial photograph information can be used in a GIS a number of factors must be considered. The first of these is scale. Scale varies across an aerial photograph, owing to the distance of the camera from the ground. The scale will be constant only at the centre of the image, and the greater the flying height, the greater the scale difference between the centre and edges of the image. Second, factors which may influence interpretation need to be considered. These include time of day and time of year. On photographs taken in winter, long shadows may assist the identification of tall building and trees, but may obscure other features on the image. Conversely, in summer, when trees are in full leaf, features which may be visible from the air in winter will be obscured. The date of the photograph may be important to ensure the data taken from it are contemporaneous with the rest of the data in the GIS.

Aerial photographs represent a versatile, relatively inexpensive and detailed data source for many GIS application. For example, local government bodies may organize aerial overage of their districts to monitor changes in the extent of quarrying of building development. At a larger scale, photographs can be used to provide data on drainage or vegetation conditions within individual fields or parcels that could not be obtained from conventional topographic map (Curran, 1989).

Satellite Image

The satellite images are collected by sensors on board a satellite and then relayed to Earth as a series of electronic signals, which are processed by computer to produce an image. These data can be processed in a variety of ways, each giving a different digital version of the image.

There are large numbers of satellites orbiting the Earth continuously, collecting data and returning them to ground stations all over the world. Some satellites are stationary with respect to the Earth (geostationary), for example Meteosat, which produces images centred over Africa along the Greenwich Meridian (Curran, 1989). Others orbit the Earth to provide full coverage over a period of a few days. Some of the well-known satellites, Landsat and SPOT, for example, operate in this way. Landsat offers repeat coverage of any area on a 16-day cycle (Mather, 1991).

Sensors on board these satellites detect radiation from the Earth for different parts of the electromagnetic spectrum, not only the portions visible to the human eye. The multispectral scanner (MSS) on board Landsat simultaneously detects radiation in four different wavebands:

near infrared, red, green and blue (Curran, 1989). After processing, the images can be used to detect features not readily apparent to the naked eye, such as subtle changes in moisture content across a field, sediment dispersal in a lake or heat escaping from roofs in urban areas.

Scanned images are stored as a collection of pixels, which have a value representing the amount of radiation received by the sensor from that portion of the earth's surface (Burrough, 1986). The size of the pixels gives a measure of the resolution of the image. The smaller the pixels the higher the resolution. The Landsat Thematic Mapper collects data for pixels of size 30 m by 30 m, but this has in the past been restricted to military use. Recent changes in US legislation and the availability of Russian military satellite data have made access to very high-resolution data easier. Resolution is a data and determines its practical value. A landsat Thematic Mapper image with a pixel size of 30 m by 30 m would be unsuitable for identifying individual houses but could be used to establish general patterns of urban and rural land use.

A new form of remotely sensed data is LIDAR (Light detection and Ranging data). LIDAR is a remote sensing system that uses aircraft-mounted lasers to collect topographic data. The lasers are capable of recording elevation measurements with a vertical precision of 15 cm, measurements wit a vertical referenced using high-precision GPS. The technology creates a highly detailed digital elevation model (DEM) that closely marches every undulation in the landscape. Even change in ground surface elevation detail caused by buildings and trees can be detected. This makes LIDER extremely useful for large-scale mapping and engineering applications.

For GIS, remotely sensed data offers many advantages. First, images are always available in digital form, so transfer to a computer is not a problem. However, some processing is usually necessary to ensure integration with other data. Processing may be necessary to reduce data volumes, adjust resolution, change pixel shape of alter the projection of the data (Burrough, 1986). Second, there is the opportunity to process images or use different wavebands for the collection of data to highlight features of particular interest, for example water of vegetation. The repeated coverage of the Earth is a further advantage, allowing the monitoring of change at regular intervals, although for some types of imagery cloud cover while the satellite passes overhead may prevent a useful image being obtained. Finally, the small scale of images provides data useful for regional studies, and applications have included mapping remote areas, geological surveys, land use monitoring and many others (Curran, 1989; Mather, 1991; Maguire, 1989). Trotter (1991) considers the advantages of remotely sensed data for GIS applications in the area of natural resource management to be:

- low cost relative to other data sources;
- currency of images;
- completeness of data; and
- uniform standards across an area of interest.

FIELD DATA SOURCES: SURVEYING AND GPS

There are several methods of collecting raw data in the field for direct input into a GIS. These are most often used when the required data do not exist in an other readily available format such as a map or satellite image. Traditional manual surveying techniques using chain, plane table, levels and theodolites are examples of direct field measurement, but the data collected need to be written down on paper first. Modern digital equivalents of these manual techniques have been adapted so that the data collected are stored in digital format ready for direct input into GIS. Examples include total stations [high-precision theodolites with electronic distance metering (EDM) and a data logger] and hand-held laser range finders.

A relatively new technique of field data collection which has found particular favour with GIS users is the use of satellite navigation systems or GPS (Global Positioning Systems. These are portable backpack or hand-held devices that use signals from GPS satellites to work out the exact location of the user on the Earth's surface in terms of *(x, y, z)* co-ordinates using trigonometry. Position fixes are obtained quickly and accurately literally at the push of a button. The accuracy obtainable from civilian GPS receivers ranges from 100 m to as 0.5 m depending on how they are used, while there are military versions that are accurate to within a few centimetres. Originally designed for real-time navigation purpose, most GPS receivers will store collected co-ordinates and associated attribute information in their internal memory so they can be downloaded directly into a GIS database. The ability to walk or drive around collecting co-ordinate information at sample points in this manner has obvious appeal for those involved in field data collection for GIS projects.

GPS BASICS

GPS is a set of satellites and control systems that allow a specially designed GPS receiver to determine its location anywhere on Earth 24 hours a day (Barnard, 1992). Two main systems exist, the American NAVSTAR system and the Russian GLONASS system. A European system is also planned for the near future. A American system consists of 24 satellites orbiting the Earth in high altitude orbits. These satellites have a 12 hours orbit time, and pass over control stations so that their orbits can be closely monitored and their positions precisely identified. Satellites and ground-based receivers transmit similarly coded radio signals, so that the time delay between transmission and receipt of the signals gives the distance between the satellites and the receiver. If a receiver can pick up signals from three or four satellites, trigonometry is used to calculate the location and height of the receiver.

A GPS will see a position 'fix' displayed on their receiver. Until recently, all readings were affected by selective available (SA), a deliberate error added to the signals by the U military. This has now been switched off and fixes of far greater accuracy can be obtained. One fix, obtained from a single receiver, will have an accuracy of about 25 m for 95 per cent of the time. When SA was in place accuracy was 100 m 95 per cent of the time. More advanced data collection methods, including the averaging of fixes and the use of two receivers in parallel (differential GPS), can be used to obtain even more accurate reading down to the sub-sentimetre level.

Differential GPS technique require two receivers, one fixed at a known location (the base station) and the other at an unknown location (the roving receiver). If both receivers are set up in exactly the same manner and use the same satellites to compute their location, then the positional error recorded will be the same for both receivers. A highly accurate positional fix can, therefore, be obtained for the roving receiver by subtracting the positional error calculated for the base station.

Despite 24-hour global coverage, GPS use can be hampered by certain factors. These include problems where the path between the satellite and the receiver is obstructed by buildings, dense tree cover or steep terrain, and in polar regions where favourable satellite configurations are not always available.

Despite practical difficulties, GPS is finding a wide range of applications, varying from navigation (air, sea and land), to mapping and surveying. Cornelius *et al.* (1994) and carver *et al.* (1995) give some examples of the applications of GPS in geomorphology and fieldwork. Recent developments in portable computing and mobile communication have opened up a whole new area of application for GPS. Knowing exactly where you are in relation to a GIS database on our portable computer, or receiving location information via your WAP phone, can be of great value and is a highly saleable commodity.

GIS DATA STANDARDS

The number of formats available for GIS data is almost as large as the number of GIS packages on the market. This makes the sharing of data difficult and means that data created on one system is not always easily read by another system. This problem has been addressed in the past by including data conversion functions in GIS software. These conversion functions adopt commonly used exchange formats such as DXF and E00. As the range of data sources for GIS has increased, the need for widely applicable data standards to facilitate exchange of data has been recognized. Some of the standards in current use are listed in Table.

Table: Current Geographical Information Standards

BS 7666	Spatial data sets for geographic referencing
CEN TC 287	European norms for geographic information
DIGEST	Digital Geographic Information Exchange Standards
DNF	Digital National Framework
GDF	Geographic Data File
Geo TIFF	Geographic Tagged Image Format File
GML	Geography Markup Language
ISO 6709	Standard representation of latitude, longitude and altitude
ISO 8211	Specification for a data descriptive file for information interchange
ISO 15046	Geographic information
NEN 1878	Netherlands transfer standards for Geographic information
NTF	Neutral Transfer Format
OGIS	Open Geodata Interoperability standard
RINEX	Receiver Independent Exchange Format
SDTS	Spatial data Transfer standard
UGDCS	Utilities Geospatial data Content standard

Source: Adapted from Harding and Wilkinson, 1997.

There is still no universally accepted GIS data standard, although the Open GIS Consortium (OGC), formed in 1994 by a group of leading GIS software and data vendors, is working to deliver spatial interface specifications that are available for global use (OGC, 2001). The OGC has proposed the Geography Markup Language (GML) as a new GIS data standard.

IN THE ABOVE CESSION

We have looked at the distinction between data and information, identified the three main dimensions of data (temporal, thematic and spatial) and looked in detail at how different spatial data sources portray the spatial dimension. The main characteristics of spatial data have been identified and a review of how the traditional map-making process has shaped these characteristics has been presented. In addition we have considered a range of other sources of spatial data. The discussion has shown that any souce of spatial data may be influenced by some, or all, of the following factors:

- The purpose for which they have been collated;
- The scale at which they have been created;
- The resoiution at which they have been captured;
- The projection which has been used to map them;
- The spatial referencing system used as locational framework;
- The nature of the spatial entities used to represent real-world features;
- The generality with which these entities have been modelled; and
- The topological structure used to represent the relationship between entities.

In some data sources, one factor will dominate; in other it will be the interplay of factor which give the data their character. Appreciating the main characteristics of spatial data is important because these characteristics will determine how the data can be used in building a GIS model. For example, data collected at different resolutions should only be integrated and analysed at the resolution of the coarsest data set. In the Zdarske Vrchy case study the 30 m by 30m resolution of the land use map generated from TM satellite data dictated the resolution of the database for analysis.

Therefore, GIS models are only as good a representation of the real world as the spatial data used to construct them. Understanding the main characteristics of spatial data is an important first step in evaluating its usefulness for GIS. The next step is to understand how these data can be stored in a form suitable for use in the computer, as this will also influence the quality of the GIS model.

REFERENCES

Aronoff, S. (1989) *Geographic Information Systems*: a *Management perspective*. WDL Publications, Ottawa.

Bonham-Carter, G.F. (1995) *Geographic Information Systems for Geoscientists:* Modelling with GIS Pergamon Press, New York.

Burrough, P.A., McDonnell, R. (1998) *Principles of Geographical Information Systems*. Oxford University Press, Oxford.

Chrisman, N.R. (1997) *Exploring Geographic Information Systems*. Wiley, New York.

Clayton, K. (1995) The land from space. In: O'Riordan, T. (ed.) *Environmental Science for Environmental Management*. Longman, London, pp. 198-222.

Curran, P. (1989) *Principles of Remote Sensing*. Longman, London..

Dale, P.F., McLaughlin, J.D. (1998) *Land Information Management. An Introduction with Special Reference to Cadastral Problems in Third World Countries*. Clarendon Press, Oxford.

Dangermond, J. (1991) The commercial setting of GIS. In: Maguire D.J., Goodchild M.F., Rhind, D.W. (eds.) *Geographic Information Systems: Principles and Application*. Longman, London, pp. 55-65, Vol. 1.

Davis, D. (1999) *GIS for Everyone*. ESRI Press, Redlands.

DeMers, M.N. (1997) *Fundamentals of Geographic Information Systems*. Wiley, New York.

Department of the Environment, (1987) *Handling Geographic Information*. Report the Committee of Enquiry chaired by Lord Chorley. HMSO, London.

Gatrell, A.C. (1991) Concept of space and geographical data. In: Maguire, D.J., Goodchild, M.F., Rhind, D. (eds) *Geographical Information Systems: Principles and Applications*. Longman, London, Vol. 1, pp. 119-143.

Goodchild, M.F. (1991b) The technological setting of GIS. In: Maguire DJ, Goodchild, M.F., Rhind, D.W. (ens.) *Geographic Information Systems: Principles and Applications*. Longman, London, pp. 45-54, Vol. 1.

Goodchild, M.F. (1997) What is Geographic Information Science? *NCGIA Core Curriculum in GI Science*. http://www..ncgia, ucsb.edu/giscc/units/u002/u002. html, posted October 7, 1997.

Grimshaw, D.J. (1994) *Bringing Geographical Information Systems into Business*. Longman, London.

Hanna, K.C. (1999) *GIS for Landscape Architects*. ESRI Press, Redlands.

Harder, C. (1997) *Arc View GIS means Business: Geographic Information Systems Solution for Business*. ESRI Press, Redlands.

Harder, C. (1998) *Serving Maps on the Internet: Geographic Information and the World Wide Wed*. ESRI Press, Redlands.

Ian Heywood, Sarah Cornelius and Steve Carver (2003) *An Introduction to Geographical Information System*, Pearson Education Ltd., New Delhi.

Keates, J.S. (1982) *Understanding Maps*. Longman, London.

Kennedy, M. (1996) *The Global Positioning System and GIS: An Introduction*. Ann Arbor Press, Ann Arobar, Michigan.ang, L. (1998) *Managing Natural Resources with GIS*. ESIR Press, Redlands.

Korte, G.B. (2000) *The GIS Book*. 5th edn. On Word Press, USA.

Longley, P.A., Goodchild, M.F., Maguire, D.J., Rhind, D.W. (eds.) (1999) *Geographical Information Systems: Principles, Techniques, Management and Applications*. John Wiley, New York.

Longley, P.A., Goodchild, M.F., Maguire, D.J., Rhind, D.W. (2001) *Geographical Information Systems and Science*. Wiley, Chichester.

Maguire, D.J. (1991) an overview and definition of GIS. In: Maguire, D.J., Goodchild, M.F., Rhind, D.W. (eds.) *Geographical Information Systems: Principles and Applications*. Longman, London, pp. 9-20, Vol. 1.

Maguire, D.J., Goodchild, M.F., Rhind, D.W. (eds.) (1991) *Geographical Information Systems: Principles and Applications*. Longman, London.

Martin, D. (1996) *Geographical Information Systems and their Socio-economic Applications*, 2nd edn. Routledge, London.

McDonnell, R., Emp, K. (1998) *International GIS Dictionary*. 2nd edn. Geo Information International, London.

Mitchell, A. (1997) *Zeroing In: Geographic Information systems at Work in the Community*. ESRI Press, Redlands.

Monmonier (1991) *How to Lie with Maps*. University of Chicago Press, Chicago.

Raper, J.F., hind, D.W., Shepherd, J.W. (1992) *Postcodes: the New Geography*. Longman, Harlow, U.K.

Robinson, A.H., Morrison, J.L., Muehrecke, P., Kimerling, A.J., Guptill, S.C. (1995) *Elements of Cartography*, 6th edn. Wiley, New York.

Seegar, H. (1999) Spatial referencing and coordinate systems. In Longley, P.A., Goodchild, M.F., Maguire, D.J., Rhind, D.W. (eds) *Geographical Information Systems*. Wiley, New York, Vol. 1, pp. 427-36.

2

NEED OF COMPUTER KNOWLEDGE IN GEOGRAPHY

S.K. PATHAN

INTRODUCTION

Land is one of the economic entities that cannot be re-produced. The growing demand for economic well being and a better quality of life has put severe stress on the management of land resources. Therefore, organization of information of land is an essential element for managing the natural resources at cadastral level. This in particular calls for the creation of a 'Land Information System (LIS)' as it is more efficient system for land management. Therefore, LIS represents a system for handling spatial data related to multivariate and multilevel data with associate attributes such as physical and environmental characteristics of the terrain, socio-economic and cultural information, revenue collection, land and property ownership, utilities, services and other infrastructure information which will help planners and administrators to carry out various tasks in an efficient manner. Thus LIS visualize integration of various databases from different registers into a new and comprehensive data input that could be analyzed with the aid of computers to draw meaningful conclusions. The scope of Land Information System until now has been confined mainly to statistical/attributes data. Therefore, the scope of LIS has to be enlarged to meet the specific need of spatial planning by including graphic data as one of the important components. This would require advanced modelling capability. The latest developments in computer hardware and software are competent of meeting these needs, i.e. handling of both alpha-numeric and graphic databases. The computer software which meets this need is 'Geographic Information System (GIS)'. It is in this context, GIS plays a major role by providing linkage between the spatial information domain (land parcel data) and non-spatial information (attributes of land parcels). So, GIS is a particular form of Information System that is applied to geographical data. Accordingly, the GIS has the capacity of full range of functions to achieve its purpose related to observations, measurement, description, explanation, forecasting and decision-making. Therefore, the objective of this lecture is to provide you the full range of capabilities of GIS with a specific reference to 'Land Information System'.

GEOGRAPHICAL INFORMATION SYSTEM (GIS)

GIS, basically refers to the science and technology dealing with the character and structure of spatial information, its methods of capture, organisation, classification, qualification, analysis, management, display and dissemination as well as the infrastructure necessary for the optimal use of the information. The integration of spatial information for the natural resources development has been done using manual methods in the past. But with the increase in the volume and dimensionality of data sets, it has become essential to use automated GIS to meet the demands in natural resources development. Hence, GIS is defined as *"an automated tool useful to capture, storage, retrieval and manipulation, display and querying of both spatial and non-spatial data to generate various planning scenarios for decision making"*.

The environment in which a GIS operates is defined by hardware (the machinery including a host computer), a digitizer or scanner for converting the input data, a plotter for presentation of processed outputs and video display unit for commanding the system by a user, the software (programs that tell the computer what to do) and the data. In this context GIS can be seen as a system of hardware, software and procedures designed to support the capture, management, manipulation, analysis, modelling and display of spatially-referenced data for solving complex planning and management problems. Although many other computer programs can use spatial data (e.g. AutoCAD and statistics packages), GIS includes the additional ability to perform spatial operations. Thus, the major components of GIS are (i) the end use or management, (ii) data acquisition, (iii) data input, (iv) data storage and retrieval, (v) analysis and (vi) information presentation.

DATA CAPTURE

The entire geographic variation on the earth can be captured into GIS using points, lines and polygons (areas). Symbols and labels are used to describe these features. Points define discrete location of geographic features which are too small to be depicted as lines or areas, i.e. telephone poles, wells, electric poles etc. Lines represent the shape of geographic objects too narrow to depict as areas, such as roads, canals, rivers, contours etc. which have length but no area. Areas are closed polygons that represent the shape and location of homogeneous, real world features such as administrative boundaries (State, District, Taluka etc.), land use types, soil classes, vegetation categories etc. However, the map scale dictates the type of map feature used to represent a real world geographic feature. For example, a polygon feature on large scale may appear as a point on small scale. Because small scale maps depict large ground areas with low spatial resolution and represent little detail. On the other hand, large scale maps depict small ground areas with high spatial resolution and represent more details.

The entire geographic data is captured in GIS environment either by digitization process or scanning the map. Data can also be captured by converting an ASCII formatted file (GPS co-ordinates, ground survey co-ordinates etc.) and by converting digital data from different sources (Fig. 2.1).

The co-ordinates after digitization will be in default units, i.e. inches. These coordinates have to be projected to real world geographic coordinates using transformation functions available in the computer system. The commonly used map projections are Polyconic and

Universal Transverse Mercator projection. Latitude and longitude terms are commonly used as a geographic reference system.

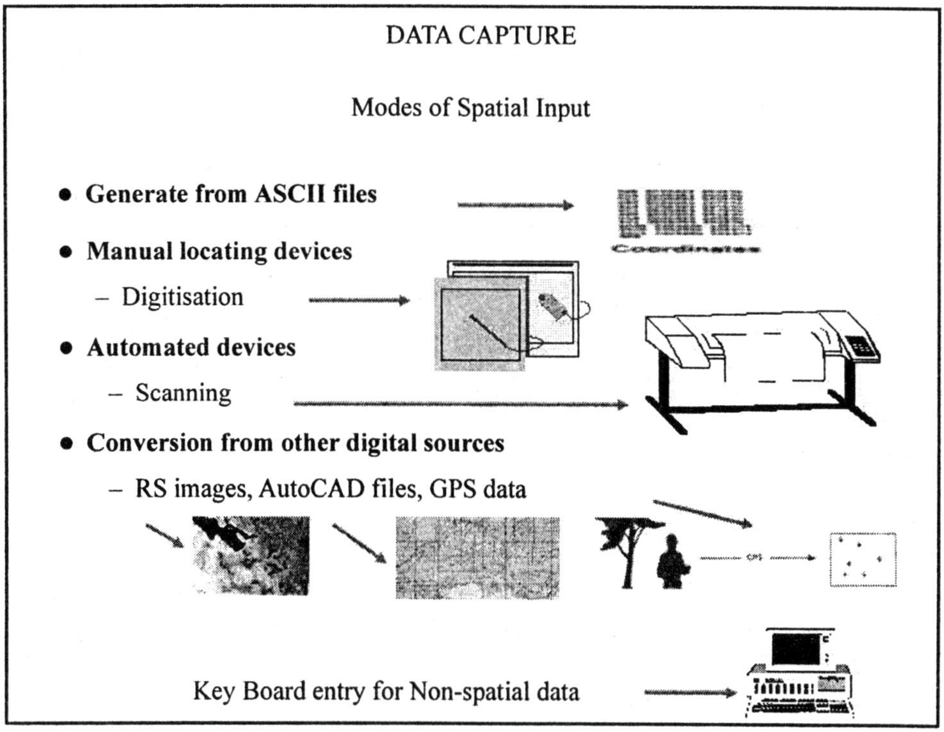

Fig. 2.1: Data Capture in GIS Environment

DATA STORAGE

As it is discussed earlier, geographic variation on the earth's surface is primarily represented on a map as points (wells, telephone poles etc.), lines (roads, canals, drainage etc.) and polygons (areas under Agriculture, Forest etc.). Current GIS differ according to the way in which they organize reality through the data model. Each model tends to fit certain types of data and applications better than others. The data model chosen for a particular project or application is also influenced by the software available, the training of the key individuals, historical precedent.

The Data Model

The procedure used to convert the geographic variation into discrete objects in GIS environment is called a data model. There are two data models available to represent this variation in GIS. They are (i) Raster data model, and (ii) Vector data model.

Raster Data Model

In raster data model, the map is divided into a regular grid of square or rectangular cells which are individually coded. The conventional sequence is row by row from the top left corner, each cell contains a single value. Simple rasters are limited by the area it can represent within the limitations of storage. The fineness of the data is limited by the cell size. The raster GIS is space-filling since every location in the study area corresponds to a cell in the raster. One set of cells and associated values is a layer. There may be many layers in a database, e.g. soil type, elevation, land use, land cover etc. Conceptually, the raster models are the simplest of the available data models. However, they occupy more storage space. This problem is now handled by resorting to coding such as run-length coding, chain coding, block coding etc. A more elegant structure in raster data model is the quad tree structure. In quad tree structure, the area of interest is recursively decomposed into smaller grids and the decomposition continues till each of the smallest grid represent a homogeneous area. Therefore, quad tree describes a class of hierarchical data structure. The resolution of decomposition depends upon the number of times the decomposition process is applied. The storage requirements of a quad tree are lower than that of the simple raster. Hence, these days, much attention is given to quad tree structures. Recent advancements in this area suggests a variety of quad tree data structures. Among them, Region quad tree and Polygon Map (PM) quad tree structures are famous.

The Vector or Object GIS

The vector model uses discrete line segments or points to identify locations. Discrete objects (boundaries, streams, cities) are formed by connecting line segments. Vector objects do not necessarily fill space, not all locations in space need to be referenced in the model. Thus, the Vector data model is based on vectors (as opposed to space-occupancy raster structures). The fundamental primitive is a point and is represented using an x, y (Cartesian) coordinate system. Each point is recorded as a single location. Lines are recorded as a series of x, y coordinates. Areas are defined by sets of lines. Areas are recorded as a series of x, y coordinates defining area that enclose the area. It is important to note that the points, i.e. x, y pairs along the arc are called vertices and the end points of the arc are called nodes in GIS terminology (Fig. 2.2). Arcs join only at nodes. The term polygon is synonymous with area in vector databases because of the use of straight-line connections between points. It is because of this, vector model tends to dominate in transportation, utility, marketing applications. However, raster and vector data models are used in resource management applications.

A coverage is the basic unit of storage in vector GIS. It is a digital version of a single map sheet layer and generally contains one type of map feature such as roads, vegetation types, soil types, land use patterns etc. A coverage contains both the locational data and the descriptive data for features in a given geographic area. A coverage is stored as a directory that contains related files describing the location and attributes of the features.

Raster Data *versus* Vector Data Models

A raster model tells what occurs everywhere, at each place in the area. A vector model tells where everything occurs, gives a location to every object. Vector data is precise and has no approximate errors for measured quantities like area, length, perimeter etc. Raster data suffers to present precise details of measured quantities due to the discretization. Generally, raster data has higher storage requirements. However, overlay and spatial analysis operations are computationally faster than vector data. Raster data is not easily amenable to association of attribute data with spatial features such as points, lines and polygons. This is primarily because of the fact that the basic entity in raster data is the grid cell and the entities such as points, lines or polygons are not recognised as objects in their own merit.

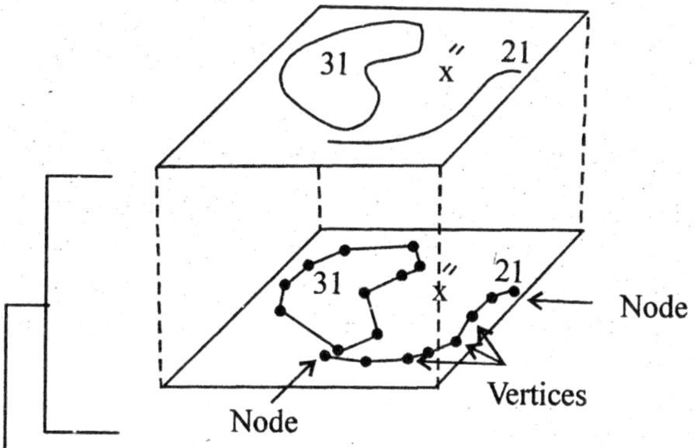

	Feature number	Location
Point		x, y (single Point)
Line	21	$x_1y_1, x_2y_2, \ldots\ldots x_ny_n$ (string)
Polygon	31	$x_1y_1, x_2y_2, \ldots\ldots x_1y_1$ (closed loop)

Fig. 2.2: Structure of Vector Coverage – Polygons, Lines and Points

Most of the GIS's in the market choose one of these two data models as the primary method of representing spatial data and provide conversion utilities from one form to the other.

Defining Spatial Relationships

Spatial relationships in GIS are defined by topology. Topology is a mathematical procedure for explicitly defining properties and spatial relationships of geograhic features which include, connectivity of lines, direction of a line, length of a line, adjacency (contiguity) of areas and definition of areas. There are two data models exist to define these relationships. They are (a) topologic model and (b) Non-topologic model. A topologic data model stores data efficiently and provides the framework for advanced geographic analysis. The model builds areas from the list of individual lines that define area borders. The system stores linear co-ordinates only once because two areas that are adjacent may share the common line between them. In contrast, the non-topologic data model stores each closed area as a single entity. The line shared by adjacent areas must be entered and stored twice, either by double digitizing or copying the line. This duplicate data makes geographic analysis difficult because of the system's inability to observe topologic relationships between areas that share a common border. The non-topologic model is a common data model supported by many computer aided drafting (CAD), mapping and graphic systems.

The topology helps to perform various types of overlaying analysis and modelling. The relationships cannot be established without building the topology.

Descriptive Data in GIS

The GIS stores aspatial data associated with geographic features similarly to the way it stores coordinates. Attributes are stored as sets of numbers (integer, real) and characters (logical or boolean). The storage is no different from the data handled by conventional data base systems. The attribute data is stored using either hierarchical, network or relational models. Most of the data base management packages available today are relational (Oracle, Ingres, Integra SQL, Sybase etc.).

Connecting Features and Attributes

The importance of GIS lies in its link between the graphic (spatial) and the tabular (aspatial) data. Basically there are three characteristics of this connection. They are (i) There is one-to-one relationship between features on the digital map and the records in the feature attribute table, (ii) The link between the feature and its record is maintained through a unique numerical identifier assigned to each feature (label points), and (iii) The unique identifier is physically stored in two places, i.e. in the file that contain the x, y coordinates and with the corresponding record in the feature attribute table. So, once this connection is established one can query the digital map to display attribute information or create a map based on the attributes stores in the feature attribute table.

DATA MANIPULATION

GIS provides number of analysis capabilities that operate on the topology or spatial aspects of geographical data and on aspatial (attribute) data. It provides a number of spatial analysis functions which signifies its importance on other information systems (P.A. Burrough, 1996). The integrated information is useful to answer queries pertaining to the parameters defined in a particular theme. For example, if a thematic map represents land use information, the associate tabular information will answer all queries related to the extent and spatial distribution of various land uses. Integrated analysis will help not only to know the land use details but also to know the soil depth, soil texture, slope etc. pertaining to a land use category. Thus GIS has many capabilities for spatial analysis and identifying spatial relationships between the parameters.

Spatial Analysis Operations

Raster GIS allows arithmetic as well as logical overlay operations. Arithmetic operations involve addition, subtraction, multiplication, division etc. between the cell values of different layers. Logical overlay operations find the areas where occurrence/non-occurrence of a set of conditions are observed. The overlay operations do not alter the input layer information but generates a new output layer combining the information of input layers (Fig. 2.3).

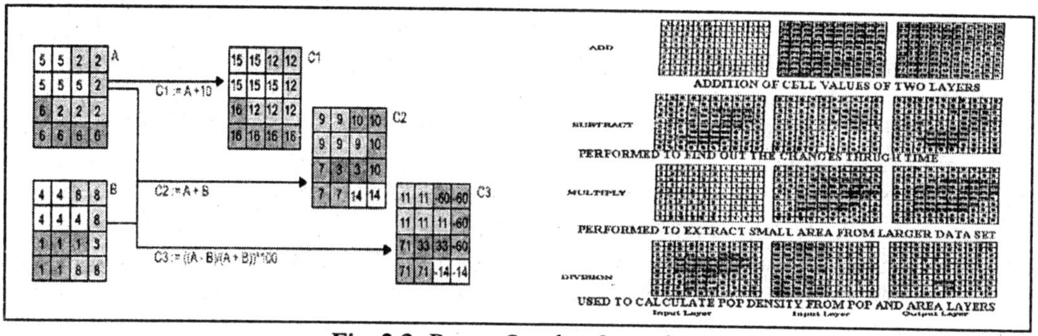

Fig. 2.3: Raster Overlay Operations

In vector based GIS, these operations are performed on two layers (maps) at a time to form a new composite map through the geometric intersection of the features. The layer on which manipulation is performed is called the input layer and the layer that controls the area of operation is called the analysis layer. These operations are classified into three categories as (a) Feature combination, (b) Feature extraction and (c) Feature extraction and combination.

Feature Combination

There are two categories identified under this operation. UNION and INTERSECT (Fig. 2.4).

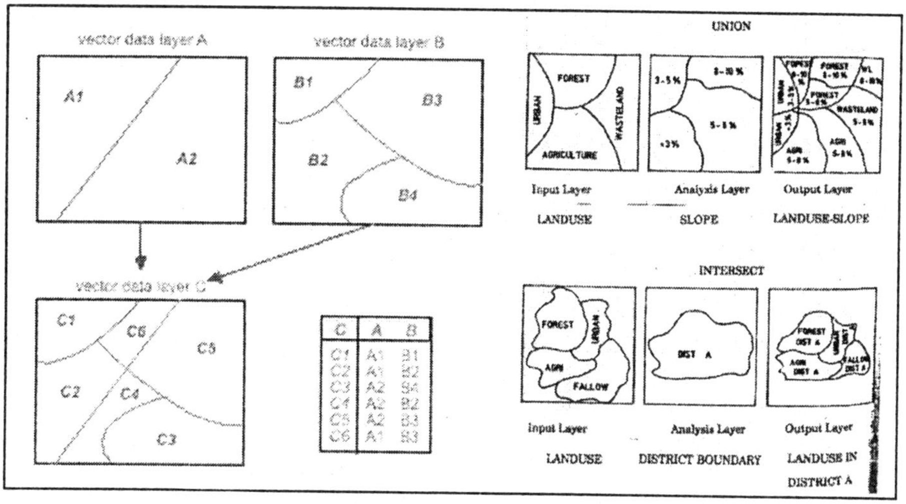

Fig. 2.4: Union and Intersect Analysis in Vector GIS

UNION process is used only on the maps having polygon features. It performs the geometric combination of the features of the two themes. Similar to mathematical operation (Ex.: A is a set of natural numbers with 1, 2, 3, 4, 5. B is a set of natural numbers with 4, 5, 6, 7. The union of A and B sets will be AUB = 1, 2, 3, 4, 5, 6, 7) to keeps all the areas from both the coverages. New polygons in the output layer are created by splitting the arcs of the input layers at the intersections. Therefore, the number of polygons in the output layer is more compared to the number of polygons in any of the input layers. This particular operation is performed, when the application demands that the combined region and combined attributes of the two input thematic layers are required for querying or for further analysis.

INTERSECT operation is performed to overlay points, lines or polygons on polygons but keep only those portions of the input coverage features falling within the overlay coverage features, i.e. A Ç B = 4, 5 from the above example. The input layer can have points, lines or polygons but the analysis layer must have polygon topology. This particular operation is normally performed to find out the number of tube wells, dug wells in a watershed, roads crossing the human settlements, land use distribution in a defined administrative boundary etc.

Feature Extraction

In this operation, two input layers are overlaid similar to feature combination analysis. However, the extent of analysis layer defines the area of interest to be retained or removed. Three operations fall under this category, viz. CLIP, ERASE and SPLIT. In CLIP operation, the features that fall within the boundary of the analysis layer are retained in the resultant layer with the attributes of the input layer only. In this operation, the input layer can be points, lines or polygons but the analysis layer should contain polygons. This function is performed to extract a smaller data set from a larger data set. ERASE is a reversal process of CLIP in which the features of the input layer within the boundary of the analysis layer are erased and the features that fall outside the boundary of analysis layer are retained. This is used to create a compliment of CLIP operation. SPLIT operation is the enhancement of CLIP operation. This operation is normally carried out to produce an output layer of various CLIP operations based on certain criteria. This operation is meant for advance users to perform a series of CLIP operations through a single command.

Feature Combination and Extraction

Two spatial operations fall under this category. They are UPDATE and IDENTITY. UPDATE operation is used to perform cut and paste analysis (Fig. 2.5). Similar to earlier operations, the analysis layer defines the area of control of the input layer that needs to be updated. Thus, the output layer will have features from the input layer in the non-overlapping area and the features from the analysis layer will be in the overlapping area. It is used to generate a time series map showing the changes in thematic information through time. For example, showing the change in land use pattern through time, Urban sprawl through time etc. IDENTITY operation is performed to create an output layer by combining the features of the overlapping areas of input and analysis layers. It is mostly used to preserve the boundaries of thematic map precisely. For example, showing the district boundaries in the state map without there being any mismatch between the district boundaries and the state boundary. This operation is also performed to overlay points, lines or polygons on polygons and keep all input coverage features.

In all the integration operations, a large number of small polygons are formed at the edges of the boundary of the output layer due to the mismatch of the boundaries of input layers. These small polygons in GIS terminology are termed as "SLIVERS". These polygons have to be removed before the output layer of integrated operation is taken for further use. The slivers are normally removed on the basis of the minimum mapping unit (MMU) area of the thematic map scale.

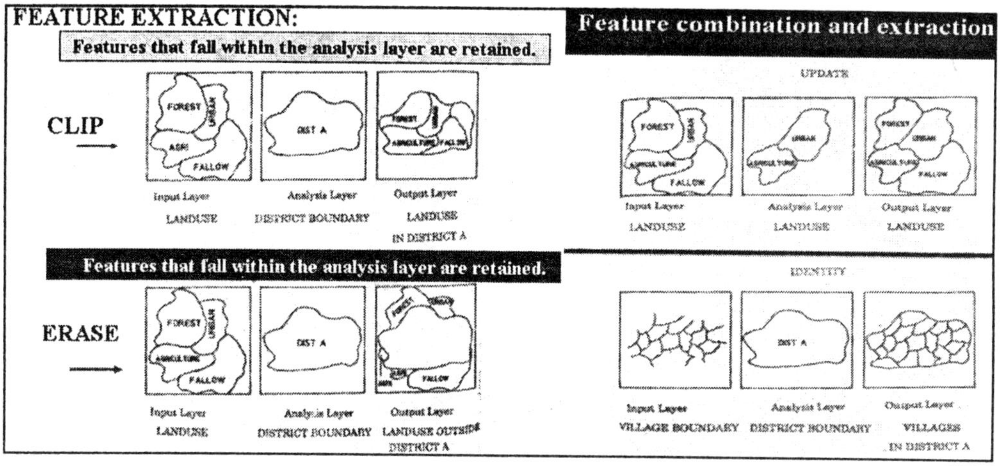

Fig. 2.5: Feature Extraction/Feature Combination and Extraction

GIS is also useful to construct the proximity boundaries for polygons at a distance specified by the user. These resultant polygons are known as proximity polygons or buffer zones (Fig. 2.6). The distance input used for the operation is called as "search radius" or "buffer distance". Buffers can be generated on points, lines and polygon (both inside and outside of polygon) features. This operation is basically carried out to find out the closeness or proximity between the features. This function is necessary in spatial data analysis to find out the distance of a road from a settlement, ownership details around a mining area, water bodies, flood hazard zones etc. The spatial analysis techniques mentioned above provide a variety of choices to user to carry out spatial modelling exercises. However, it all depends upon the technical understanding and experience of the user to identify the need of a particular function.

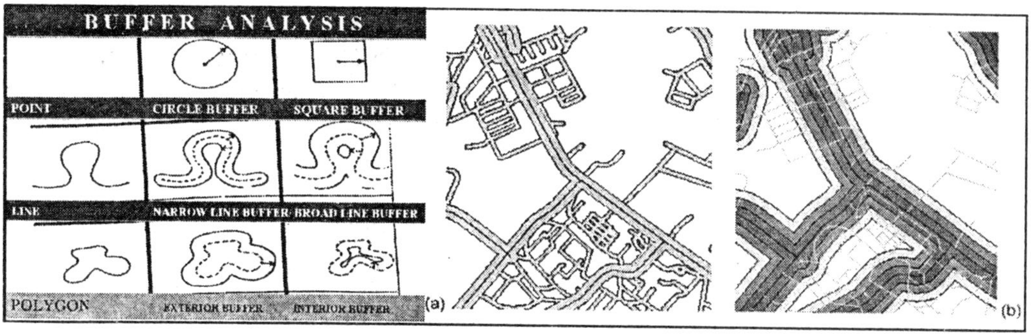

Fig. 2.6: Buffer Analysis

DATA DISPLAY AND QUERYING

Both spatial and attribute data can be displayed on the computer screen. It is known that the geographic features are stored as points, lines and polygons in GIS, they can be displayed as separate layers or in combination with each other. The displays can be sent to a hard copy output device, i.e. plotter or printer.

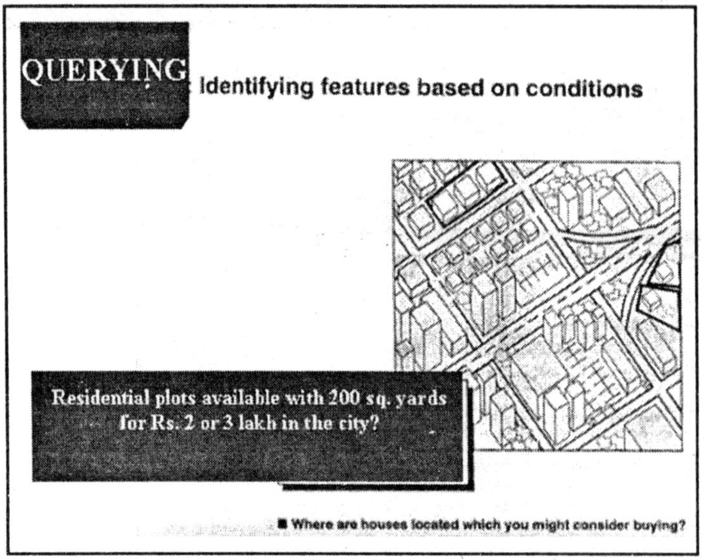

Fig. 2.7: Identification of Features Based on Conditions

In GIS, one can determine what exists at a particular location. For this, one must first specify the location of an object or region for which the information is needed. Commonly used methods of specification are: Pointing to the object or region, typing in a particular address or typing in a co-ordinate location. After the specification of the object or location, one can obtain either all of its characteristics or some of its characteristics, viz. Location address, current owner, existing land use, assessed value etc. (Fig. 2.7) The second function in GIS is that one can obtain information on locations satisfying certain conditions. For this, one can specify the condition as: making selections from a set of predefined options, writing logical expressions and filling out forms interactively on a terminal. After specifying the logical condition, one can obtain the list of all objects meeting the specified conditions and also display highlighting all features meeting the specified conditions.

THE FUTURE OF GIS

The GIS is originated in the mid-1960s and it is continuous history since then. However, many see GIS as a phenomenon of the late 1980s. This is primarily because of developments in software, cost-effectiveness of hardware. The expansion has also been fuelled by continuing advances in computing technology, increasing awareness of digital databases, new application areas etc. The fundamental question now is that how long can this growth continue ? The growth of GIS will continue, if the institutions provide focus on programs, journals, magazines, books, conferences and raise awareness of GIS technology and its applications in education and training (the effective use of spatial information requires higher levels of training than the word processing software's). The prospects for the future of GIS lies in automated geography, i.e. to create "paperless map library" on the similar lines of "paperless office". The potential of automated geography may lead to much greater levels of use if people have better access to the database, easy procedures to use the database and usage of geographical database more frequently by planners, administrators etc. for various decision making processes.

REFEERENCES

Burrough, P.A., 1986. *Principles of Geographical Information Systems for Land Resources Assessment*, Clarendon, Oxford. Chapter 3 reviews alternative methods of terrain representation.

Dutton-Marion, K.E., 1988. *Principles of Interpolation Procedures in the Display and Analysis of Spatial Data: A Comparative Analysis of Conceptual and Computer Contouring*, unpublished Ph.D. Thesis, Department of Geography, University of Calgary, Calgary, Alberta.

Elements of Cartography by Robinson, H and Randall, D. Sale.

Everett, J.E. and D.S. Simonett, 1976. "Principles, concepts and philosophical problems in remote sensing," in J. Lintz and D.S. Simonett, Editors, *Remote Sensing of Environment*, Addison-Wesley, Reading, MA, pp 85-127. A review of remote sensing from the mid-1970s with striking parallels with current debates within GIS.

ESRI, 1990. *Understanding GIS: The ARC/INFO Way*, ESRI, Redlands, CA. A general introductory tutorial for ARC/INFO, a well-known contemporary GIS.

Evans, I.S., 1980. "An integrated system for terrain analysis and slope mapping," *Zeitschrift fur Geomorphologie* 36:274-95.

Fisher, P.F., and R. Lindenberg, 1989. "On distinctions among Cartography, Remote Sensing, and Geographic Information Systems," *Photogrammetric Engineering and Remote Sensing* 55(10):1431-1434. Reviews definitions of each of the three and shows how the disciplines are interrelated.

Goodchild, M.F., 1988. "Stepping over the line: technological constraints and the new cartography," *American Cartographer* 15:311-9. Argues that cartography's traditions are derived from its reliance on pen and paper, and looks at how these constraints are removed by automation.

Lecture notes—Training course on GIS for resource management and development planning, Volume-I, II and III, 1993, SAC (ISRO), Ahmedabad.

NCGIA Core Curriculum, Volumes-1, 2 and 3, Michael F Goodchild and Karen K. Kemp, d (1990), A NCGIA, University of California, Santa Barbara, USA publication.

Pathan, S.K. and Navalgund, R.R., "The Role of Geomatics in Natural Resources Development and Planning", 1996, Proceedings of Indian Society of Geomatics.

Raper, Jonathan. Three Dimensional Applications in GIS, Taylor and Francis, New York. A collection of papers on the developing technology of 3D GIS.

Tutorial notes on Geographic Information System, GIS—The Basics, Indian Society of Geomatics, 1996, Space Applications Centre (ISRO).

3

USE OF REMOTE SENSING IN MATHEMATICAL MODELLING

S.K. PATHAN

INTRODUCTION

Remote Sensing is the science of making inferences about material objects from measurements, made at a distance, without coming into physical contact with the objects under study. Visual perception of objects is the best example of remote sensing. We see an object by the light reflected from the object falling on the human eye. Here, eye is the sensor and the nervous system carries information to the brain, which interprets the information in terms of the identification and location of the objects seen. Modern remote sensing is an extension of this natural phenomenon. However, currently the term remote sensing is used more commonly to denote identification of earth features by detecting the characteristic electromagnetic radiation that is reflected and or emitted by the earth surface. A remote sensing system consists of a sensor to collect the radiation and a platform—an aircraft, balloon, rocket, satellite or even a ground-based sensor-supporting stand-on which a sensor can be mounted. The information received by the sensor is suitably manipulated and transported back to the earth—may be telemetered as in the case of unmanned spacecraft, or brought back through films, magnetic tapes, etc. as in aircraft or manned spacecraft systems. The data are reformatted and processed on the ground to produce either photographs, or computer compatible magnetic tapes (CCT). The photographs/CCTs are interpreted visually/digitally to produce thematic maps and other resources information. This lecture notes deal with all the aspects of remote sensing for various natural resources development and planning.

What is Remote Sensing?

"Remote sensing is the science and art of acquiring information about the Earth's surface without actually being in contact with it". This is done by sensing and recording reflected or emitted energy and processing, analyzing, and applying that information.

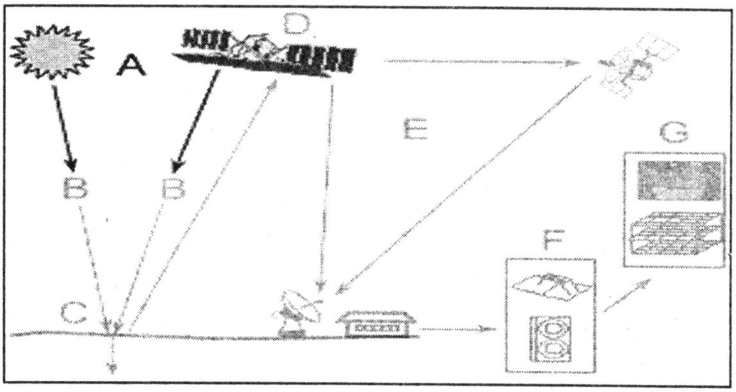

Fig. 3.1: Seven Elements of Remote Sensing *(Adopted from CCRS tutorials)*

In much of remote sensing, the process involves an interaction between incident radiation and the targets of interest. This is represented by the use of imaging systems, in which the following seven elements are involved (Fig. 3.1). (Note, however, that remote sensing also involves the sensing of emitted energy and the use of non-imaging sensors).

1. *Energy Source or Illumination (A)* – the first requirement for remote sensing is to have an energy source, which illuminates or provides electromagnetic energy to the target of interest.
2. *Radiation and the Atmosphere (B)* – as the energy travels from its source to the target, it will come in contact with and interact with the atmosphere it passes through. This interaction may take place a second time as the energy travels from the target to the sensor.
3. *Interaction with the Target (C)* – once the energy makes its way to the target through the atmosphere, it interacts with the target depending on the properties of both the target and the radiation.
4. *Recording of Energy by the Sensor (D)* – after the energy has been scattered by, or emitted from the target, we require a sensor (remote—not in contact with the target) to collect and record the electromagnetic radiation.
5. *Transmission, Reception, and Processing (E)* – the energy recorded by the sensor has to be transmitted, often in electronic form, to a receiving and processing station where the data are processed into an image (hardcopy and/or digital).
6. *Interpretation and Analysis (F)* – the processed image is interpreted, visually and/or digitally or electronically, to extract information about the target which was illuminated.
7. *Application (G)* – the final element of the remote sensing process is achieved when we apply the information we have been able to extract from the imagery about the target in order to better understand it, reveal some new information, or assist in solving a particular problem.

These seven elements comprise the remote sensing process from beginning to end.

Electromagnetic Radiation

As was noted in the previous section, the first requirement for remote sensing is to have an *energy source to illuminate the target* (unless the sensed energy is being emitted by the target). This energy is in the form of electromagnetic radiation.

All electromagnetic radiation has fundamental properties and behaves in predictable ways according to the basics of wave theory. *Electromagnetic radiation* consists of an electrical field (E) which varies in magnitude in a direction perpendicular to the direction in which the radiation is travelling, and a magnetic field (M) oriented at right angles to the electrical field. Both these fields travel at the speed of light (c).

Two characteristics of electromagnetic radiation are particularly important for understanding remote sensing. These are the *wavelength and frequency.*

The wavelength is the length of one wave cycle, which can be measured as the distance between successive wave crests. Wavelength is usually represented by the Greek letter lambda (l). Wavelength is measured in metres (m) or some factor of metres such as **nanometres** (nm, 10^{-9} metres), **micrometres** (μm, 10^{-6} metres) or centimetres (cm, 10^{-2} metres). Frequency refers to the number of cycles of a wave passing a fixed point per unit of time. Frequency is normally measured in **hertz** (Hz), equivalent to one cycle per second, and various multiples of hertz.

Wavelength and frequency are related by the following formula:

$$c = \gamma v$$

where:

γ = wavelength (m)
v = frequency (cycles per second, Hz)
c = speed of light (3×10^8m/s)

Therefore, the two are inversely related to each other. The shorter the wavelength, the higher the frequency. The longer the wavelength, the lower the frequency. Understanding the characteristics of electromagnetic radiation in terms of their wavelength and frequency is crucial to understanding the information to be extracted from remote sensing data. Next we will be examining the way in which we categorize electromagnetic radiation for just that purpose.

The Electromagnetic Spectrum

The *electromagnetic spectrum* ranges from the shorter wavelengths (including gamma and x-rays) to the longer wavelengths (including microwaves and broadcast radio waves). There are several regions of the electromagnetic spectrum which are useful for remote sensing. For most purposes, the *ultraviolet or UV* portion of the spectrum has the shortest wavelengths which are practical for remote sensing. This radiation is just beyond the violet portion of the visible wavelengths, hence its name. Some Earth surface materials, primarily rocks and minerals, fluoresce or emit visible light when illuminated by UV radiation. The light which our eyes—

our "remote sensors"—can detect is part of the *visible spectrum.* It is important to recognize how small the visible portion is relative to the rest of the spectrum. There is a lot of radiation around us which is "invisible" to our eyes, but can be detected by other remote sensing instruments and used to our advantage. The visible wavelengths cover a range from approximately 0.4 to 0.7 μm. The longest visible wavelength is red and the shortest is violet. Common wavelengths of what we perceive as particular colours from the visible portion of the spectrum are listed below. It is important to note that this is the only portion of the spectrum we can associate with the concept of colours.

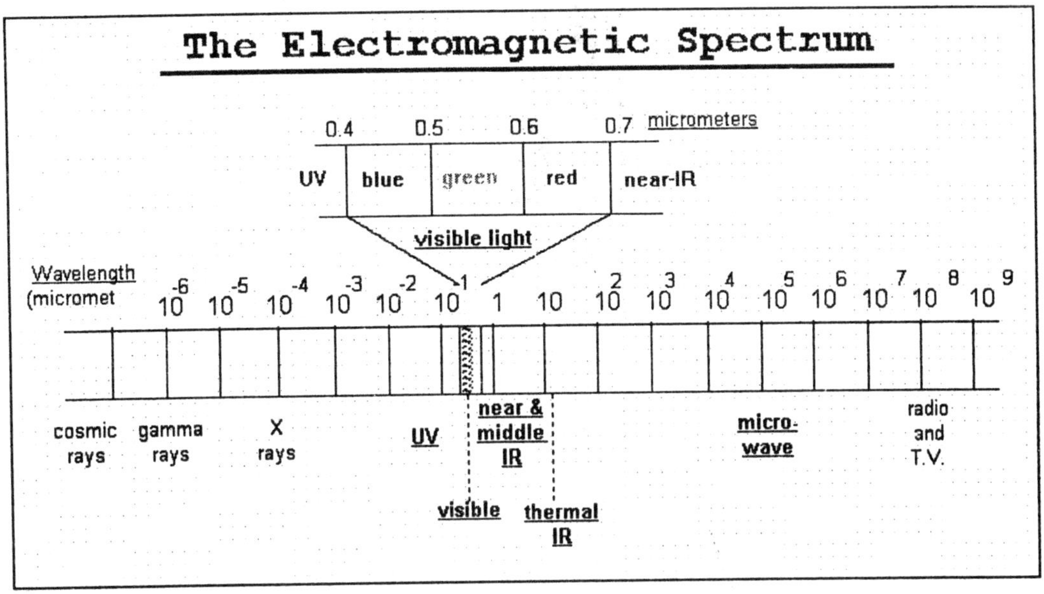

Violet: 0.4-0.446 μm
Blue: 0.446-0.500 μm
Green: 0.500-0.578 μm
Yellow: 0.578-0.592 μm
Orange: 0.592-0.620 μm
Red: 0.620-0.7 μm

Blue, **green**, and **red** are the **primary colours** or wavelengths of the visible spectrum. They are defined as such because no single primary colour can be created from the other two, but all other colours can be formed by combining blue, green, and red in various proportions. Although we see sunlight as a uniform or homogeneous colour, it is actually composed of various wavelengths of radiation in primarily the ultraviolet, visible and infrared portions of the spectrum. The visible portion of this radiation can be shown in its component colours when sunlight is passed through a *prism*, which bends the light in differing amounts according to wavelength.

The next portion of the spectrum of interest is the *infrared (IR) region* which covers the wavelength range from approximately 0.7 μm to 100 μm—more than 100 times as wide as the visible portion. The infrared region can be divided into two categories based on their radiation properties—the **reflected IR**, and the emitted or **thermal IR**. Radiation in the reflected IR region is used for remote sensing purposes in ways very similar to radiation in the visible portion. The reflected IR covers wavelengths from approximately 0.7 μm to 3.0 μm. The thermal IR region is quite different than the visible and reflected IR portions, as this energy is essentially the radiation that is emitted from the Earth's surface in the form of heat. The thermal IR covers wavelengths from approximately 3.0 μm to 100 μm.

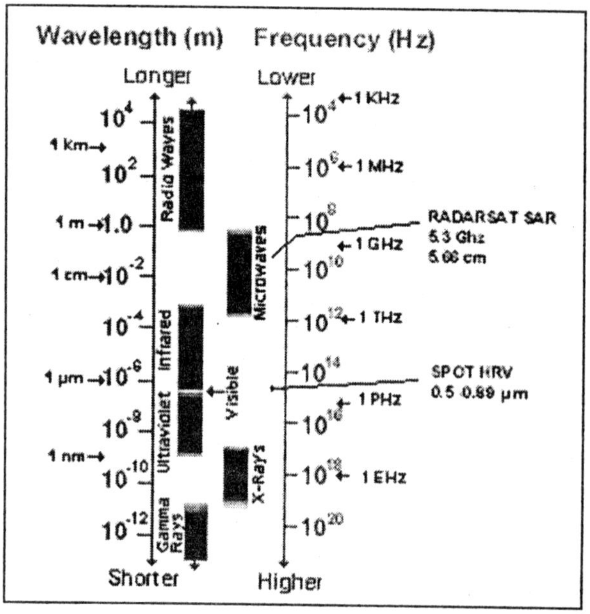

The portion of the spectrum of more recent interest to remote sensing is the *microwave region* from about 1 mm to 1 m. This covers the longest wavelengths used for remote sensing. The shorter wavelengths have properties similar to the thermal infrared region while the longer wavelengths approach the wavelengths used for radio broadcasts.

On the Ground, In the Air, In Space

In order for a sensor to collect and record energy reflected or emitted from a target or surface, it must reside on a stable *platform* removed from the target or surface being observed. Platforms for remote sensors may be situated on the ground, on an aircraft or balloon (or some other platform within the Earth's atmosphere), or on a spacecraft or satellite outside of the Earth's atmosphere.

Ground-based sensors are often used to record detailed information about the surface which is compared with information collected from aircraft or satellite sensors. In some cases, this can be used to better characterize the target which is being imaged by these other sensors, making it possible to better understand the information in the imagery. Sensors may be placed on a ladder, scaffolding, tall building, cherry-picker, crane, etc. Aerial platforms are primarily stable wing *aircraft*, although helicopters are occasionally used. Aircraft are often used to collect very detailed images and facilitate the collection of data over virtually any portion of the Earth's surface at any time. In space, remote sensing is sometimes conducted from the *space shuttle* or, more commonly, from satellites. *Satellites* are objects which revolve around another object—in this case, the Earth. For example, the moon is a natural satellite, whereas man-made satellites include those platforms launched for remote sensing, communication, and telemetry (location and navigation) purposes. Because of their orbits, satellites permit repetitive coverage of the Earth's surface on a continuing basis.

Satellite Characteristics: Orbits and Swaths

As is evident from in the previous section that remote sensing instruments can be placed on a variety of platforms to view and image targets. Although ground-based and aircraft platforms may be used, satellites provide a great deal of the remote sensing imagery commonly used today. Satellites have several unique characteristics which make them particularly useful for remote sensing of the Earth's surface.

The path followed by a satellite is referred to as its **orbit**. Satellite orbits are matched to the capability and objective of the sensor (s) they carry. Orbit selection can vary in terms of altitude (their height above the Earth's surface) and their orientation and rotation relative to the Earth. Satellites at very high altitudes, which view the same portion of the Earth's surface at all times have *geostationary orbits*. These geostationary satellites, at altitudes of approximately 36,000 kilometres, revolve at speeds which match the rotation of the Earth so they seem stationary, relative to the Earth's surface. This allows the satellites to observe and collect information continuously over specific areas. Weather and communications satellites commonly have these types of orbits. Due to their high altitude, some geostationary weather satellites can monitor weather and cloud patterns covering an entire hemisphere of the Earth.

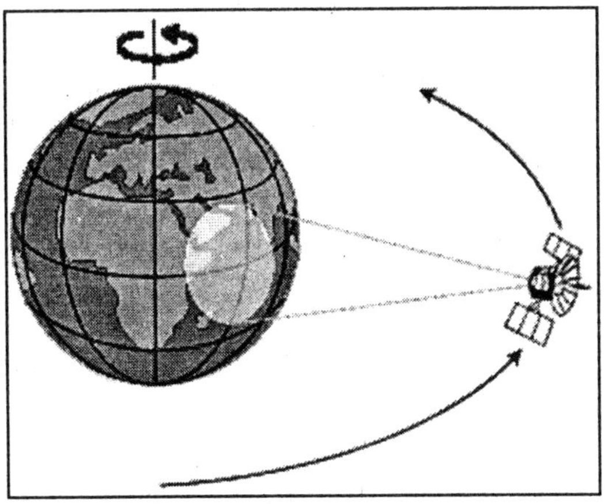

Geostationary orbit *(Adopted from CCRS tutorials)*

Many remote sensing platforms are designed to follow an orbit (basically north-south) which, in conjunction with the Earth's rotation (west-east), allows them to cover most of the Earth's surface over a certain period of time. These are *near-polar orbits*, so named for the inclination of the orbit relative to a line running between the North and South poles. Many of these satellite orbits are also *sun-synchronous* such that they cover each area of the world at a constant local time of day called **local sun time**. At any given latitude, the position of the sun in the sky as the satellite passes overhead will be the same within the same season. This ensures consistent illumination conditions when acquiring images in a specific season over successive years, or over a particular area over a series of days. This is an important factor for monitoring changes between images or for mosaicking adjacent images together, as they do not have to be corrected for different illumination conditions.

Most of the remote sensing satellite platforms today are in near-polar orbits, which means that the satellite travels northwards on one side of the Earth and then toward the southern pole on the second half of its orbit. These are called **ascending and descending passes**, respectively. If the orbit is also sun-synchronous, the ascending pass is most likely on the shadowed side of the Earth while the descending pass is on the sunlit side. Sensors recording reflected solar energy only image the surface on a descending pass, when solar illumination is available. Active sensors which provide their own illumination or passive sensors that record emitted (e.g. thermal) radiation can also image the surface on ascending passes.

As a satellite revolves around the Earth, the sensor "sees" a certain portion of the Earth's surface. The area imaged on the surface, is referred to as the **swath**. Imaging swaths for space borne sensors generally vary between tens and hundreds of kilometres wide. As the satellite orbits the Earth from pole to pole, its east-west position wouldn't change if the Earth didn't rotate. However, as seen from the Earth, it seems that the satellite is shifting westward because the Earth is rotating (from west to east) beneath it. This apparent movement allows the satellite swath to cover a **new area with each consecutive pass**. The satellite's orbit and the rotation of the Earth work together to allow complete coverage of the Earth's surface, after it has completed one complete cycle of orbits.

If one start with any randomly selected pass in a satellite's orbit, an orbit cycle will be completed when the satellite retraces its path, passing over the same point on the Earth's surface directly below the satellite (called the **nadir** point) for a second time. The exact length of time of the orbital cycle will vary with each satellite. The interval of time required for the satellite to complete its orbit cycle is not the same as the "**revisit period**". Using steerable sensors, a satellite-borne instrument can view an area (off-nadir) before and after the orbit passes over a target, thus making the 'revisit' time less than the orbit cycle time. The revisit period is an important consideration for a number of monitoring applications, especially when frequent imaging is required (for example, to monitor the spread of an oil spill, or the extent of flooding). In near-polar orbits, areas at high latitudes will be imaged more frequently than the equatorial zone due to the increasing *overlap in adjacent swaths* as the orbit paths come closer together near the poles.

Spatial Resolution, Pixel Size, and Scale

For some remote sensing instruments, the distance between the target being imaged and the platform, plays a large role in determining the detail of information obtained and the total area imaged by the sensor. Sensors onboard platforms far away from their targets, typically view a larger area, but cannot provide great detail. Compare what an astronaut onboard the space shuttle sees of the Earth to what you can see from an airplane. The astronaut might see your whole province or country in one glance, but couldn't distinguish individual houses. Flying over a city or town, you would be able to see individual buildings and cars, but you would be viewing a much smaller area than the astronaut. There is a similar difference between satellite images and air photos.

The detail discernible in an image is dependent on the **spatial resolution** of the sensor and refers to the size of the smallest possible feature that can be detected. Spatial resolution of passive sensors depends primarily on their *Instantaneous Field of View (IFOV)*. The IFOV is the angular cone of visibility of the sensor (a) and determines the area on the Earth's surface which is "seen" from a given altitude at one particular moment in time (b). The size of the area viewed is determined by multiplying the IFOV by the distance from the ground to the sensor (c). This area on the ground is called the **resolution cell** and determines a sensor's maximum spatial resolution. For a homogeneous feature to be detected, its size generally has to be equal to or larger than the resolution cell. If the feature is smaller than this, it may not be detectable as the average brightness of all features in that resolution cell will be recorded. However, smaller features may sometimes be detectable if their reflectance dominates within a Particular resolution cell allowing sub-pixel or resolution cell detection.

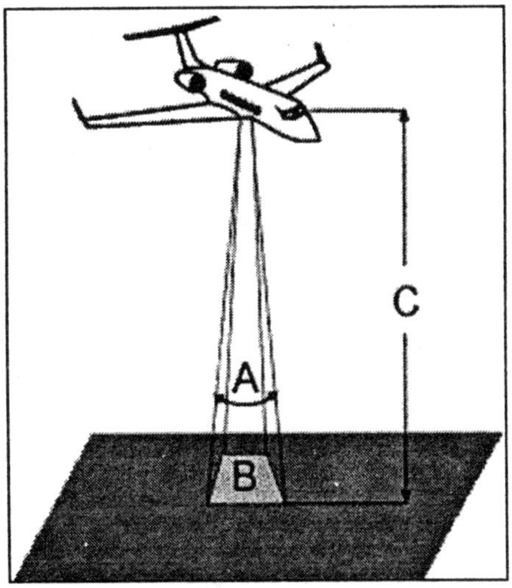

Instantaneous Field of View *(Adopted from CCRS tutorials)*

Remote Sensing images are composed of a matrix of picture elements, or **pixels**, which are the smallest units of an image. Image pixels are normally square and represent a certain area on an image. It is important to distinguish between pixel size and spatial resolution—they are not interchangeable. If a sensor has a spatial resolution of 20 metres and an image from that sensor is displayed at full resolution, each pixel represents an area of 20m × 20m on the ground. In this case the pixel size and resolution are the same. However, it is possible to display an image with a pixel size different than the resolution. Many posters of satellite images of the Earth have their pixels averaged to represent larger areas, although the original spatial resolution of the sensor that collected the imagery remains the same.

Images where only large features are visible are said to have *coarse or low resolution.* In *fine or high resolution* images, small objects can be detected. Military sensors for example, are designed to view as much detail as possible, and therefore, have very fine resolution. Commercial satellites provide imagery with resolutions varying from a few metres to several kilometres. Generally speaking, the finer the resolution, the less total ground area can be seen.

The ratio of distance on an image or map, to actual ground distance is referred to as scale. If you one a map with a scale of 1:100,000, an object of 1cm length on the map would actually be an object 100,000cm (1km) long on the ground. Maps or images with small "map-to-ground ratios" are referred to as small scale (e.g. 1:100,000), and those with larger ratios (e.g. 1:5,000) are called large scale.

Spectral Resolution

Different classes of features and details in an image can often be distinguished by comparing their responses over distinct wavelength ranges. Broad classes, such as water and vegetation, can usually be separated using very broad wavelength ranges—the visible and near infrared. Other more specific classes, such as different rock types, may not be easily distinguishable using either of these broad wavelength ranges and would require comparison at much finer wavelength ranges to separate them. Thus, We would require a sensor with higher **spectral resolution**. Spectral resolution describes the ability of a sensor to define fine wavelength intervals. The finer the spectral resolution, the narrower the wavelength range for a particular channel or band.

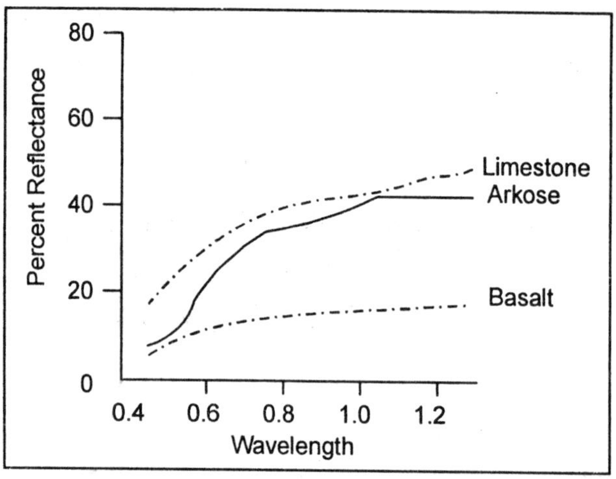

Black and white film records wavelengths extending over much, or all of the visible portion of the electromagnetic spectrum. Its *spectral resolution* is fairly coarse, as the various wavelengths of the visible spectrum are not individually distinguished and the overall reflectance in the entire visible portion is recorded. Colour film is also sensitive to the reflected energy over the visible portion of the spectrum, but has higher spectral resolution, as it is individually sensitive to the reflected energy at the blue, green, and red wavelengths of the spectrum. Thus, it can represent features of various colours based on their reflectance in each of these distinct wavelength ranges.

Many remote sensing systems record energy over several separate wavelength ranges at various spectral resolutions. These are referred to as **multi-spectral sensors** and will be described in some detail in following sections. Advanced multi-spectral sensors called **hyperspectral** sensors, detect hundreds of very narrow spectral bands throughout the visible, near-infrared, and mid-infrared portions of the electromagnetic spectrum. Their very high spectral resolution facilitates fine discrimination between different targets based on their spectral response in each of the narrow bands.

Radiometric Resolution

While the arrangement of pixels describes the spatial structure of an image, the radiometric characteristics describe the actual information content in an image. Every time an image is acquired on film or by a sensor, its sensitivity to the magnitude of the electromagnetic energy determines the **radiometric resolution**. The radiometric resolution of an imaging system describes its ability to discriminate very slight differences in energy. The finer the radiometric resolution of a sensor, the more sensitive it is to detecting small differences in reflected or emitted energy.

Imagery data are represented by positive digital numbers which vary from 0 to (one less than) a selected power of 2. This range corresponds to the number of bits used for coding numbers in binary format. Each bit records an exponent of power 2 (e.g. 1 bit = 2^1 = 2). The maximum number of brightness levels available depends on the number of bits used in representing the energy recorded. Thus, if a sensor used 8 bits to record the data, there would be 2^8 = 256 digital values available, ranging from 0 to 255. However, if only 4 bits were used, then only 2^4 = 16 values ranging from 0 to 15 would be available. Thus, the radiometric resolution would be much less. Image data are generally displayed in a range of grey tones, with black representing a digital number of 0 and white representing the maximum value (for example, 255 in 8-bit data). *By comparing a 2-bit image with an 8-bit image*, one see that there is a large difference in the level of detail discernible depending on their radiometric resolutions. (Visualise the following figures.)

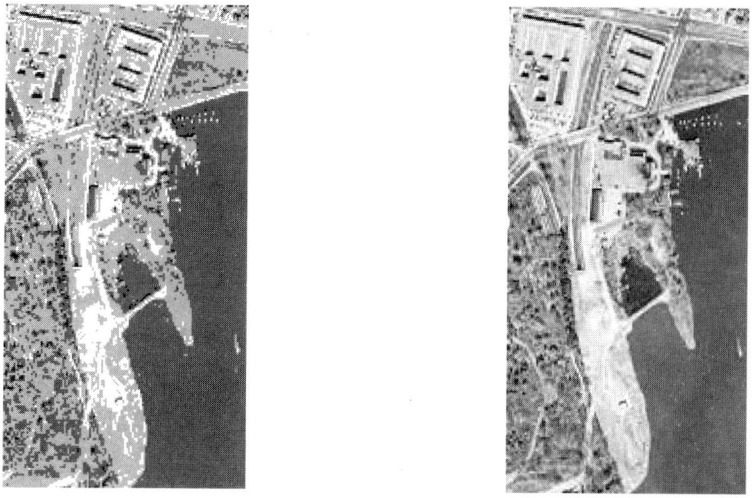

Comparision of a 2 bit image with an 8 bit image *(Adopted from CCRS tutorials)*

Temporal Resolution

In addition to spatial, spectral, and radiometric resolution, the concept of **temporal resolutions** also important to consider in a remote sensing system. As we discussed the concept of revisit period, which refers to the length of time it takes for a satellite to complete one entire orbit cycle. The revisit period of a satellite sensor is usually several days. Therefore, the absolute temporal resolution of a remote sensing system to image the exact same area at the same viewing angle a second time is equal to this period. However, because of some degree of overlap in the imaging swaths of adjacent orbits for most satellites and the increase in this overlap with increasing latitude, some areas of the Earth tend to be re-imaged more frequently. Also, some satellite systems are able to *point their sensors to image the same area* between different satellite passes separated by periods from one to five days. Thus, the actual temporal resolution of a sensor depends on a variety of factors, including the satellite/sensor capabilities, the swath overlap, and latitude.

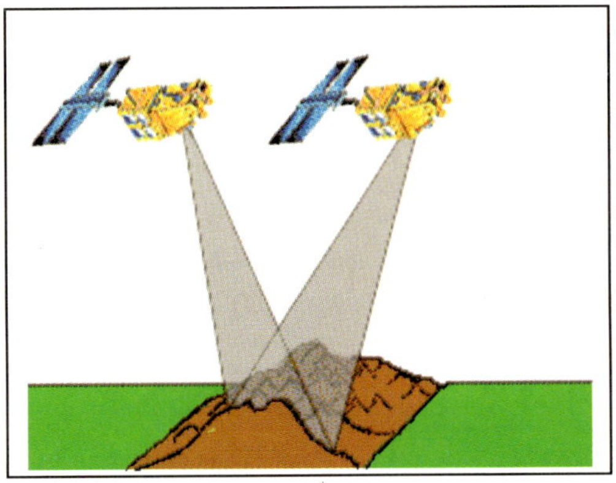

Imaging the same area *(Adopted from CCRS tutorials)*

The ability to collect imagery of the same area of the Earth's surface at different periods of time is one of the most important elements for applying remote sensing data. Spectral characteristics of features may change over time and these changes can be detected by collecting and comparing **multi-temporal** imagery. For example, during the growing season, most species of vegetation are in a continual state of change and the ability to monitor those subtle changes using remote sensing is dependent on when and how frequently one collect imagery. By imaging on a continuing basis at different times it as possible to monitor the changes that take place on the Earth's surface, whether they are naturally occurring (such as changes in natural vegetation cover or flooding) or induced by humans (such as urban development or deforestation).

Geometric Distortion in Imagery

Any remote sensing image, regardless of whether it is acquired by a multispectral scanner on board a satellite, a photographic system in an aircraft, or any other platform/sensor combination, will have various **geometric distortions.** This problem is inherent in remote sensing, as we attempt to accurately represent the three-dimensional surface of the Earth as a two-dimensional image. All remote sensing images are subject to some form of geometric distortions, depending on the manner in which the data are acquired. These errors may be due to a variety of factors, including one or more of the following, to name only a few:

- the perspective of the sensor optics,
- the motion of the scanning system,
- the motion and (in) stability of the platform,
- the platform altitude, attitude, and velocity,
- the terrain relief, and
- the curvature and rotation of the Earth.

Framing systems, such as cameras used for aerial photography, provide an instantaneous "snapshot" view of the Earth from directly overhead. The primary geometric distortion in vertical aerial photographs is due to *relief displacement*. Objects directly below the centre of the camera lens (i.e. at the **nadir**) will have only their tops visible, while all other objects will appear to lean away from the centre of the photo such that their tops and sides are visible. If the objects are tall or are far away from the centre of the photo, the distortion and positional error will be larger.

The geometry of along-track scanner imagery is similar to that of an aerial photograph for each scan line as each detector essentially takes a "snapshot" of each ground resolution cell. Geometric variations between lines are caused by random variations in platform altitude and attitude along the direction of flight.

Images from across-track scanning systems exhibit two main types of *geometric distortion*. They too exhibit relief displacement (a), similar to aerial photographs, but in only one direction parallel to the direction of scan. There is no displacement directly below the sensor, at nadir. As the sensor scans across the swath, the top and side of objects are imaged and appear to lean away from the nadir point in each scan line. Again, the displacement increases, moving towards the edges of the swath. Another distortion (b) occurs due to the rotation of the scanning optics. As the sensor scans across each line, the distance from the sensor to the ground increases further away from the centre of the swath. Although the scanning mirror rotates at a constant speed, the IFOV of the sensor moves faster (relative to the ground) and scans a larger area as it moves closer to the edges. This effect results in the compression of image features at points away from the nadir and is called **tangential scale distortion**.

All images are susceptible to geometric distortions caused by variations in platform stability including changes in their speed, altitude, and attitude (angular orientation with respect to the ground) during data acquisition. These effects are most pronounced when using aircraft platforms and are alleviated to a large degree with the use of satellite platforms, as their orbits are relatively stable, particularly in relation to their distance from the Earth. However, the eastward rotation of the Earth, during a satellite orbit causes the sweep of scanning systems to cover an area slightly to the west of each previous scan. The resultant imagery is thus skewed across the image. This is known as **skew distortion** and is common in imagery obtained from satellite multispectral scanners.

The sources of geometric distortion and positional error vary with each specific situation, but are inherent in remote sensing imagery. In most instances, we may be able to remove, or at least reduce these errors but they *must* be taken into account in each instance before attempting to make measurements or extract further information.

Land Observation Satellites/Sensors

Landsat

Although many of the weather satellite systems (such as those described in the previous section) are also used for monitoring the Earth's surface, they are not optimized for detailed mapping of the land surface. Driven by the exciting views from, and great success of the early meteorological satellites in the 1960s, as well as from images taken during manned spacecraft missions, the first satellite designed specifically to monitor the Earth's surface, Landsat-1, was launched by NASA in 1972. Initially referred to as ERTS-1, (Earth Resources Technology Satellite), Landsat was designed as an experiment to test the feasibility of collecting multi-spectral Earth observation data from an unmanned satellite platform. Since that time, this highly successful program has collected an abundance of data from around the world from several Landsat satellites.

TM Bands

Channel	*Wavelength Range (mm)*	*Application*
TM 1	0.45-0.52 (blue)	soil/vegetation discrimination; bathymetry/coastal mapping; cultural/urban feature identification
TM 2	0.52-0.60 (green)	green vegetation mapping (measures reflectance peak); cultural/urban feature identification
TM 3	0.63-0.69 (red)	vegetated *vs.* non-vegetated and plant species discrimination (plant chlorophyll absorption); cultural/urban feature identification
TM 4	0.76-0.90 (near IR)	identification of plant/vegetation types, health, and biomass content; water body delineation; soil moisture
TM 5	1.55-1.75 (short wave IR)	sensitive to moisture in soil and vegetation; discriminating snow and cloud-covered areas
TM 6	10.4-12.5 (thermal IR)	vegetation stress and soil moisture discrimination related to thermal radiation; thermal mapping (urban, water)
TM 7	2.08-2.35 (short wave IR)	discrimination of mineral and rock types; sensitive to vegetation moisture content

Routine collection of MSS data ceased in 1992, as the use of TM data, starting on Landsat 4, superseded the MSS. The **TM** sensor provides several improvements over the MSS sensor including: higher spatial and radiometric resolution; finer spectral bands; seven as opposed to four spectral bands; and an increase in the number of detectors per band (16 for the non-thermal channels versus six for MSS). Sixteen scan lines are captured simultaneously for each non-thermal spectral band (four for thermal band), using an oscillating mirror which scans during both the forward (west-to-east) and reverse (east-to-west) sweeps of the scanning mirror. This difference from the MSS increases the **dwell time** and improves the geometric and radiometric integrity of the data. Spatial resolution of TM is 30 m for all but the thermal infrared band which is 120 m. All channels are recorded over a range of 256 digital numbers (8 bits). The accompanying table outlines the spectral resolution of the individual TM bands and some useful applications of each.

Data from both the TM and MSS sensors are used for a wide variety of applications, including resource management, mapping, environmental monitoring, and change detection (e.g. *monitoring forest encroachments and tree felling)*.

Indian Remote Sensing (IRS) Satellites

The Indian Remote Sensing (IRS) satellite series, combines features from both the Landsat MSS/TM sensors and the SPOT HRV sensor. The third satellite in the series, IRS-1C, launched in December, 1995 has three sensors: a single-channel panchromatic (PAN) high resolution camera, a medium resolution four-channel Linear Imaging Self-scanning Sensor (LISS-III), and a coarse resolution two-channel Wide Field Sensor (WiFS). The accompanying table outlines the specific characteristics of each sensor.

IRS Sensors

Sensor	*Wavelength Range (mm)*	*Spatial Resolution*	*Swath Width*	*Revisit Period (at equator)*
PAN	0.5-0.75	5.8 m	70 km	24 days
LISS-II				
Green	0.52-0.59	23 m	142 km	24 days
Red	0.62-0.68	23 m	142 km	24 days
Near IR	0.77-0.86	23 m	142 km	24 days
Shortwave IR	1.55-1.70	70 m	148 km	24 days
WiFS				
Red	0.62-0.68	188 m	774 km	5 days
Near IR	0.77-0.86	188 m	774 km	5 days

In addition to its high spatial resolution, the panchromatic sensor can be steered up to 26° across-track, enabling stereoscopic imaging and increased revisit capabilities (as few as five days), similar to SPOT. This high resolution data is useful for urban planning and mapping applications. The four LISS-III multispectral bands are similar to Landsat's TM bands 1 to 4 and are excellent for vegetation discrimination, land-cover mapping, and natural resource planning. The WiFS sensor is similar to NOAA AVHRR bands and the spatial resolution and coverage is useful for regional scale vegetation monitoring. While the Government-sponsored earth observation systems such as LANDSAT, SPOT, IRS, JERS/MOS have been extensively used for the survey of renewable and non-renewable resources, commercial earth observation satellites in the private sector providing data of very high spatial resolutions in the range of 1-2 m in the panchromatic and 3-5 m in multi-spectral mode are revolutionising the scenario

(Fritz, 1996). These systems are distinguished by their flexible pointing ability, high geometric fidelity and very rapid image-collection to customer delivery. Major specifications of some of the commercial satellites are given in the following Table.

Broad Specifications of Commercial High Resolution Satellites

Corporation system	*Earth Watch*				*Orbital Sciences*		*Space Imaging*	
	Early Bird		*Quick Bird*		*OrbView-1*		*EOSAT*	
Mode	*Pan*	*MS*	*Pan*	*MS*	*Pan*	*MS*	*Pan*	*MS*
Resolution (nadir GSD)	3m	15m	1m, 2m	4m	1m, 2m	8m	1m	4m
Spectral Bandwidths (μm)	.45-.80	.50-.59 .61-.68 .79-.89	.45-.90	.45-.52 .53-.59 .63-.69 .77-.90	.45-.90	.45-.52 .52-.60 .63-.69 .76-.90	.45-.90	.45-.52 .52-.60 .63-.69 .76-.90
Swath width (km)	6 × 6	30 × 30	36	36	8	8	11	11
Scene Size (at nadir)	36 km^2	900 km^2	36 × 36 km	36 × 36 km	8 × 8 km	8 × 8 km	60 × 60 km	60 × 60 km
Accuracy	Horiz	Vert	Horiz	Vert	Horiz	Vert	Horiz	Vert
With GCPs	6m	4m	2m	3m	2m	3m	2m	3m
Without GCPs	...	...	...	...	12m	8m	12m	8m
Stereo	In track		In track		In track		In and cross track	

Source: L. Fritz, PE & RS, January 1996, p. 43.

Data Reception, Transmission and Processing

Data obtained during airborne remote sensing missions can be retrieved once the aircraft lands. It can then be processed and delivered to the end user. However, data acquired from satellite platforms need to be electronically transmitted to Earth, since the satellite continues to stay in orbit during its operational lifetime. The technologies designed to accomplish this can also be used by an aerial platform if the data are urgently needed on the surface.

There are three main options for **transmitting data** acquired by satellites to the surface. The data can be directly transmitted to Earth if a Ground Receiving Station (GRS) is in the line of sight of the satellite (A). If this is not the case, the data can be recorded on board the satellite (B) for transmission to a GRS at a later time. Data can also be relayed to the GRS through the Tracking and Data Relay Satellite System (TDRSS) (C), which consists of a series of communications satellites in geosynchronous orbit. The data are transmitted from one satellite to another until they reach the appropriate GRS.

In India, National Remote Sensing Agency (NRSA) operates one ground receiving station-located at Shadnagar, just outside of Hyderabad city. The coverage circle of GRS enable the potential for reception of real-time data from satellites passing over almost any part of India's land mass.

The data are received at the GRS in a raw digital format. They may then, if required, be processed to correct systematic, geometric and atmospheric distortions to the imagery, and be translated into a standardized format. The data are written to some form of storage medium such as tape, disk or CD. For many sensors it is possible to provide customers with **quick-turnaround** imagery when they need data as quickly as possible after it is collected. Near real-time processing systems are used to produce low resolution imagery in hard copy or soft copy (digital) format within hours of data acquisition. Such imagery can then be faxed or transmitted digitally to end users. Low resolution **quick-look** imagery is used to preview archived imagery prior to purchase. The spatial and radiometric quality of these types of data products is degraded, but they are useful for ensuring that the overall quality, coverage and cloud cover of the data is appropriate.

APPLICATIONS OF REMOTE SENSING

Over the past two and a half decades, DOS has conducted various end-to-end application projects with various user agencies. Many of these have been taken up as national level projects encompassing large areas and various agencies. In the eighties and early nineties RS data was extensively used to prepare thematic maps on various themes/resources such as ground water, geology, flood inundation areas, forest cover, land use/cover, wastelands, snow clad areas, watersheds, coastal, and inland wetlands, etc. Crop production forecasting using digital data is another major area of application. Subsequent to these efforts, the thrust is now towards integrating information related to various resources conjunctively with socio-economic, demographic and infrastructure data of the region to arrive at sustainable development plans at the local level on watershed basis. Salient outcome of some of the major projects are given below.

As part of the Scientific Source Finding for the Technology Mission on Drinking Water, hydrogeomorphological maps on 1:250,000 scale for the entire country (447 districts) have been prepared showing demarcation of potential zones/locales where there is high probability of finding ground water. Prospective sites for tube wells within 1.6 km radius of a village have also been identified on 1:50,000 scale maps for many of the problem villages. Feedback received from various sources indicates that scientific source finding has resulted in success rates of more than 90 per cent in drilling tube wells in hardrock areas.

Over 2800 wasteland maps, showing 13 fold wasteland categories, for 146 critically affected districts of the country (districts which have more than 15 per cent area under wastelands) have been prepared on 1:50,000 scale at the request of the National Wasteland Development Board (NWDB). Village boundaries have been superposed on these maps to facilitate taking up development/restoration of these wastelands. About hundred additional districts have been mapped as part of Phase-II. Land use/cover maps at 1:250,000 scale for the entire country are prepared to help in agroclimatic zoning of the country. More than sixteen hundred maps showing inland wetlands have been prepared for the entire country.

The project Vasundhara, covering Indian peninsula south of 17 degree latitude, aimed at delineating potential target areas for mineral search has led to the location of a lead-zinc occurrence in parts of Andhra Pradesh. A fairly comprehensive geographic information system package has also been developed for mineral information, facilitating storage, analysis and retrieval of thematic information. The National Agricultural Drought Assessment and Monitoring System (N-ADAMS) Project provides fortnightly bulletins indicating the prevalence, severity level and persistence of drought conditions taking district as a unit (Thiruvengatachari, 1988), using NOAA-AVHRR data with meteorological and other collateral information.

Acreage estimation procedures involving various sampling and digital image processing techniques have been developed for major crops like wheat, rice, sorghum, groundnut, cotton and mustard and the project is being implemented in fifteen states of the country. Yield-spectral index relationships derived using multi-year data are being used to make trial yield predictions. Realising the need to have more comprehensive, multiple forecasts for major crops of the country at national, state and districts levels, a conceptual framework for Forecasting Agricultural Output using Space, Agrometeorology and Land based observations (FASAL) has been formulated. Setting up a National Centre for Crop Forecasting (NCCF) as part of this has been established. Microwave data available from ERS and RADARSAT are used for rice estimation (Sahai and Dadhwal, 1989; Navalgund *et al.*, 1991. NNRMS Bulletin, 1997).

Forest cover map at 1:250,000 scale for the entire country is being prepared by Forest Survey of India once in every two years. Besides this, procedures have been developed and demonstrated for mapping forest types, density, and monitoring afforestation/deforestation. Using multi-year satellite data, it has been possible to identify areas of illegal encroachment in the forest area. Use of satellite data in biodiversity studies is being investigated.

Snow cover mapping in the Sutlej basin is being regularly done. Forecasts of the seasonal inflow of snow melt run-off into Bhakra reservoir are regularly being made available (Ramamoorthy, 1983). Glacier inventory in the Beas and the Sutlej basins has been done using two-season data. Flooded-area mapping in near real-time has been done for flooded areas of the Brahmputra and Ganga basins.

The Integrated Mission for Sustainable Development (IMSD) aimed at preparing locale-specific action plans has been taken up in 174 districts of the country. Action plans for development of land and water resources have been generated on priority basis for 92 blocks in the country. In many of the watersheds, action plans are being implemented by the local government agencies and non-government agencies with active participation of the local people. Satellite data is also being used to monitor the changes in vegetation cover, cropping intensity and afforestation to see the impact of development measures.

Besides these, several application projects and R&D studies in various areas of natural resources and oceanography are being conducted by different central and state departments, research institutes in the country. For details reader may refer to special issues of Current Science, 1991 and 1996.

CONCLUDING REMARKS

Remote sensing has made rapid advances since its inception a few decades ago and has established itself as a powerful tool in providing information in many resource areas. Availability of data with very high ground and spectral resolutions is opening up new application areas hitherto unknown. Scope of many application areas will be enhanced with data from microwave sensors becoming operationally available. Integration of remote sensing (map) data, non-spatial (socio-economic) data and other collateral information using a Geographic Information System

(GIS) is providing an efficient and powerful tool for resource managers and decision makers, Development of user-friendly and expert systems are breaking the barrier between the technologists and common users.

REFERENCES

Colwell (Ed.),1983. Manual of RS two volumes. American Society of Photogrammetry Falls Church, Va.

Current Science (1996), Special Section: IRS-1C, Vol. 70 (No. 7).

Current Science (1991), Special Issue, 61 (3 & 4).

Earth Observing System (1984). NASA Technical Memorandum 86129, Vols. I and II.

http://CCRS Remote Senisng Tutorial.htm

Joseph, G., (1996), Imaging sensors for remote sensing, *Remote Sensing Reviews*, Vol. 13, pp. 257-342.

Joseph, G., *et al.*, (1996), Cameras for Indian remote sensing satellite IRS-1C, *Current Science*, Vol. 70.

Navalgund, R.R., and Kasturirangan, (1983). Indian remote sensing satellite—a programme over view, Proc. Indl. Acad. of Sci., Engg. Sciences, Vol. 6, pp. 313.336.

Navalgund, R.R., Parihar, J.S., Ajai and Nageswar Rao, P.P. (1991), Crop inventory using remotely sensed data, *Current Science*, speical issue, 61 (3 & 4), pp. 162-171.

NNRMS Bulletins (1996), 20, 21, published by NNRMS, DOS, Bangalore.

Ramamoorthy, A.S., (1983). Forecasting of snow melt run off (1982) of Sutlej river using NOAA satellite imageries, Proc. NNRMS Seminar, pp 11-25.1 to 11-25.8.

Slater, P.N. (1980). Remote sensing: optics and optical systems, Pub. Addison, Wesley, London.

Swain, P.H., and Davis, S.M. (1978). *Remote sensing: the quantitative approach* (McGraw Hill Inc. 1978).

4

DATA PREPARATION FOR SPATIAL ANALYSIS

SUBHAN K. PATHAN

INTRODUCTION

The analysis of remote sensing imagery involves the identification of various targets in an image and those targets may be environmental or artificial features which consist of points, lines or areas. Targets may be defined in terms of the way they reflect or emit radiation. This radiation is measured and recorded by a sensor, and ultimately is depicted as an image product such as an air photo or a satellite image.

Visual interpretation and digital image processing techniques are two important techniques of data analysis to extract resource related information either independently or in combination with other data. Visual interpretation methods have been the traditional methods for extracting information, based on the target characteristics as seen on the imagery. A human interpreter uses not only tone (spectral pattern) but also the spatial pattern at different scales which are not easily detected by digital methods.

IMAGE INTERPRETATION

Photography is defined as the art or process of producing images on a sensitized surface by the action of light or other radiant energy. Similarly an image is the reproduction or imitation of the form of a person or thing. The optical counter-part of an object produced by a lens or mirror or other optical system. Therefore, image is a pictorial representation of the energy reflected and/or emitted from the scene in different parts of the electromagnetic spectrum.

Image interpretation is defined as 'the art of examining images for the purpose of identifying objects and judging their significance.' Interpreters study remotely sensed data and attempt through logical processes in detecting, identifying, classifying, measuring and evaluating the significance of physical and cultural object, their patterns and spatial relationships.

Aspects of Image Interpretation

Image interpretation is a complex process of physical, physiological activities occurring in a sequence begins with the detection and identification of images and later by their measurements.

Different aspects of image interpretation (Vink 1964) are listed below in a simpler form most of these have overlapping function.

Detection and Identification

The primary task of the interpreter is the detection and identification of objects, features, phenomena and processes. This is primarily a stimulus and response activity. The interpreter conveys his or her response by labelling. These labels are often expressed in qualitative terms, e.g. likely, possible, probable or certain. Therefore, it is a process of 'picking out' object element from the photo or image through interpretation techniques. It may be detection of point, line or polygons, viz. Forest land, agricultural fields, settlement patterns, road network, tube wells, dug wells etc.

Recognition

It is a process of classification or trying to distinguish an object by its characteristics or patterns which are familiar on the image. It preceded the process of detection. Sometimes it is also termed as photo reading, e.g. vegetation, water bodies, built-up land etc.

Analysis

It is a process of resolving or separating a set of objects or features having similar set of characteristics. In analysis 'lines of separation' are drawn between groups of objects and the degree of reliability of these lines can also be indicated, e.g. vegetation as that of built-up land, water bodies with that of waste lands etc.

Classification

It is a process of identification and grouping of objects or features resolved by analysis. It arranges 'features of recurrence' in the same class or group to which the feature belongs. Any wrong identification and analysis may often lead to misclassification.

Deduction

Deduction may be directed to the separation of different groups of objects or elements and deducing their significance based on covering evidence. The evidence is derived from mainly visible objects or from invisible elements, which give only partial information on the nature of certain correlative indications. Deduction as regard to the identify of objects made without proper pre-interpretation checks in the field, may often be misleading and result in wrong classification. For complicated interpretation, therefore, it is advisable to effect the separation under this process and leave the deduction of the identity till after the clarification.

Idealization

It is a process of drawing ideal or standard representation from what is actually identified and interpreted from the image or map with standard symbols and colours.

Elements of Visual Interpretation

Recognizing targets is the key to interpretation and information extraction. Observing the differences between targets and their backgrounds involves comparing different targets based on any, or all, of the visual elements of **tone, shape, size, pattern, texture, shadow, location, association and height**. Some of these elements are interconnected and are classified into three orders, viz. Basic or First order elements (Tone), Second order (Texture, shape, size, pattern) and Third order (Location, association, shadow and height).

Visual interpretation using these elements is often a part of our daily lives, whether we are conscious of it or not. Examining satellite images on the weather report, or following high speed chases by views from a helicopter are all familiar examples of visual image interpretation. Identifying targets in remotely sensed images based on these visual elements allows us to further interpret and analyze. The nature of each of these interpretation elements is described below, along with an image example of each.

Basic, First Order Elements of Image Interpretation

Tone/Colour

Tone can be defined as each distinguishable variation from white to black. Colour may be defined as each distinguishable variation on an image produced by a multitude of combinations of hue, value and chroma. Therefore, tone refers to the relative brightness or colour of objects on an image. Hence, tone is the fundamental element for distinguishing between targets or features. Variations in tone also allow the elements of shape, texture and pattern of objects to be distinguished.

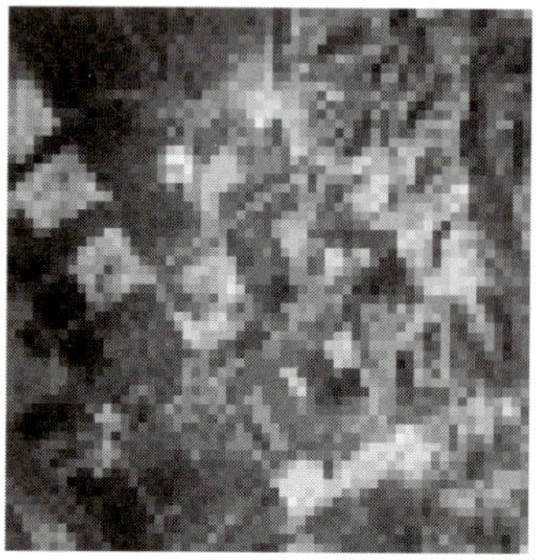

Tonal Variations

Colour Variations

Many factors influence the tone or color of objects or features recorded on photographic emulsions. But, if there is not sufficient contrast between an object and its background to permit at least detection, there can be no identification. While a human interpreter may only be able to distinguish between ten and twenty shades of grey; interpreters can distinguish many more colours. Some authors state that interpreters can distinguish at least 100 times more variations of colour on colour photography than shades of gray on black and white photography.

Second order—Geometric Arrangements of Objects

Size

The size of objects can be important in discrimination of objects and features (cars *vs.* trucks or buses; single family *vs.* multi-family residences, scrubs vs. trees, etc.). In the use of size as a diagnostic characteristic both the relative and absolute sizes of objects can be important. Size can also be used in judging the significance of objects and features (size of trees related to board feet which may be cut; size of agricultural fields related to water use in arid areas, or amount of fertilizers used; size of runways gives an indication of the types of aircraft that can be accommodated). The size of objects in an image is a function of scale. It is important to assess the size of a target relative to other objects in a scene, as well as the absolute size, to aid in the interpretation of that target.

Shape

The shape of objects/features can provide diagnostic clues that aid identification. Therefore, the shape of the object refers to the general form, structure, or outline of individual objects. Shape can be a very distinctive clue for interpretation. Straight edge shapes typically represent urban or agricultural (field) targets, while natural features, such as forest edges, are generally more irregular in shape, except where man has created a road or clear cuts. Farm or crop land irrigated by rotating sprinkler systems would appear as circular shapes. Similarly roads can have right angle turns, railroads do not.

Road Network

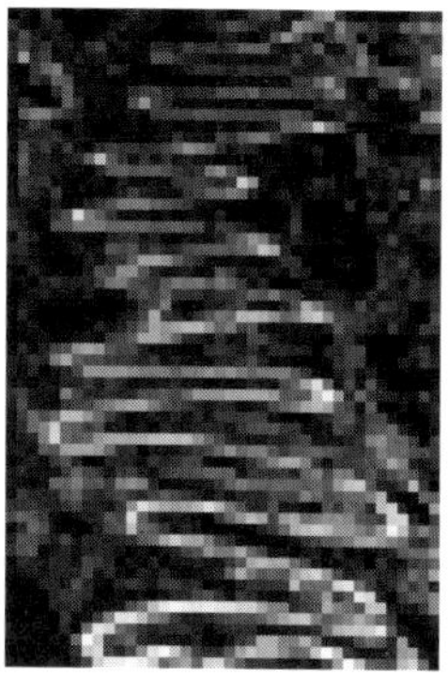

River

Texture

It refers to the arrangement and frequency of tonal variation in particular areas of an image. Rough textures would consist of a mottled tone where the grey levels change abruptly in a small area, whereas smooth textures would have very little tonal variation. Smooth textures are most often the result of uniform, even surfaces, such as fields, asphalt, or grasslands. A target with a rough surface and irregular structure, such as a forest canopy, results in a rough textured appearance.

Textural Variations

The visual impression of smoothness or roughness of an area can often be a valuable clue in image interpretation. Water bodies are typically fine textured, while grass is medium, and brush is rough, although there are always exceptions.

Pattern

Pattern is the spatial arrangement of objects. Pattern can be either man-made or natural. Pattern is a macro image characteristic. It is the regular arrangement of objects that can be diagnostic of features on the landscape. An orchard has a particular pattern. Likewise, the network or grid of streets in a subdivision or urban area can aid identification and aid in problem solving such as the growth patterns of a city. Pattern can also be very important in geologic or geomorphologic analysis. Drainage pattern can tell the trained observer a great deal about the lithology and structural patterns in an area. Dendritic drainage patterns develop on flat bedded sediments; radial on/over domes; linear or trellis in areas with faults or other structural controls.

Therefore, the Pattern and Texture are thought of being scale dependent.

Third Order—Locational or Positional Elements

Shadow

It is useful in interpretation as it may provide an idea of the profile and relative height of a target or targets which may make identification easier. However, shadows can also reduce or eliminate interpretation in their area of influence, since targets within shadows are much less (or not at all) discernible from their surroundings. Shadow is also useful for enhancing or identifying topography and landforms, particularly in radar imagery.

Location

How objects are arranged with respect to one another; or with respect to various terrain features, can be an aid in interpretation. Aspect, topography, geology, soil, vegetation and cultural features on the landscape are distinctive factors that the interpreter should use when examining a site. The relative importance of each of these factors will vary with local conditions, but all are nportant. Just as some vegetation grows in swamps others grow on sandy ridges. Agricultural rops may like certain conditions. Man-made features may also be found on rivers (e.g. power lant) or on a hill top (observatory or radar facility).

Association

t takes into account the relationship between other recognizable objects or features in proximity o the target of interest. The identification of features that one would expect to associate with ther features may provide information to facilitate identification. Some objects are so commonly ssociated with one another that identification of one tends to indicate or confirm the existence f another. Smoke stacks, step buildings, cooling ponds, transformer yards, coal piles, railroad acks = coal fired power plant. Arid terrain, basin bottom location, highly reflective surface, parse vegetation = playa. Association is one of the most helpful clues in identifying man made nstallations. Aluminium manufacture requires large amounts of electrical energy. Absence of power supply may rule out this industry. Cement plants have rotary kilns. Schools at different evels typically have characteristic playing fields, parking lots, and clusters of buildings in rban areas. Large farm silos typically indicate the presence of livestock.

Height

Height can add significant information in many types of interpretation tasks; particularly, those hat deal with the analysis of man-made features. How tall a tree is can tell something about board feet. How deep an excavation is can tell something about the amount of material that was removed (in some mining operations excavators are paid on the basis of material removed s determined by photogrammetric analysis).

IMAGE CHARACTERISTICS FOR IDENTIFICATION OF LAND USE/LAND COVER USING SATELLITE IMAGERY

The information on land use/land cover pattern is essential as it is the prime requisite in most of the planning and development exercises. Land use refers to "man's activities on land, which are directly related to the land" (Clawson 1965). Land cover denotes "the vegetational and artificial constructions covering the land surface" (Burley 1961). The terms land use and land cover are closely related and are interchangeable. The remote sensing data record, information essentially on land cover from which the information on land use has to be deduced or inferred. For example, if remotely sensed data identifies the land cover as 'vegetation', it leads to the information that the land is put either for agricultural produce, timer production or recreational use etc. Therefore, it is very important to correctly identify the land cover and deduce the information related to its use. The final map prepared using remote sensing data should satisfy the following criteria:

- A minimum level of accuracy (about 85 to 90 per cent or better) should be tried in the interpretation of the data. This level of accuracy is comparable with the level of accuracy attained by various planning departments using conventional surveys.

- A fair level of reliability of interpretation for the several categories included in the classification system should be attained.
- Reproducible results should be obtained by different interpretations as it is necessary that many persons are involved in the interpretative process in a given project. Hence It is very important to have a classification system that can yield comparable results each time the monitoring exercise is repeated for a given area.
- While adopting the classification system, a care should be taken that it is suitable for use with imagery taken at different times during the year. Because the most effective interpretation depends upon the use of all three seasons data.

Thus, by understanding the image characteristics of each of the thematic classes, an image interpretation key is evolved. The visual interpretation key enables an interpreter having the minimum of knowledge and interpretation skill to identify different categories of land use land cover categories on the satellite imagery. In other words, it simplifies and facilitate interpretation of the imagery in a systematic way. However, it may be mentioned here that visual interpretation is subjective and differs from individual to individual and also upon the season, scale, spectral bands and spatial resolution of the imagery. The interpretation key is to be prepared and later finalised after field checks.

A typical interpretation key developed for the identification of various land use/land cover categories is given in Table 4.1 and a comprehensive land use/land cover classification system developed by National Remote Sensing Agency (NRSA) for use with different spatial and spectral resolutions of remote sensing data is presented in Appendix I. Similarly the vegetation type and crown density classification system developed by NRSA for use with satellite is also presented in Appendix II. The definitions of different land use/cover categories are presented in Appendix III.

STEPS TO PREPARE THEMATIC MAPS USING VISUAL INTERPRETATION

Step 1: Acquisition of Data

(i) Standard false colour composite made of green, red and infrared bands of RS data of two seasons Kharif (June-September) and Rabi (October-March).

(ii) Ancillary data – Topographical maps (As per the requirement, i.e. 1:250,000 or 1:50,000 scale).

Step 2: Preliminary Interpretation

(i) Check the area to be interpreted is covered on RS data.

(ii) Spread the satellite imagery on a light table and fix it firmly.

(iii) Orient the imagery with respect to SOI map corresponding to the scene, with reference to latitudes and longitudes marked on the imagery and physical and cultural features.

(iv) Cut to the size of the imagery the artian tracing sheet or polyester tracing sheet (rough surface facing top) or acetate base film and fix it on the scene with transparent tape.

(v) For interpretation use ordinary pencil (HB or H).

Table 4.1: Interpretation Key for Use with RS Data (False Colour Composite)

Sr. No.	Land use/land cover	Image characteristics						
		Tone	Texture	Size	Shape	Shadow	Pattern	Association
I.	BUILT-UP LAND Residential	Dark cyan	Coarse	Variable	Variable	Shadows seen	Regular roads/streets	Mixed with different rise buildings
	(a) High rise	White	Coarse	-do-	-do-	Small Shadows	Regular, streets	A group of uniform houses
	(b) Low rise	White	Medium	-do-	-do-	–	-do-	-do-
	(c) Detached	White	Coarse	-do-	-do-	–	-do-	-do-
	(d) Slums	Light cyan	Smooth	-do-	-do-	–	Irregular	Surrounded by a marsh &
	(e) Mixed	Mottled gray	Coarse	-do-	-do-	–	Unorganised	Along roads, & railways
	Residential area without vegetation	White and cyan mixed with pinkish specks	Coase	Variable	Variable	–	–	–
	Industrial	White	Medium	Variable	Rectangle	–	Regular	Interspersed with vegetation and open lands
	Recreational							
	(a) Gardens/parks	Red	Medium	Variable	Variable	–	Dispersed	–
	(b) Play grounds	Pinkish white	-do-	-do-	-do-	–	-do-	–
	(c) Stadiums	Reddish	-do-	-do-	Definite	–	–	–
	Transportational							
	(a) Roads	Dark grey	Medium	Variable	Straight with bends	–	Linear	Interspersed by pink tone
	(b) Railways	Light grey	-do-	Fixed	Straight	–	-do-	–
	(c) Runways	Light blue	Smooth	Variable	Straight streeps	–	-do-	Interspersed by pink tone
II.	AGRICULTURE							
	(a) Cropped land	Pinkish red	Fine	Variable	Variable	–	Scattered	–
	(b) Fallow land	Whitist blue	Coarse	-do-	-do-	–	-do-	–
	(c) Plantations	Dark pink	Smooth	-do-	-do-	–	-do-	–
III.	FOREST LAND	Dark red	Medium	Variable	Variable	–	Scattered	–
IV.	WASTELAND							
	Scrub land	Light pink to grey	Medium to coarse	Variable	Variable	–	Scattered	–
	Quarries	Whitish cyan	coarse	Variable	Variable	–	Scattered	–
V.	WATER BODIES							
	(a) Clear water	Dark blue	Smooth	Variable	Variable	–	Scattered	–
	(b) Turbid water	Light blue	Medium	-do-	-do-	–	-do-	–

(vi) At the bottom corner, note latitude, longitude from scene, date and year of scene.
(vii) Scene path-row number, name of district, state etc.

Step 3: Final Interpretation

(i) Identify and recognise the different land use/land cover, on the imagery based on image characteristics, for correct classification of land use, refer classification system.
(ii) Considering the general interpretation key for land-use/land cover, develop a separate interpretation key for the specific area under study.
(iii) After identification of a category, delineate and map it on transparent overlay, for better clarity and discernibility, use magnifying lens.
(iv) To minimize doubtful area, use other satellite images, information from topographical maps, forest maps or any other maps available.
(v) Use numerical or alphabetical notation or colours to each of the land use/land cover classes delineated.
(vi) Land use/land cover categories having similar spectral signature, appearing side by side require detail ground verification.
(vii) Minimum mapping unit is 3 mm × 3 mm, all details less than the minimum mapping unit are excluded from mapping, for better visual separation and cartographic representation.

Step 4: Transfer of Details on to the Base Map

(i) Transfer the interpreted details on to a base map prepared from topographic map. Transfer notified forest boundaries on to the base map.
(ii) Finalise the map after incorporating post field data.

Step 5: Area Calculation

(i) Calculate area of different land use/land cover categories either using a millimetre Polythene graph sheet and/or a planimeter for calculating areas.
(ii) For better consistency and accuracy of area calculation, repeat the operation at least thrice and select the average of the two nearest values.
(iii) Compute the area both in square km and in hectares, also calculate percentage.

Step 6: Generation of Cartographic Quality Map

(i) Retrace the map on a new transparent sheet using appropriate rotting pens.
(ii) Prepare the fair drawing original, with proper legend, scale, north arrow etc.

INTERPRETATION AIDS

On the basis of tasks to be carried out, visual interpretation aids can be classified into three categories, viz. Aids for viewing including enhancement transfer of detail and measurement.

Interpretation aids can also be classified as equipment for monoscopic viewing and stereo viewing. It should, however, be pointed out that the categorisation is not rigid as certain equipment can be classified into more than one category particularly some of the projection devices which can be used for both viewing and transfer of details. Some of the equipments have become obsolete due to availability of high quality and cost effective FCC data products. In context of land use/cover the aids used only for transferring of details have been discussed here. Basically the equipment in this category are meant for delineating the details from image to a map on the same scale or different scale. These devices employ the means for simultaneous viewing of both the image and the map. There are following three types of equipments which are commonly used for mapping, i.e. HME, LFOE, PROCOM and light tables. First two types come under the optical projection types instruments.

High Magnification Enlarger (HME)

HME is developed at SAC is a versatile aid for visual interpretation of remotely sensed data in the form of transparencies, i.e. B/W or FCC dia-positives. Enlargements upto 20 times are possible in this instrument thus enabling a 1:1million dia-positives such as LANDSAT TM transparency to be enlarged up to the scale of 1:50,000 corresponding to SOI topographic sheets. Even high magnifications have also been attempted with this equipment comparison. Projection of the image is carried out on a table attached to the equipment on which a map can be fixed for registration with image and also a tracing film for the preparation of a thematic map.

Large Format Optical Enlarger (LFOE)

Large format optical enlarger has also been developed at SAC. It is used for two and four times enlargement of 240mm dia-positives.1:1M scale images such as LANDSAT TM and MSS data can be enlarged up to the scale of 1:250,000 scale corresponding to SOI topographical scale at this scale. IRS LISS II images at 1:500,000 scale can be enlarged to a scale of 1:250,000 using a LFOE having capability of two times enlargements. It can also project multi-spectral images in 70 mm format for easy comparison.

Light Tables

In addition to the instruments used for optical projection an interpretation aid used for delineation of features from hardcopy paper prints are light tables. Interpretation corresponding to the scale of paper print can be carried out. Handling of images for interpretation is relatively easier using light tables.

APPENDIX I

REMOTE SENSING BASED COMPREHENSIVE LAND USE/LAND COVER CLASSIFICATION SYSTEM FOR USE WITH SATELLITE DATA

Level *Detail* *Scale*	*Level-I* *Major class* *1:1 million*	*Level-II* *Type* *1:250,000*	*Level-III* *Sub-class/status* *1:50,000*
1.	• [@1]Built Up	Built Up (Urban)[+1]	• [@1]Residential with vegetation
			• Residential without Vegetation
			• Industrial
			• Recreational (Parks/Gardens)
			• [@2]Mixed Built Up Area
			• Open/Vacant Land
		Built Up (Rural)[+2]	• Built Up Area
		Transportation	• Transportation
2.	• Agricultural Land	Crop Land[+3]	• [@3] Kharif Crop
			• [@4] Rabi Crop
			• Kharif + Rabi [@5]
			• (Double Cropped)
			• [@6] Zaid Crop
		• Fallow	• [@7] Current Fallow
			• [@8] Other Fallow
			Agricultural I } [@9]
		• Plantation/Orchards	• Horticulture I } [@9]
			• Forest I } [@9]
			• Agri Horticulture I } [@9]
3.	• [@2]Forest	• Evergreen/Semi Evergreen	• [@10] Dense/Closed
			• [@11] Open
		• Deciduous (Dry/Moist)	• [@12] Degraded
			• Dense/Closed
		• Scrub/Shrub Forest	• Open
		• Forest Blank	• Degraded
		• Grassland/Grazing Land	• Dense/Closed
			• Open
		• Forest Plantations	• Forest Blank
		• Mangrove	• Grassland/Grazing Land
		• [+5]Non-Forest activities	• Forest Plantations [@13]
			• Dense
			• Open
			• Agriculture
			• Mining
			• Reservoir/Tank
			• Built Up

(*Contd.*)

Level *Detail* *Scale*	*Level-I* *Major class* *1:1 million*	*Level-II* *Type* *1:250,000*	*Level-III* *Sub-class/status* *1:50,000*
4.	• [@3]Natural/ Semi-Natural Grassland & Grazing land	• [+6]Alpine/Sub-Alpine • [+7]Temperate/Sub Tropical • [+8]Tropical [+9]Manmade Grasslands	• Moist/Dry Alpine pastures • Sub-Alpine pastures • Grasslands with tree cover • Man-made Grasslands
5.	• Wastelands	• Salt-affected land • Gullied/Ravinous land • Scrub Land • Sandy area • Mine dumps/Industrial Waste or dumps • Barren rocky/Stony waste/Sheet rock	• Strong • Moderate • Slight • [@14] Shallow • [@15] Medium • [@16] Deep • Land with scrub • Land without scrub • Desertic • Coastal • Riverine • Mine dumps/Industrial Waste or dumps • Barren rocky/Stony waste/Sheet rock
6.	• Wetlands	• Inland Wetland • Coastal Wetland	Natural • Ox-bow lakes/ Cut-off meanders • [@17] Playas • Marshy/Swampy • Waterlogged (seasonal) Man-made • [@18]Pisciculture ponds Natural • Estuary • [@19] Lagoon • Creek • Backwaters (Kayals) • Tidal flat/Mud flat • Marshy/Swampy • [@20] Coral Reefs Man-made • [@21]Aquaculture ponds • Salt-pans/Salt Dumps
7.	• Water Bodies	• [+7]River/ Stream • Canal/Drain • Lakes/Ponds • Reservoir/Tanks	• [@22] River/stream • Lined/Unlined Canal and Drain • Perennial/Dry • Reservoir/Tanks • Ash pond/Cooling pond or Reservoir

(*Contd.*)

Level	*Level-I*	*Level-II*	*Level-III*
Detail	*Major class*	*Type*	*Sub-class/status*
Scale	*1:1 million*	*1:250,000*	*1:50,000*
	•	• +8Ashpond/Cooling pond or reservoir • Abandoned quarry pits with water	• Abandoned quarry pits with or without water
8.	• Others	• Shifting cultivation • +9Snow covered / Glacial area • Mining • Quarry • Brick kilns	• @23Current • Abandoned • Snow covered/Glacial area • Active • Abandoned • Rock/Mud quarry • Active • Abandoned

Notes:

- This is a general-purpose comprehensive classification scheme whose land use classes are comparable and compatible with the Nine-fold Revenue Land use classification.
- This classification is amenable for use with remotely sensed IRS satellite data at various scales.
- Level-I: IRS (LISS-I); Level-II: IRS (LISS-II); Level-III: IRS (LISS-III) for general land use; IRS (LISS-III+PAN) for Urban Land use.
- Level-I on need basis may be mapped using IRS (LISS-I)/IRS (WiFS)/data on compatible scale on 1:1 Million and small.
- Specific land use/cover classes can be grouped to suite the region/local area requirements like for Desert region, Coastal region and hilly region, considering the terrain, climatic and cultural practices.
- Inputs from ACZP National land use/land cover classification, Wasteland classification and Nation-wide Grassland/Grazing land classification also referred.

Level-I	*Level-II*	*Level-III*
• [@1]Non-Agricul-ture Use • [@2]Reserve/Protected/Classified within Notified areas • [@3] Outside forest on saline areas, river plains/islands, hill slopes; open and with tree cover, natural or man-made	• [+1] Towns/Cities • [+2] Village/Hamlets • [+3] Include Mono/Double and Mixed/Multiple crops area • [+4] Miscellaneous trees (excluding sacred groves & Wood Lands] • [+5] Non-related forest activities within forest area • [+6] High altitude and above tree line zone. Sub-alpine is below tree line • [@7] Those which occur in temperate/sub-tropical zone • [@8] Those which occur in tropical zone • [+9] Those raised and managed scientifically • [+10] Include Perennial/Seasonal or Dry • [+11] Associated with thermal plants • [+12] Permanent/Seasonal	• [@1] Include slums • [@2] Mixed with dominant residential/commercial • [@3] Rainfed or Non-irrigated • [@4] Winter season • [@5] More than once/Irrigated • [@6] Summer season • [@7] Seasonal (less than one year) • [@8] Long (1 to 5 years) • [@9] Indicate type of plantation wherever possible • [@10] Canopy 40% and above • [@11] Canopy 20-40% • [@12] Less than 20% canopy cover • [@13] Indicate type of plantation wherever possible • [@14] Qualitative depth • [@15] Qualitative depth • [@16] Qualitative depth • [@17] Salt lake • [@18] Fish ponds • [@19] Coastal lake • [@20] Accumulated dead polyp material • [@21] Shrimp Ponds • [@22] Stream network of different orders • [@23]Freshly burnt and under use/Abandoned (not in use for more than one year)

APPENDIX II

VEGETATION TYPE AND CANOPY DENSITY CLASSIFICATION SCHEME

Sl. No.	*Level-I*	*Level-II*	*Level-III*	*Canopy density*
1.	Forest	Evergreen	Tropical Wet Evergreen Forest Tropical Dry Evergreen Forest Sub-tropical Broad-leaved Hill Forest Sub-tropical Pine Forest Sub-tropical Dry Evergreen Forest Temperate Broad-leaved forest Temperate Coniferous Forest Sub-alpine Forest	10 – 40%, 40 – 70%, >70%[1]
		Semi-Evergreen	Tropical Semi-evergreen Forest	
		Moist Deciduous	Tropical Moist Mixed Deciduous Forest Gregarious formations	
		Dry Deciduous	Tropical Dry Mixed Deciduous Forest Gregarious formations	
		Tropical Thorn Forest	Local Specific formations	
		Scrub	Moist Alpine Scrub Dry Alpine Scrub Edaphic/Degraded/Seral types	
		Littoral/Swamp Forest	Tidal/Fresh Water Swamp	10-40%, >40%
		Forest Plantation	Teak/Bamboo Sal/Pine/Acacia/Eucalyptus etc	
2.	Natural/Semi-Natural Grassland[2]	Alpine/Sub alpine	Moist Alpine pasture Dry Alpine pasture	
		Temperate/Sub-Tropical		
		Tropical/Desertic	Bhabars, Sub-Tropical Wet Grasslands (shoals), Reeds, Bamboo mixed forest, Riverine/Swamps, Open Grasslands and Grasslands on Saline areas	
		Man Made Grassland		
3.	Agriculture[3]			
4.	Village Woodlots[3]			
5.	Orchards[3]			

1. Canopy Density > 70% need to be experimentally tested in prototype studies.
2. Density Levels in grasslands could be attempted in prototype studies on experimental basis.
3. Classes to be prepared under Land Use/Land Cover theme:
   - All forest land use classes would be covered under land use/land cover map.
   - Level I to Level II classes would be integrated with Land Use/Land Cover map.

APPENDIX III

DEFINITIONS OF LAND USE-LAND COVER CLASSES

Built-up Land

It is defined as an area of human habitation developed due to non-agricultural use and that which has a cover of buildings, transport, communication, utilities in association with water, vegetation and vacant lands.

Agricultural Land

It is defined as the land primarily used for farming and for production of food, fibre, other commercial and horticultural crops. It includes land under crops (irrigated and unirrigated), fallow, plantations etc.

Crop Land

It includes those lands with standing crop (*per se*) as on the date of the satellite imagery. The crops may be of either Kharif or Rabi or Kharif + Rabi seasons.

Fallow Land

It is described as agricultural land which is taken up for cultivation but is temporarily allowed to rest, un-cropped for one or more seasons, but not less than one year. These lands are particularly those which are seen devoid of crops at the time when the imagery is taken of both seasons.

Plantations

It is described as an area under agricultural tree crops, planted adopting certain agricultural management techniques. It includes tea, coffee, rubber, coconut, areca nut, citrus, orchards and other horticultural nurseries.

Forest

It is an area (within the notified forest boundary) bearing an association predominantly of trees and other vegetation types capable of producing timber and other forest produce.

Evergreen/Semi-evergreen Forest

It is described as a forest which comprises of thick and dense canopy of tall trees which predominantly remain green throughout the year. It includes both coniferous and tropical broad leaved evergreen trees. Semi-evergreen forest is a mixture of both deciduous and evergreen trees but the latter predominate.

Deciduous Forest

It is described as a forest which predominantly comprises of deciduous species and where the trees shed their leaves once in a year.

Degraded Forest or Scrub

It is described as a forest where the vegetative (crown) density is less than 20 per cent of the canopy cover. It is the result of both biotic and a-biotic influences. Scrub is a stunted tree or bush/shrub.

Forest Blank

These occur on forest hill tops in association with areas accessible for tree felling (for timber or cultivation) etc. it includes openings of assorted size and shapes as seen on the imagery.

Forest Plantations

It is described as an area of trees of species of forestry importance and raised on notified forest lands. It includes, eucalyptus, Casuarina, bamboo etc.

Mangrove

It is described as a dense thicket or woody aquatic vegetation or forest cover occurring in tidal waters, near estuaries and along the confluence of delta in coastal areas, it includes species of the genera Rhizophora and Avicennia.

Waste Lands

It is described as degraded land which can be brought under vegetative cover with reasonable effort, and which is currently under-utilised and land which is deteriorating due to lack of appropriate water and soil management or on account of natural causes. Wastelands can result from inherent/imposed constraints such as, by location, environment, chemical and physical properties of the soil or financial or management constraints. (NWDB, 1987).

Salt Affected Land

The salt affected land is generally characterised as the land that has adverse effects on the growth of most plants due to the action or presence of excess soluble or high exchangeable sodium. Alkaline land has exchangeable sodium percentage (ESP) of about 15, which is generally considered as the limit between normal and alkali soils. The predominant salts are carbonates and bicarbonates of sodium. Coastal saline soils may be with or without ingress or inundation by sea water.

Waterlogged Land

Waterlogged land is that land where the water is at/or near the surface and water stands for most of the year. Such lands usually occupy topographically low lying areas.

Marshy/Swampy Land

Marshy land is that which is permanently or periodically inundated by water and is characterised by vegetation which includes grasses and weeds.

Gullied/Ravinous Land

The gullies are formed as a result of localised surface runoff affecting the friable unconsolidated material in the formation of perceptible channels resulting in undulating terrain. The gullies are the first stage of excessive land dissection followed by their networking which leads to the development of ravinous land. The word 'ravine' is usually associated not with an isolated gully but a network of deep gullies formed generally in thick alluvium and entering a nearby river, flowing much lower than the surrounding high grounds. The ravines, are extensive systems of gullies developed along river courses.

Land With or Without Scrub

They occupy (relatively) higher topography like uplands or high grounds with or without scrub. These lands are generally prone to degradation or erosion. These exclude hilly and mountainous terrain.

Sandy Area (Coastal and Desertic)

These are the areas which have stabilised accumulations of sand *in situ* or transported in coastal, riverine or inland (desert) areas. These occur either in the form of sand dunes, beaches, channel (river/stream) islands etc.

Barren Rocky/Stony Waste/Sheet Rock

It is defined as the rock exposures of varying lithology often barren and devoid of soil cover and vegetation. They occur amidst hill forests as openings or scattered as isolated exposures or loose fragments of boulders or as sheet rocks on plateau and plains.

Water Bodies

It is an area of impounded water, area in extent and often with a regulated flow of water. It includes man-made reservoirs/lakes/tanks/canals, besides natural lakes, rivers/streams and creeks.

River/Stream

It is a natural course of flowing water on the land along definite channels. It includes from a small stream to a big river and branches. It may be perennial or non-perennial.

Reservoir/Lakes/Tanks/Canal

It is a natural or man-made enclosed water body with a regulated flow of water. Reservoirs are larger than tanks/lakes and are used for generating electricity, irrigation and for flood control. Tanks are smaller in area extent with limited use than the former. Canals are inland waterways used for irrigation and sometimes for navigation.

Others

It includes all those which can be treated as miscellaneous because of their nature of occurrence, physical appearance and other characteristics.

Shifting Cultivation

It is the result of cyclic land use practice of felling of trees and burning of forest areas for growing crops. Such lands are also known as Jhum lands. This results in extensive loss of soil leading to land degradation.

Grassland/Grazing Land

It is an area of land covered with natural grass along with other vegetation, often grown for fodder to feed cattle and other animals. Such lands are found in river beds, on uplands, hill slopes etc. Such lands can also be called as permanent pastures or meadows. Grazing lands are those where certain pockets of land are fenced for allowing cattle to graze.

Snow Covered/Glacial Area

It is defined as a solid form of water consisting of minute particles of ice. It includes permanently snow covered areas as on the Himalayas. Glacier is a mass of accumulated ice occluding amidst permanently snow covered areas.

5

MATHEMATICAL MODELLING, APPLICATION AND AUTOMATION DEVELOPMENT TO FURTHER EXPLOIT EXISTING SPACIALISED SOFTWARE

S.K. PATHAN

1. Image Extractions and Enhancement

(a) Loading Image
- Viewer
- File
- Open
- Raster layer
- Select file **liss3.img**
- Raster option Select band 321 fits to frame
- OK

(b) Enhancement
- Raster (from viewer)
- Contrast
- Histogram equalize or general contrast

(c) Extraction of Image
- Utility (from viewer)
- Inquire box <
- Define box to select area of interest
- Apply
- Dataprep
- Subset image
- Input image file
- Select file **liss3.img**
- Enter output file **subJiss3.img**
- From inquire box
- OK

(d) Load image **subJiss3.img** using steps shown in (a)

2. Image Classification

2.1 Unsupervised Classification

- Classifier
- Unsuperwise Classification
- Enter input file liss3.img
- Enter output file liss3_uc.img
- Enter output signature file liss3_uc.sig
- Enter number of classes (5)
- Ok
- Load classified image liss3_uc.sig
- Raster option (band 1, pseudo colour)
- Raster
- Tools
- Select display edit raster attributes
- Edit color for each class
- File
- Save

2.2 Supervised Classification

- Load image liss3.img in viewer
- AOI
- Tools
- Classifier
- Signature editor
- Select icon create polygon AOI from AOI tools
- Draw polygon in viewer for a class signature
- Select create new signature from AOI icon from signature editor tools
- Enter class name in place of class and change color if require
- Repeat for other class signatures
- File
- Save
- Enter output signature file name (liss31.sig)
- OK
- Select super wise classification from classification menu
- Enter input file name liss3.img
- Enter input signature file liss31.sig
- Enter output file liss3_sup.img
- OK

3. Raster to Vector Conversion

- Select vector from top level menu bar

- Raster to vector (from vector utilities)
- Enter input file (classified image liss3_sup.img)
- Enter output vector file name l3coverage (Arc/Info format)
- OK

4. On Screen Vector Tracing from Image

(a) Load image (pan.img)
(b) Open new vector layer
  - File
  - New
  - Vector layer
  - Enter name of new vector layer (road)
  - Ok
  - Trace line using vector tools
  - Change line symbol using vector viewing properties
  - Save vector using save top layer

(c) Building topology
  - Vector (from main menu bar)
  - Clean vector layer
  - Build vector layer topology

(d) Listing attributes
  - Vector (from viewer)
  - Attributes (listing attributes)

5. Projection

(a) Change projection information and pixel value
  - Load image
  - Utility
  - Layer info
  - Edit map projection
  - Edit map model

(b) Reprojection of image
  - Display image
  - Raster
  - Geometric correction
  - Reproject
  - Change projection
  - OK
  - Apply
  - Run 'Display resample image dialog'Icon from geometric correction tools
  - Enter out put file
  - OK

(c) Image Coordinates
- Display image
- Inquire cursor
- Display in full resolution
- Locate cursor
- Map (for geocoded coordinates)
- File (No. of pixel and scan lines)
- Long/lat (geographic value)

6. Map Composition

(a) Open a map file and setting up of parameters
- Display image
- Composer
- New map composition
- Enter output file name
- Define page size
- View (from map composer tools)
- Scale
- Map to window

(b) Selection of image area
- Select 'create map frame' icon from annotation tool bar
- Define a box in map viewer

EXERCISE ON GIS

7. Creation and Organization of GIS Database

(a) Spatial
- Identification of spatial extent of the study area and registration points (GCPs).
- Create spatial framework using registration points and specific required map projection.
- Digitization of thematic layer.
- Editing and error removal.
- Building topology.

(b) Non-spatial
- Defining data structure.
- Adding tabular data.

(c) Spatial and non-spatial database linkages
- Defining of key field (link field).
- Joining of two datasets.
- Defining relationships between two datasets.

8. Displaying and Querying on GIS Database

(a) Displaying of thematic layer.
(b) Querying on displayed layer.
- What lies here ? (point and ask query).
- Where lies this category ? (conditional query).
- Statistics (summation of area for each category within theme).

9. Spatial Analysis

(a) Integrated analysis.
- Integrate more than one layers.
- View the characteristics of combined layers.
- Apply logical condition and display the areas, which are satisfying specific criteria.
- Generate buffer around a feature by applying distance criteria.
- Clip and select one of the layer falling inside this buffer polygon.
- Erase one of the layers falling inside the buffer polygon.

10. Map Layout in GIS

(a) Designing of map layout
(b) Plotting different map elements

Demonstration on 3D Analysis

- Generate 3D perspective view using elevation data.
- Displaying 3D view and draping of thematic layers.
- Volumetric analysis.
- Slope analysis.

Demonstration on Network Analysis

(a) Routing analysis
- Find out shortest path using distance criteria.
- Find out optimum path using time criteria.

6

INDIAN SPACE PROGRAMME TOWARDS NATIONAL DEVELOPMENT

J.R. Sharma

The Indian Government established the Department of Space in 1972 to promote development and application of space science and technology for socio-economic benefits. Indian Space Research Organisation (ISRO), is the primary agency under Department of Space, for executing space programmes. During the seventies, India undertook demonstration of space applications for communication, broadcasting and remote sensing; designing and building experimental satellites – Aryabhata, Bhaskara, APPLE and Rohini and experimental Satellite Launch Vehicles – SLV-3 and ASLV. Today, India has established space systems that form an important element of the national infrastructure.

NATIONAL SYSTEMS USING SPACE TECHNOLOGY

INSAT	IRS
* Telecommuications	* Natural Resources Management
* Broadcasting (TV, Radio)	* Natural Resources Information System
* Business Communication	– Water
* Mobile Communication	– Forest
* Telemedicine	– Crop
* Search and Rescue	– Land
* Satellite Navigation	– Minerals
* Meteorology	– Environment
	– Ocean
	* Disaster Management Support
	* Village Resource Centres

Indian National Satellite (Insat) System

Commissioned in 1983, INSAT is a multipurpose satellite system for telecommunications, television broadcasting, meteorology, disaster warning and search and rescue. Besides telecommunications and other regular broadcasting services, INSAT is widely used for interactive educational television, developmental communication and telemedicine. Meteorological imaging and direct-to-community broadcast capabilities of INSAT help in issuing warnings on impending cyclones. INSAT also carries transponders dedicated to search and rescue operations as part of international COSPAS/SARSAT programme.

Indian Remote Sensing Satellite (IRS) System

Commissioned in 1988, India now has the world's largest constellation of remote sensing satellites. The system offers space-based data in a range of spectral bands, spatial resolutions and swaths. The data is used for several applications covering agriculture, water resources, urban development, mineral prospecting, environment, forestry, drought and flood forecasting and ocean resources.

Launch Vehicles

India has developed and commissioned Polar Satellites Launch Vehicle (PSLV) and Geosynchronous Satellite Launch Vehicle (GSLV). PSLV can launch 1400 kg. Class remote sensing satellites into an 820 km polar Sun Synchronous Orbit (SSO). It can also place a satellite weighing about 1100 kg in Geosynchronous Transfer Orbit (GTO) or a 3500 kg class satellite in Low Earth Orbit (LEO). GLSV is capable of launching 2000 kg class satellites into GTO. GSLV Mark-III, to place 4000 kg class satellites in GTO, is under development.

A variety of sounding rockets have also been developed for conducting scientific and technological experiments.

SPACE SCIENCE

India has flown Gamma–Ray and Retarding Potential Analyser payloads on two of its SROSS satellites launched in 1992 and 1994. IRS-P3, launched in 1996 carries an X-ray astronomy payload. Chandrayaan-1, which envisages orbiting a spacecraft at a height of 100 km in the polar orbit around the moon for physical and chemical mapping of the moon's surface, is planned in 2007-08. An exclusive astronomical satellite, ASTROSAT, is also planned. Several ground-based facilities for space sciences, including a Mesosphere Stratosphere Troposphere (MST) Radar, have been set up in India.

INFRASTRUCTURE FOR SPACE DEVELOPMENT

India has established a strong infrastructure for executing its space programme. They include facilities for the development of satellite and launch vehicles and their testing; launch infrastructure for sounding rockets and satellite launch vehicles; telemetry, tracking and command network; and data reception and processing systems for remote sensing. A number of academic and research institutions as well as industries participate in the Indian Space Programme. Several Indian industries have the expertise to undertake sophisticated jobs required for space systems.

SPACE SERVICES FROM INDIA

Indian space capabilities are available to international customers through Antrix Corporation of the Department of Space. Presently, these include the reception of Indian Remote Sensing Satellite data, launch services, lease of transponders on INSAT and providing TTC support.

INTERNATIONAL CO-OPERATION

International co-operation has been the hallmark of Indian space programme. India participates in major space fora including the UN, IAF, COSPAR and CEOS. India has set up the Centre for Space Science and Technology Education in Asia and the Pacific (CSSTE-AP), which is sponsored by the United Nations. India offers training in space applications to personnel from developing countries under the programme Sharing of Experience in Space (SHARES). ISRO and the French Space Agency CNES have undertaken a joint atmospheric satellite mission Megha-Tropiques to be launched in 2008-09.

SPACE CENTRES IN INDIA

New Delhi

- DOS Branch Secretariat
- ISRO Branch Office
- Delhi Earth Station

Jodhpur (Rajasthan)

- Western RRSSC

Udaipur (Rajasthan)

- Solar Observatory

MT. ABU (Rajasthan)

- Infrared Observatory

Lucknow (Uttar Pradesh)

- ISTRAC Ground Station

Dehra Dun (Uttaranchal)

- Indian Institute of Remote Sensing
- Northern RRSSC

Ahmedabad (Gujarat)

- Space Applications Centre
- Physical Research Laboratory
- Development and Education Communication Unit

Nagpur (Maharashtra)

- Central RRSSC

Mumbai (Maharashtra)

- ISRO Liaison Office

Aluva (Kerala)

- Ammonium Perchlorate Experimental Plant

Thiruvananthapuram (Kerala)

- Vikram Sarabhai Space Centre
- Liquid Propulsion System Centre
- ISRO Inertial System Unit

Mahendragiri (Tamil Nadu)

- Liquid Propulsion Facilities

Bangalore (Karnataka)

- Space Commission
- Department of Space & ISRO Headquarters
- INSAT Programme Office
- NNRMS Secretariat
- Civil Engineering Division
- Antrix Corporation
- ISRO Satellite Centre
- Laboratory for Electro-Optic Systems
- ISTRAC
- Southern RRSSC
- Liquid Propulsion Systems Centre

Hassan (Karnataka)

- Master Control Facility (MCF)

Sriharikota (Andhra Pradesh)

- Satish Dhawan Space Centre, SHAR

Tirupati (Andhra Pradesh)

- NMRF

Hyderabad (Andhra Pradesh)

- National Remote Sensing Agency

Port Blair

- Down Range Station

Kharagpur (West Bengal)

- Eastern RRSSC

Shillong

- North Eastern Space Applications Centre

INDIAN SATELLITES LAUNCHED SO FAR

S. No.	*Name*	*Date of Launch*	*Weight (Kg)*	*Launch Vehicle*
1.	Aryabhata	19.04.1975	358	C-1 Intercosmos, USSR
2.	Bhaskara-1	07.06.1979	444	C-1 Intercosmos, USSR
3.	RTP	10.08.1979	35	SLV-3, India
4.	RS-1	18.07.1980	35	SLV-3, India
5.	RS-D1	31.05.1981	38	SLV-3, India
6.	Apple	19.06.1981	670	Ariane, Europe
7.	Bhaskara-2	20.11.1981	436	C-1 Intercosmos, USSR
8.	INSAT-1A	10.04.1982	1150	DELTA, USA
9.	RS-D2	17.04.1983	41.5	SLV-3, India
10.	INSAT-1B	30.08.1983	1194	Space Shuttle, USA
11.	SROSS-1	24.03.1987	150	ASLV, India
12.	IRS-1A	17.03.1988	980	Vostok, USSR
13.	SROSS-2	13.07.1988	150	ASLV, India
14.	INSAT-1C	22.07.1988	1190	Ariane, Europe
15.	INSAT-1D	12.06.1990	1293	Delta, USA
16.	IRS-1B	29.08.1991	990	Vostok, USSR
17.	SROSS-C	20.05.1992	106	ASLV, India
18.	INSAT-2A	10.07.1992	1906	Ariane, Europe

(*Contd.*)

S. No.	Name	Date of Launch	Weight (Kg)	Launch Vehicle
19.	INSAT-2B	23.07.1993	1932	Ariane, Europe
20.	IRS-1E	20.09.1993	845	PSLV, India
21.	SROSS-C2	04.05.1994	113	ASLV, India
22.	IRS-P2	15.10.1994	904	PSLV, India
23.	INSAT-2C	07.12.1995	2050	Ariane, Europe
24.	IRS-1C	28.12.1995	1250	Molniya, Russia
25.	IRS-P3	21.03.1996	920	PSLV, India
26.	INSAT-2D	04.06.1997	2070	Ariane, Europe
27.	IRS-1D	29.09.1997	1200	PSLV-C1, India
28.	INSAT-2DT	Procured in Orbit from Arabsat in January, 1998		
29.	INSAT-2E	03.04.1999	2550	Ariane, Europe
30.	IRS-P4	26.05.1999	1050	PSLV-C2, India
31.	INSAT-3B	22.03.2000	2070	Ariane, Europe
32.	GSAT-1	18.04.2001	1530	GSLV-D1, India
33.	TES	22.10.2001	1108	PSLV-C3, India
34.	INSAT-3C	24.01.2002	2750	Ariane, Europe
35.	Kalpana-1	12.09.2002	1060	PSLV-C4, India
36.	INSAT-3A	10.04.2003	2950	Ariane, Europe
37.	GSAT-2	08.05.2003	1825	GSLV-D2, India
38.	INSAT-3E	28.09.2003	2775	Ariane, Europe
39.	Resourcesat-1	17.10.2003	1360	PSLV-C5, India
40.	Edusat	20.09.2004	1950	GSLV-F01, India

INTERNATIONAL SATELLITES LAUNCHED BY INDIA

S. No.	Name	Date of Launch	Weight (Kg)	Launch Vehicle
1.	DLR-TUBSAT, Germany	26.05.1999	45	PSLV-C2
2.	KITSAT-3, Republic of Korea	26.05.1999	110	PSLV-C2
3.	BIRD, Germany	22.10.2001	92	PSLV-C3
4.	PROBA, Belgium	22.10.2001	94	PSLV-C3

INDIAN SATELLITE LAUNCH VEHICLES

1. SLV

Height	:	22 m
Lift-off weight	:	17 t
Propulsion	:	All Solid
Payload mass	:	40 kg
Orbit	:	Low Earth

2. ASLV

Height	:	23.5 m
Lift-off weight	:	39 t
Propulsion	:	All Solid
Payload mass	:	150 kg
Orbit	:	Low Earth

3. PSLV

Height	:	44 m
Lift-off weight	:	295 t
Propulsion	:	Solid & Liquid
Payload mass	:	1400 kg
Orbit	:	820 km Sunsynchronous Polar (1050 kg in Geosynchronous Transfer Orbit)

4. GSLV

Height	:	49 m
Lift-off weight	:	414 t
Propulsion	:	Solid, Liquid & Cryogenic
Payload mass	:	2000 kg
Orbit	:	Geosynchronous Transfer

7

SPATIAL MODELLING IN RASTER AND VECTOR DATA

RAJASHREE V. BOTHALE

GIS – A TOOL

A GIS is a specific Information System applied to geographical data and is mainly referred to as system of hardware, software and procedures designed to support the capture, storage, management, manipulation, analysis, modelling and display of spatially-referenced data for solving complex planning and management problems.

While GIS allows the handling and manipulation of data in both spatial and non-spatial format, the power of the GIS is in the organization of Spatial Information Systems and orienting it towards decision-making or resource management. All of this is based on the concepts of technology – Spatial Science. The organization of applications around GIS requires a high-level of GIS-knowledge and visualization of integrated analysis. The design and organization of GIS is an involved activity and requires a good understating of the concepts of geography, cartography, database management, resource modeling and management etc.

WHY GIS?

Efficient means for data distribution and handling
Elimination of redundant database
Integration of data from various sources
Maximize the efficiency of planning and decision making

WHICH GENERIC QUERIES GIS CAN ANSWER?

GIS can answer queries like pertaining to:

Location

What exist at a particular location by name of the place, geographic co-ordinate or by interactive pointing. Thereby it can answer queries like What lies here?

Condition

It can find out areas where certain conditions exist. Criteria for the query can be specified. Thereby it can answer queries like Where lies this?

Neighbourhood

GIS allows to find out neighbourhood information from a specified place, path of area. Thereby it answers queries like What lies within this distance?

Proximity and Route

Nearest feature on the map from the particular feature can also be located. Thereby it can answer queries like Where is nearest entity from here?

Trends

GIS provide the changes occurring over the period of time and thereby allow to find trends.

Modelling

Modelling allows to solve complex problem of any application domain. Modelling for generation of land resources development plan will provide viable action plans for sustainable development of the area. Forest fire prediction models will show all probable zones, which are prone to fire. Flood hazard modeling will highlight areas, which could be inundated during flood. Network models allow planning of resource location based on demand and supply criteria. What happens if kind of scenarios can be generated using network models.

Patterns

GIS also help in finding out spatial patterns in the data which may lead to make some conclusions and the finding the cause for occurrence of these patterns.

DATA STRUCTURES IN GIS

Hierarchical Database Structure

When the data have a parent/child or one-to-many relations, such as areas for various levels of administration, soil series within a soil family, or pixels within a polygon, hierarchical methods provide quick and convenient means of data access. The main advantages of the model are its simplicity and ease of access via keys that define the hierarchy. They are easy to understand and to update and expand and can be useful for organizing data in mass storage systems.

Network Database Structure

In hierarchical structures, travel within the database is restricted to the path up and down the taxonomic pathways. In many situations much more rapid linkage is required, particularly in data structures for graphics features where adjacent items in a map need to be linked together even though the actual data about their coordinates may be written in very different parts of the database. Both redundancy and linkage problems are avoided by the compact network structure. Network database structures are very useful when the relations or linkages can be specified beforehand.

Relational Database Structure

In its simplest form the relational database structure stores no pointers and has no hierarchy. Instead, the data are stored in simple records, known as tuples, which are sets of fields, each containing an attribute; tuples are grouped together in two-dimensional tables, known as relations, somewhat in the manner of a spreadsheet programme. Each table or relation is usually a separate file. The pointer structures in network structures and the keys in hierarchical structures are replaced by data redundancy in the form of identification codes that are used a unique keys to identify the records in each file.

Data Models In GIS

Data input is the first step in GIS operation. Before we do this, we should know different ways of representing the data in GIS. The model in which we represent the data in GIS is termed as data model. Two basic data modeis are raster and vector. Application requirement will decide the representation to chose. Each data model has its advantages and disadvantages. In most of the GIS software, vector is given as a basic data model and the procedures to convert from vector to raster data model are provided.

Raster Model

Representation of real world data in any of the model is nothing but the problem of explaining/representing/indicating the geographic space in the computer system. Raster method serves to quantize or divide space as a series of packets or units or grid cells of fixed or uniform size. Converting or explaining the real world data in raster is essentially storing the data at each position (grid cell). Finer the grid cell higher the amount of details in the raster data. Size of the grid cell is nothing but the resolution of the data. The grid-cell or raster model is a relatively simple approach to data representation, both conceptually and operationally. The representation of attributes is by specifying values to each grid cell. Rasters are limited by the area they can represent and also the limits of storage space. Also the fineness of the data is limited by the cell size. Thus the area coverage is trade-off with the resolution of the coverage. The storage problems are handled by resorting to compression techniques like run-length coding, block coding etc.

Vector Data Model

The vector data structure represents each geographical feature by a set of coordinates. Vectors as xy coordinates define points, lines and polygons. The basic premise of the vector based structuring is to define a 2-dimensional space where features are represented by coordinates on the two-axes. Generally, representing points and lines is straight forward. Points are characterized by a x-y coordinate pair and lines by a set of x-y coordinate pairs with a specific beginning and ending vector. However, representing polygons in vector storage poses a challenge.

Raster *Versus* Vector

As there are two characteristically different ways of spatial data representation for GIS applications, questions of preferring one method over the other arise. Some of the issues that need consideration are as follows:

(a) Vector data is precise and has no approximation errors for measured quantities like area, length, perimeter, etc. (subject, of course, to the representational accuracy of the hardware). Due to the discretization raster data suffers from such approximation errors.

(b) Generally, raster data has higher storage requirements though compression techniques can be adopted. Vector storage is relatively less.

(c) GIS manipulation operations on raster data are computationally faster and less complex than their equivalents on vector data. Imposing a structure on 2D vector data that is sufficiently formal and supports efficient searching is fairly complex.

(d) Raster data is not easily amenable to association of attribute data with spatial features such as points, lines or polygons. This is due to the fact that the basic entity in the raster approach is the grid cell and entites such as points, lines and polygons are not recognized as objects in their own merit.

(e) Raster is limited by the cell-size and sampling or dis-aggregation to a lower size is not possible. Vector structure does not impose this limitations and thus analysis, plotting etc. is more accurate in Vector structure.

GIS DATABASE CREATION

Spatial Database Creation

Based on the design, the steps of database creation are worked out and a procedure laid down. The steps for the spatial database creation include following:

(a) Creating spatial framework
(b) Master template creation
(c) Thematic map manuscript preparation

(d) Digitisation of features
(e) Coverage editing
(f) Attribute coding verification

Non-spatial Database Organization

Non-spatial data elements are mostly available in analog mode. Towards converting it into a digital mode, a suitable application package could be used to configure a data entry system.

8

CARTOGRAPHIC DATA EXCHANGE IN DIFFERENT FILE FORMAT

A.K. BERA

Spatial databases contain many files with data on related aspects of the same entities, or of data on entities that because of their spatial proximity or connectivity have to be linked or grouped together. It is essential to organize the way these files are stored and linked in the computer to model real world phenomena and to ensure efficient storage and retrieval of data.

DATABASE ORGANIZATION

The GIS has two distinct utilization capabilities—the first pertaining to querying and obtaining information and the second pertaining to integrated analytical modelling. However, both these capabilities depend upon the core of the GIS database that has been organized. Generally, a proper database organization needs to ensure the following:

(a) A flexibility in the design to adapt to the needs of different users.
(b) A controlled and standardized approach to data input and updation.
(c) A system of validation checks to maintain the integrity and consistency of the data elements.
(d) A level of security for minimising damage to the data.
(e) Minimizing redundancy in data storage.

The design of the GIS database will include three major elements:

(i) Conceptual design, basically laying down the application requirements and specifying the end-utilisation of the database. The conceptual design is independent of hardware and software and could be a wish-list of utilisation goals.
(ii) Logical design, which is the specification of the database *vis-à-vis* a particular GIS package. This design sets out the logical structure of the database elements and is determined by the GIS package.
(iii) Physical design, which pertains to the hardware and software characteristics and requires consideration of file structure, memory and disk space, access and speed etc.

DATABASE CREATION

Based on the design, the steps of database creation are worked out and a procedure laid down. The procedure for the spatial database creation is described below:

(a) *Creating of spatial framework*: The spatial framework of the GIS database can be organized in the GIS by specifying the registration points for the total database and specifying the coordinate system of the database. Registration Points for the total area can be entered through key-board.

(b) *Master template creation:* As discussed earlier, a master template is created as a reference layer and consisting of the district boundary, rivers etc. This template is then used for the component themes digitization.

(c) *Thematic map manuscript preparation:* Based on the spatial the different theme oriented information is transferred from the base map to a mylar/transparent sheet. Spatial data manuscripts are mylars consisting features that are to be digitized. These manuscripts are prepared on a sheet-by-sheet basis for digitization. These manuscripts consist "instructions" for digitization or scanning-which include:

- Registration point locations and identifiers.
- Feature codes as per the dictionary defined earlier.
- Feature boundaries.
- Tolerance specifications.
- Any other digitization/scanning instructions.

(d) *Digitisation of features:* The theme features of the spatial dataset are then digitized/ scanned using the GIS PACKAGE. The digitization/scanning is done for each map-sheet of the spatial reference. The master registration-point reference are used for the digitisaion. The theme digitization is done as a component into a copy of the master template layer.

(e) *Coverage editing:* The digitized coverage is processed for digitization errors such as dangles, constituting the overshoots or undershoots, and labels for polygons. This constitutes obtaining a report of these errors and then a manual editing of these features. The editing of these features is also best carried out on a map-sheet basis. Finally, the coverage is processed for topology creation. In the case of raster GIS packages, the topology construction may not be relevant. However, a clumping process to identify the clump of rasters having similar characteristics would be essential. The tools for topology organization or the raster-polygon generation is generally available in the GIS package.

(f) *Attribute coding verification:* The attribute codes for the different categories need to be then verified and additional attributes—feature-name, description etc. are added into the feature database.

CONCLUSIONS

Some errors might occur as a result of computational miscalculations and rounding error in the GIS software. However, most database errors result from improper input.

The errors which we normally encounter are:

(a) *Entity error (positional error)*—This is primarily encountered in vector systems. Entity error can come in three different forms: missing entities, incorrectly placed entities, and disordered entities.
(b) *Attribute error*—Attribute error occurs in both vector and raster systems, often with the same frequency. Most often attributes are typed in, and it constitutes a major source of error. In vector systems attribute errors include using the wrong code for an attribute, as well as misspellings, which make an attribute impossible to retrieve if a query uses the correct spelling. In raster, the input most often consists of attributes, so the result of typing a wrong code number or placing it in the wrong grid location is a map that displays these incorrectly coded cells in the wrong place.
(c) *Entity-attribute agreement error*—Incorrectly placed attribute data gives rise to this error (logical consistency), which also occurs in vector systems when correctly typed codes are attached to the wrong entities.

Data input is the major activity in GIS and is very crucial as the end results are totally based on the correctness of the data. The wrong or insufficient data will lead to erroneous results. Hence data input need to be followed by strict quality checks.

9

DIGITAL CARTOGRAPHY IS THE BASIC REQUIREMENT FOR GEOGRAPHERS

M.P. PUNIA

GEOGRAPHICAL INFORMATION SYSTEM

Geographical Intelligent System (GIS): The system which explain the geographical phenomena with the help of software supported Intelligent power.

Geographical Information System (GIS)

Geographical Information System (GIS) is an organized collection of computer hardware, software and geographic data designed to efficiently capture, store, manipulate, analyse and display all forms of geographically referenced information. GIS is an interdisciplinary tool, which has application in various fields such as Geography, Geology, Cartography, Engineering, Surveying, Rural and Urban Planning, Agriculture, Water Resources, etc.

The GIS is much more than simple coding, storing and retrieving data about aspects of the earth surface. In a very real sense the data can be accessed, transformed and manipulated interactively, and they can serve as a test bed for studying possible results of planning decisions.

Capabilities of GIS

Uses of geographic information technology vary widely. There has been an explosion of GIS applications in spatial data analysis over the past few years. There are very good example to solve geo-scientific problems. Three major capabilities of GIS are:

Cartographic Capability

Cartographic Capability allows accurate maps and engineering drawing to be produced efficiently. This capability includes digitizing (converting analog products to digital form), graphic display generation, interactive graphic manipulation (e.g. add, modify, delete, create window) and plotting.

Data Management Capability

Data Management Capability enables the efficient storage and manipulation of geographic data, both graphic and non-graphic. Storage and retrieval of non-geographic data is linked to graphic images. It is sometimes called Attribute Processing. Attribute processing can select data and produce graphic and reports on the basis of attribute value.

Analytical Capability

Analytical Capability permits sophisticated processing and interpretation of spatial data.

Collectively, these capabilities give managers an enhanced ability to manipulate and use data more effectively. Graphic representations are especially powerful for conveying information.

1. GIS as a Set of Interrelated Sub-Systems

GIS is a combination of various sub-systems. They are as follows:

Data Processing System

- Data acquisition – From maps, image or field surveys.
- Data Input – Data must be input from source material to the Digital database.
- Data Storage – Storage of data based on the frequency of utility.

Date Analysis Subsystem

- Retrieval and analysis – may be simple responses to queries or complex statistical analysis of large sets of data.
- Information output – Display of the result as maps or table or digital output.

Information Use Subsystem

- Users may be researcher, planners and managers.
- Interaction needed between GIS group and users to plan analytical procedures and data structures.

Management Subsystem

- Organizational role – GIS section is organized as a separate unit within a resource management agency offering spatial data base and analysis services. Extensive interaction is needed between the GIS and the rest of the organization if the system is to function effectively.

Database Management

A data base here refer to a computerized collection of related information stored in such a way that retrieval can be performed by linking various pieces of information together. It consists of one or more data files, which are collection of related information treated as one unit on a computer. Databases are managed and accessed via software termed Database Management System (DBMS). Data base management system (DBMS) is a set of computer programs for organizing the information in a database. Typically, a DBMS contains routines for data input, verification, storage, retrieval and combination. The combination of hardware, software and the database itself is referred to as a data base system.

The main characteristic of Geographical database is its spatial nature. A spatial database is a collection of spatially referenced data that act as a model of reality. All the basic data types in geography/geology are spatially distributed such as geomorphological feature, rock type, well site, lineaments, roads etc. Hence Geographic Information System provides an excellent tool to design, implement and manage geographical data in a most efficient manner.

Database Design

As in normal activity, GIS database needs to be properly designed to cater to the needs of specific application. The design should define a comprehensive framework of database, identification of essential and correct data elements, updating procedure etc. Generally, the database design include:

Conceptual Design

It is independent of software and hardware and defines the application needs and the objective of GIS database:

- Specific to the ultimate use of GIS database, e.g., GIS database for natural resource management, ground water management etc.
- Defining level of database indicates the scale or level of data contents of database.
- Spatial elements of database – defining the spatial database (primary and derived) that will populate the database.
- Non-spatial elements of the database – defining the non-spatial datasets (primary and derived) that will populate the database.
- Sources of spatial and non-spatial data-identifying the data collection and data generation activity.

Logical Design

It pertains to the logical definition of the database and is specific to a GIS package. It includes:

- Defining the coordinate system of the database – All spatial elements can be referenced to uniform coordinate system.
- Defining spatial framework – Latitude/Longitude graticules, spatial files design, identification of registration points.
- Defining attribute codes and their description.

- Spatial database normalization—
  - Identification of master templates.
  - Ensuring that the features of various elements are coordinate coincident.
- Tolerance definitions—
  - Coordinate movement tolerance – Defines the position and is a function of scale.
  - Weed Tolerance-Minimum separation between coordinates while digitization.
  - MSU—Maximum spatial units, indicating the smallest representative area. Feature having less area than MSU can be aggregated.
  - Defining the linkage between spatial and non-spatial database through a code.

Physical Design

It is based on experience and pertains to:

- Disk space requirement.
- Load of database.
- Access and speed requirement.
- Platform related aspects.

Database Characteristics

- It should be contemporaneous – should contain information of the same vintages for the entire measured variable.
- It should be positionally accurate.
- The category of information and sub-categories within them should contain all the data needed to analyse or model the behaviour of the resource using conventional methods and model.
- Exactly compatible with other information that may be compared with it.
- Internally accurate, portraying the nature of phenomena without error requires clear definition of phenomena that are included.
- Readily updated on a regular schedule.
- Accessible to whoever needs it.

Spatial Objects and Database Models

The objects in a spatial database are representative of real world entities with associated attributes. The power of GIS comes from its ability to look at entities in their geographical context and examine relationships between entities. The different types of spatial objects in GIS are:

Point Feature

It is the simplest type of spatial object. Co-ordinate of each point can be stored as two additional attributes. Information on a set of points can be viewed as an extended attribute table. Each point is independent of every other point, represented as separate row in the database model.

Line Feature

It is a composed of vertices, nodes and arcs. It includes a set of ordered coordinates that when connected represent a linear feature. Examples of such data are infrastructure networks, transport network, drainage network, etc.

Area Feature

An area feature is a closed figure whose boundary encloses a homogenous area. These are represented by polygons defined by any natural boundaries. Examples of such features are water body, aquifer units, rock units, geomorphic units, etc.

Representation of Continuous Surface

Various examples of continuous surface are topography, hydrogeochemical patterns, etc. Any digital representation of the continuous variation of relief over space is known as a digital election model (DEM), because the term terrain often implies attributes of a landscape (other than the attitude) of the land surface. DEMs were originally developed for modelling relief. They can, of course, be used to model the continuous variation of any other attributes 'Z' over a two dimensional surface.

Tin as Model of Surface Representation

The Triangulation Irregular Network (TIN) is a system designed by Peuker and his co-worker (Peuker *et al.* 1978) for digital elevation modelling that avoids the redundancies of the altitude matrix and which at the same time would also be more efficient for many types of computation (such as slope) than systems that are based on digitized contours. TIN is a terrain model that uses a sheet of continuous, connected triangular facets based on a Debunay triangulation of irregularly spaced nodes or observation points.

The TIN model is a vector topological structure similar to the fully topologically defined structures for representing polygon networks with exception that the TIN doesn't have to make provision for islands or holes. The TIN model records the nodes of the network as primary entities in the database. The topological relations are built into the database by constructing pointers from each node to each of its neighboring nodes. The neighbour list is sorted clockwise around each node starting at north. A dummy node on the "reverse side" of the topological sphere on to which TIN is projected represents the world outside the area modelled by the TIN. These dummy nodes assist in describing the topology of the border points and simplify their processing.

Need for TIN

- For computing slope profile that can be used to prepare shaded relief maps, assisting hydrogeochemical for estimating ground water quality using geochemical data.
- By replacing 'Z' by any other continuously varying attribute, it can represent ground water levels and fluctuation.

Feature Represented by GIS

- **Point:** Label, Tics and Centre of the area feature
  - X-Direction
  - Y-Direction
  - Z-value (Attributes)
- **Line:** Arc Line Segments or any linear phenomena
  - Node Topology
    - First-Node (For Node)
    - X-Direction
    - Y-Direction
    - Last Node (To Node)
    - X-Direction
    - Y-Direction
- **Arc:**
  - Topology
    - Arc-Ids
    - Right arc
      - Vertex X-Direction
        - Y-Direction
      - Left arc
        - Vertex X-Direction
        - Y-Direction
      - Length
      - Z-value (attribute)
  - Connectivity
      - Node Connectivity
      - Arc Connectivity
      - Arc-Node connectivity
      - Route/Network Topology
- **Polygon:** Area, Region and Island or any area features
  - Node Topology
    - First-Node (For Node)
      - X-Direction
      - Y-Direction
    - Last Node (To Node)
      - Y-Direction
      - X-Direction
- **Arc:**
  - Topology
    - Arc-Ids
    - Right arc
      - Vertex X-Direction
        - Y-Direction
    - Left arc
      - Vertex X-Direction
        - Y-Direction
  - Polygon Topology
    - Polygon Ids
    - Area
    - Perimeters

	Left Poly
	Label X-Direction
	Y-Direction
	Right Poly
	Label X-Direction
	Y-Direction
	Label Z-value (Attribute)
Connectivity	Node Connectivity
	Arc Connectivity
	Arc-Node Connectivity
	Label – Arc Connectivity
	Label – Island Connectivity
	Envelop label Connectivity
	Label – Label Connectivity
	Route/Network Topology

Working with Point Feature

Example: Well Location (Point) Co-ordinate location (Tic) Label mark (Area or Point)

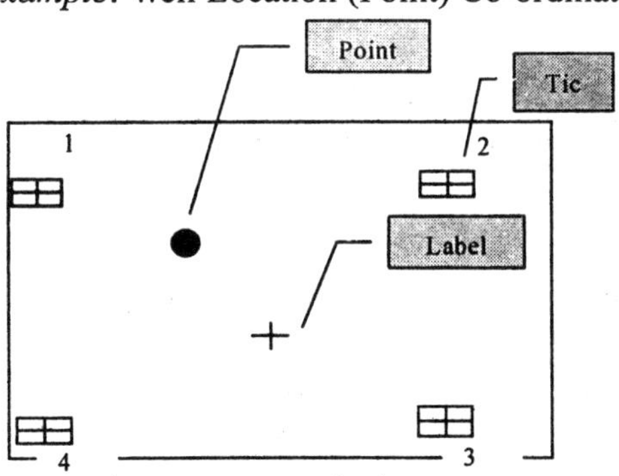

GIS Utility: *Building Point Topology,*
Label-error Finding
Creating Labels
Updating Tics with Respecct to Earth

Important GIS Analysis

1. Spatial Anlysis
2. Visibility Analysis
3. Point Buffer
4. Surface Analysis (Tin or Grid)
5. Thiessen Polygon

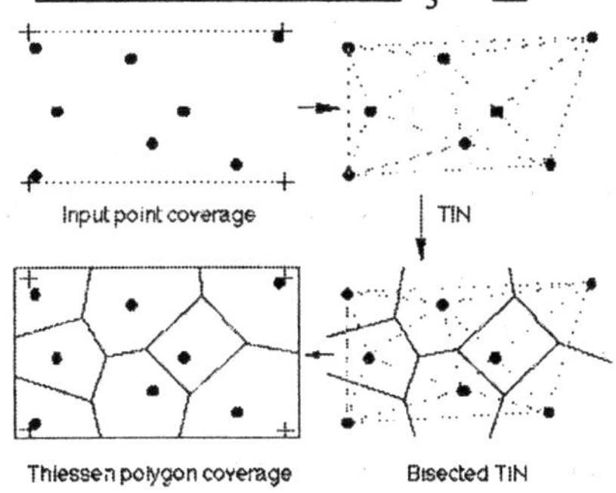

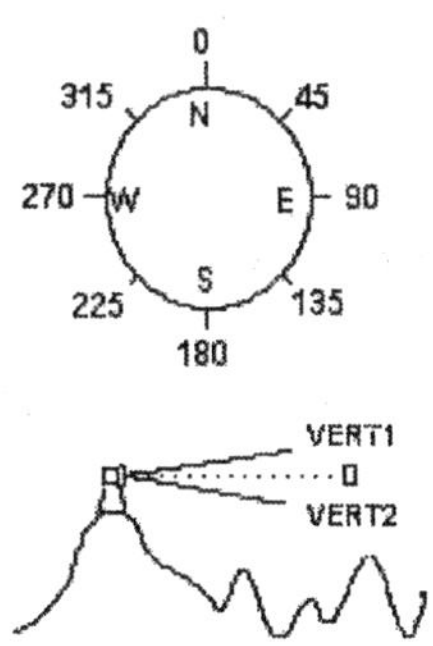

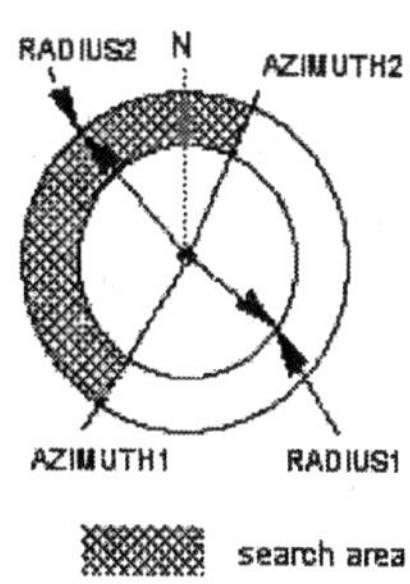

Working with Line Feature

Arcs represent both linear features and the borders of areal features. Each linear feature may be made up of many arcs. Arcs also form the basic linear infrastructure from which route-systems are built.

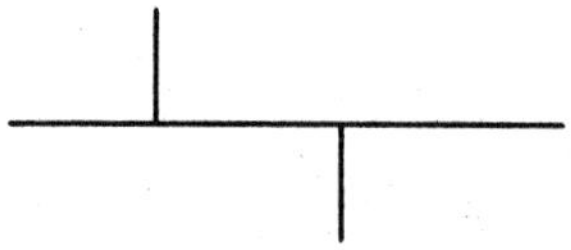

Road Network

Drainage Network

An arc is a continuous string of x, y coordinate pairs (vertices) beginning at one location and ending at another.

A node is an endpoint of an arc. The from-node is the first vertex in the arc; the to-node is the last vertex. Together, they define the direction of the arc. A node is shared by the set of arcs, which connect to each other at the node.

Topological Relationships

Arc-node topology defines connectivity-arcs are connected to each other if they share a common node. This is the basis for many network tracing and path finding operations.

Node Types

A dangling node, represented by a square symbol, refers to the unconnected node of a dangling arc.

Pseudo nodes drawn with a diamond symbol occur where a single line connects with itself (an island) or where only two arcs intersect.

At a minimum, the following items are contained in the AAT:

FNODE#	Internal sequence number of the from-node.
TNODE#	Internal sequence number of the to-node.
LPOLY#	Internal sequence number of the left polygon; set to 0 if the coverage does not contain polygons.
RPOLY#	Internal sequence number of the right polygon; set to 0 if the coverage does not contain polygons.
LENGTH	Length in coverage units.
<cover>#	Internal sequence number (i.e., the record number) of the arc in the ARC file.
<cover>-ID	User-assigned feature ID.

How Networks Fits Together in Coverage

The RAT and SEC of a route-system are stored within the coverage which they reference. The integrity of routes and sections are automatically maintained as arcs are edited or updated. Events are not part of ARC/INFO coverage, but simply reference a route-system from an external source.

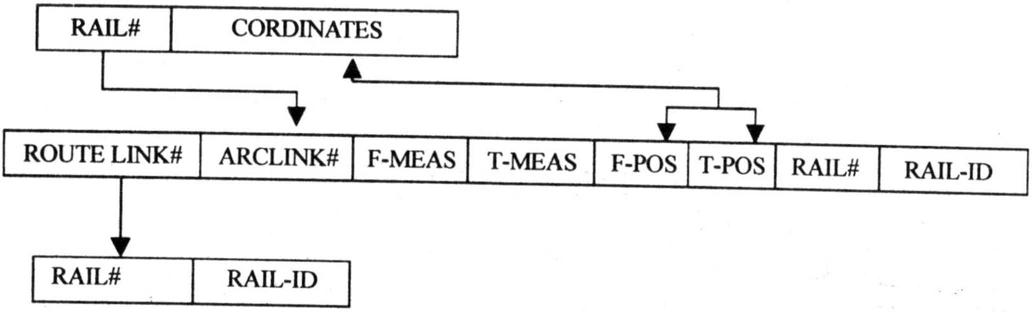

This figure shows how RAT, SEC, and ARC files in RAIL coverage fit together to implement a route-system named RAIL. The ARCLINK# item in the SEC points to the arc internal number in the ARC file. The items F-POS and T-POS in the SEC file reference the coordinates of the arc. The RAT and the SEC are related by the ROUTELINK# in the SEC and the <subclass># in the RAT.

Example:

Any linea feature (arc)
Railway/Roads/Canal
Structure/Directional linear feature
Morphomatric linear feature (Drain-system)
Route Networks (Rail/Road/Sewage/Telephone or cable line

GIS Utility

Building line topology,
Label-error finding
Creating labels (cover-id)
Creating stream-order(drainage/canal)
Creating stream-link(drainage/canal)
Flip-editing (FLIP can be useful for editing network coverages. The from/to direction of selected arcs can be changed to make sure that the flow is in the required direction.)
Trace (To determine upstream and downstream, the directionality of the network must be known. Directionality is best understood by the concept of a directed network.)
Preparing the network coverage AAT

Example

Ap:>calculateUsage: CALCULATE <cover> <feature_class> <target_item> = <arithmetic_expression>Arcplot: calc streets arc from_to_imp = (60 * (length/5280))/ speed_limit

FNODE# = 3

TNODE# = 4
LPOLY# = 1
RPOLY# = 4
LENGTH = 10000.00
STREETS# = 3
STREETS-ID = 2
SPEED_LIMIT = 35
FROM_TO_IMP = 3.25
TO_FROM_IMP = 3.25

Important GIS Analysis

1. Spatial Anlysis (Lineament Direction or Filter)
2. Route/Network Analysis
3. Line Buffer
4. Surface Analysis (Tin or Grid)
5. Identity Analysis (Line in Polygon)
6. Intersect Analysis (Line in Polygon)
7. Nearest Analysis
8. Append
9. Clip
10. Erase
11. Split
12. Line of Sight
13. Visibility Analysis
14. Contour Generation
15. Profile

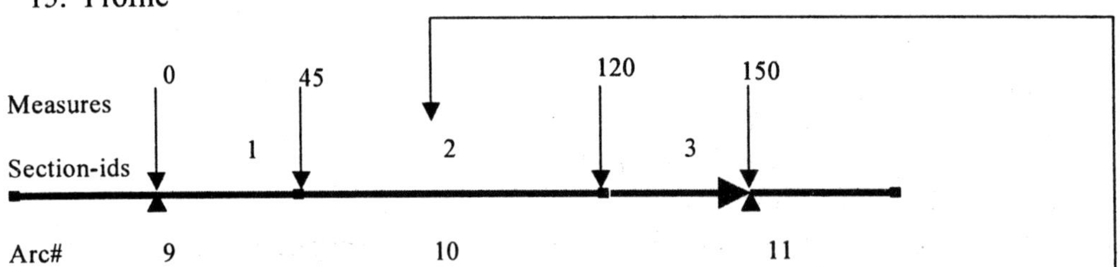

Route Link#	*Arc Link#*	*F-Meas*	*T-meas*	*F-pos*	*T-pos*	*Rail#*	*Rail-id*
2401	9	0	45	35	100	401	1
2401	10	45	120	0	100	402	2
2401	11	120	150	0	36	405	3

(This section table shows how sections for a route – stem called NR are coded. The table contains three sections belonging to one route. Section 401 is define between 35 per cent and

100 per cent along arc 9. Section 402 is defined on all of arc 10. Section 403 is define between 0 per cent and 30 per cent along arc 11.)

The position of a section on an arc is stored in a SEC as from and to values. These values represent percentage distance along the arc, starting from the FROM-NODE of the arc. For example, in the figure above, the first section is coded as 35 to 100 for the F-POS and T-POS position items since it is located on arc 9, from 35 per cent along the arc to 100 per cent along the arc (i.e., the end of the arc). The above figure assumes that the arcs are in the same direction as the route. If any arc is in the opposite direction of the route crossing it, the values of FROM and TO position items in the SEC are reversed.

Working with Area/Polygon Feature

Polygons represent area features. A polygon is made up of arcs which define the boundary and a label point which links the polygon feature to an attribute record in the polygon attribute tables (.pat).

The aim of the polygon data structure is to be able to describe the topological properties of the areas (that is their shapes neighbours and hirarchy) in such a way that the associated properties of these basic spatial building blocks can be displayed and manipulated as thematic map.

If polygon topology is not current or has never been built, run *CLEAN* or *BUILD* in either ARCEDIT or ARC. These commands work basically the same in both modules. The commands in ARCEDIT create only polygon topology, but have additional options to correct label errors.

Working with Grid Feature

This feature represent by number of cells in X & Y direction with equal size. Z-direction represent the attribute of spatial feature.

Important GIS Analysis

1. Digital elevation model.
2. Topographical elevation model.
3. Slope direction.
4. Run off analysis (flow direction).
5. Flow accumulation.
6. Darcy flow vector.
7. Watershed definition.

10

HARDWARE AND SOFTWARE OF GEOGRAPHICAL INFORMATION SYSTEM

M.P. PUNIA

COMPONENT OF GIS

- Software component
- Hardware component
- Management factor

Hardware

Used to store, process and display data. Hardware capabilities affect processing speed, ease of use and type of outputs available.

Software

Perform GIS operations. It contains procedures for performing various tasks.

Expertise

People who provide the intelligence to use the system, develop procedures and define the tasks of GIS.

Software Component

Efficient Operating System: To process large volume of data.
GIS Software (Raster & Vector): To create user oriented/define queries.
Image Processing Software: To use old scan data or Remote Sensing data.
Other programming software (Window or Command): To create object-oriented programme for different department requirements.

Hardware Component

Basic computer component:
Scanner: To scan the maps and other geographical information.
Plotter/Printer: To print the Map/Information or query about Geographical Phenomena.
Digitizer: To convert hardcopy maps/information in digital files.

Management Component

To get efficient work.
To get maximum outputs.
To get proper maintenance of hardware and software components.

GIS OBJECTIVES

Maximize the efficiency of planning and decision making.
Provide efficient means for data distribution and handling.
Elimination of redundant database—minimize duplication.
Capacity to integrate information from many sources.
Complex analysis/query involving geographical referenced data to generate new information.
For any application there are five generic questions that a GIS can answer:

Location – Location of a particular object
Condition – Identify locations where certain conditions/phenomena occur.
Trends – What change occurs?
Patterns – What spatial pattern exists?
Modelling – What type of model will fit for a phenomena.

GIS INFORMATION

Spatial Information

Represents geographic features (Location and shape) associated with the real world locations and their relationship to other features.

Non-spatial Information

Descriptive Information about the characteristics of the feature.

Representations of the Spatial Information

Geographical features are depicted on map by **Point, Line Polygon**.

- POINT feature—A discrete location depicted by a special symbol or label. A single x, y coordinates.

- LINE feature—Represents a linear feature. A set of ordered x, y coordinates.
- POLYGON feature—An area feature where boundary encloses a homogeneous area.

Representation of Non-spatial (Attribute) Information

Consists of textural description on the properties associated with geographical entities. Attributes are stored as a set of numbers and characters in the form of a table. Many attribute data files can be linked together through the use of common identifier code.

Organizing Map Data

Map features are logically organized into a set of layers or themes of information. Some of the most common considerations for organizing layers are:

Feature type – Typically, layers are organized so that points, lines, and polygons are stored in separate layers.
Logical group – Features are organized logically by what they represent.
Intended Use – Application specific.

TOPOLOGY

Geographic data describes objects in terms of location, their attributes and spatial relationship with each other (Topology). It is a mathematical procedure that determines the spatial relationships of features.

Area Definition

Polygons are stored as set of lines, rather than defining polygon as a loop of coordinates. Line coordinates are stored only once, hence reduces the storage space.

Contiguity

Identification of polygons, which touch each other. It is implemented by finding left and right polygon of each arc and takes care of shared arcs.

Connectivity

Identification of interconnected arcs. From and To node indicate the direction. Node coordinates are stored only once.

Advantages of Topology

Polygon network is fully integrated.
Optimal storage and free from excessive amount of redundant information.
Neighbours are identified.
Polygon in polygon can be represented.

11

GEOMETRICAL RECTIFICATION OF DIGITAL CARTOGRAPHIC DATA

M.P. PUNIA

DATA MODELS

Geographical variations are infinitely complex and must be represented in terms of discrete objects. Conversion of real world geographical variation into discrete objects is done through data models. It represents the linkage between the real world domain of geographic data and computer representation of these features.

Raster Data Model

Divides the entire area into rectangular grid cells, where x = y distance. Each cell contains a single value and every location corresponds to a cell. One set of cells and associated values is a *LAYER/CHANNEL*.

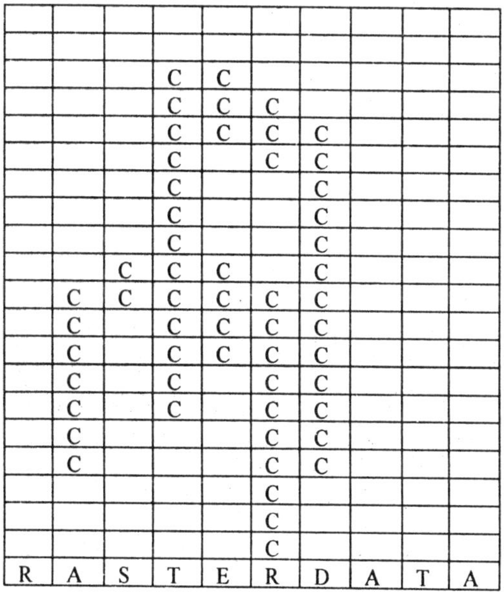

VECTOR DATA

Capabilities

Displaying layers—Each cell can be assigned a unique colour on the display.

Local Operations: Cell by cell basis.

Recoding – Assigning cells to different classes.

Overlaying layers – Layers can be overlaid/combined on the basis of arithmetic, logical criteria or unique combinations.

Focal Operations: Considering the neighbourhood cells.

Filtering – Performed for smoothening or enhancement by moving the window across the entire raster.

Buffer – Spread of objects spatially by a given distance.

Slope/Aspect

Zonal Operations: Group of homogenous cells

Identification of zones – By identifying all patches of same value and giving each zone a unique number.

Area measurement – Measure the area of each zone.

Surface analysis

Visibility analysis

Hydrologic modeling etc.

Advantages

Simple data structure, Sampling is done uniformly (Uniform resolution).

Spatial analysis is easier and faster.

Data are acquired in that form—Remote Sensing, Scanning.

Most common format (for data interchange).

Disadvantages

Huge volume of data (Though may be reduced by coding).

Fitness of data is limited by cell size.

Less appreciable output.

Projection transformations are time-consuming.

Vector Data Model

Uses discrete line segments or points represented by their explicit x, y coordinates to identify locations.

Discrete objects (boundaries, streams) are formed by connecting line segments.

Area is defined by set of line segments.

Capabilities

Display – Entity type and attributes can be displayed by different colours, line patterns or symbols.

Reclassify areas by single attributes. For example, reclassify soil area by soil type only.
Dissolve boundaries between areas of same type by deleting the arc between them.
Merge polygons into large objects.
Topological overlays—Point in polygon, Line on polygon, Polygon on polygon.

Advantages

Less storage.

Editing is faster and convenient.
Appreciable outputs.
Estimation of area/perimeter is accurate.
Network analysis is faster.
Projection transformations are easier.

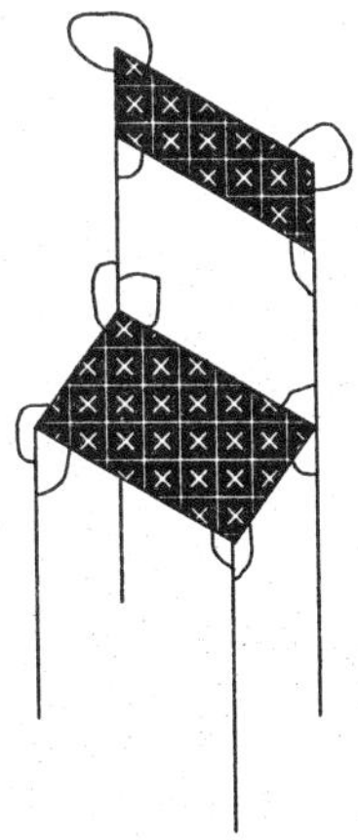

Disadvantages

Overlay based on criteria difficult.
Spatial analysis is cumbersome.

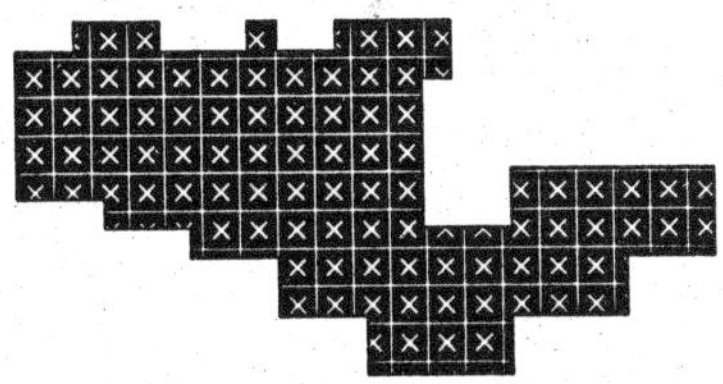

Raster Data Structure

Chain Code
Block Code
Quadra tree
Run length

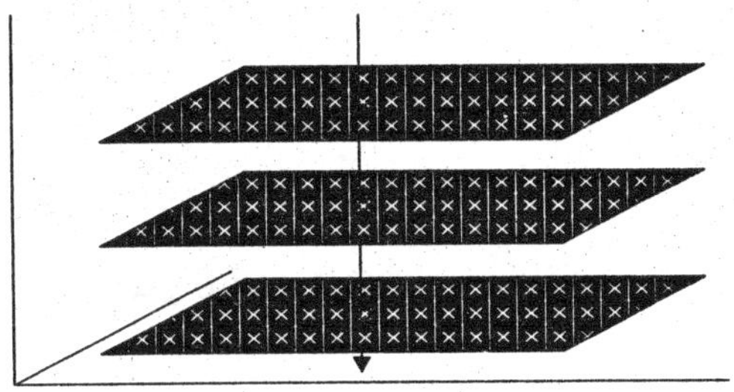

Chain-code

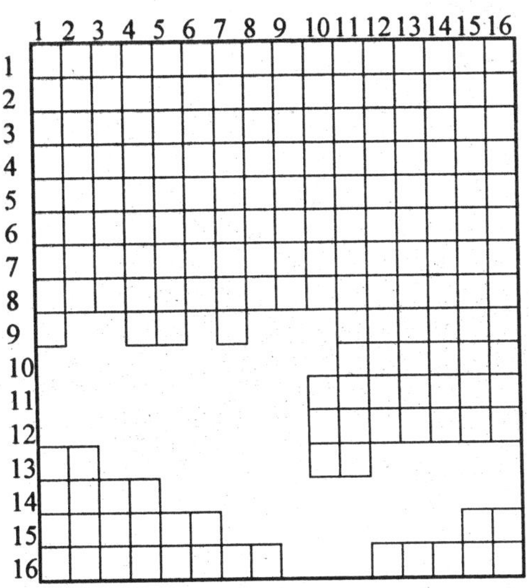

Block-code

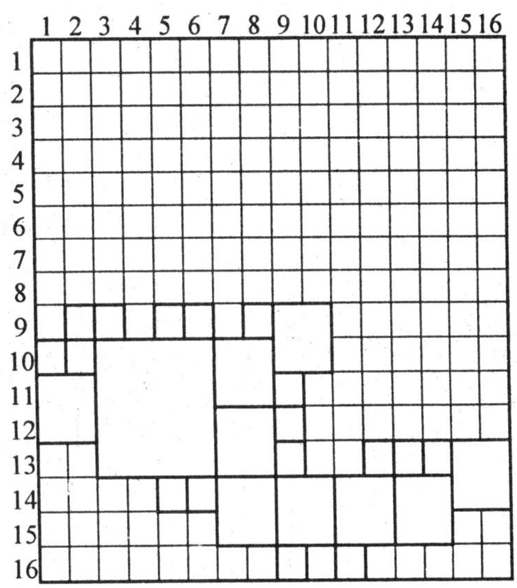

Quadra Tree

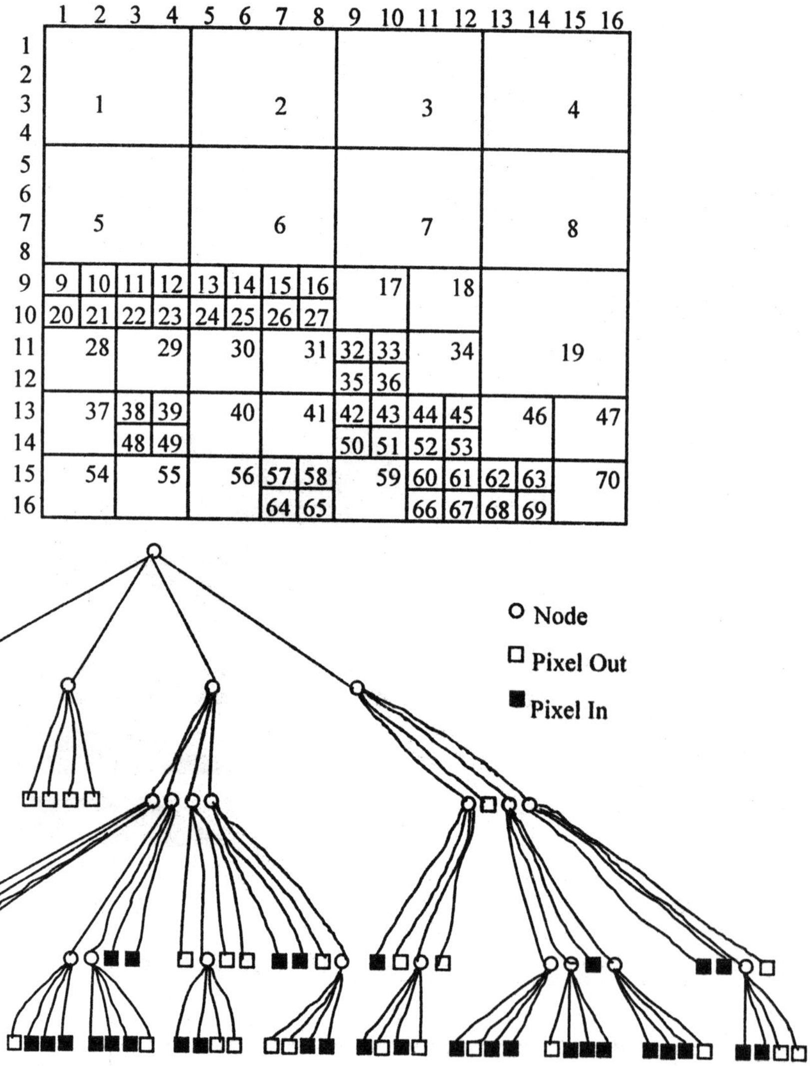

Vector Data Structure

Point

Line

Polygon/Area

Point feature ⊕

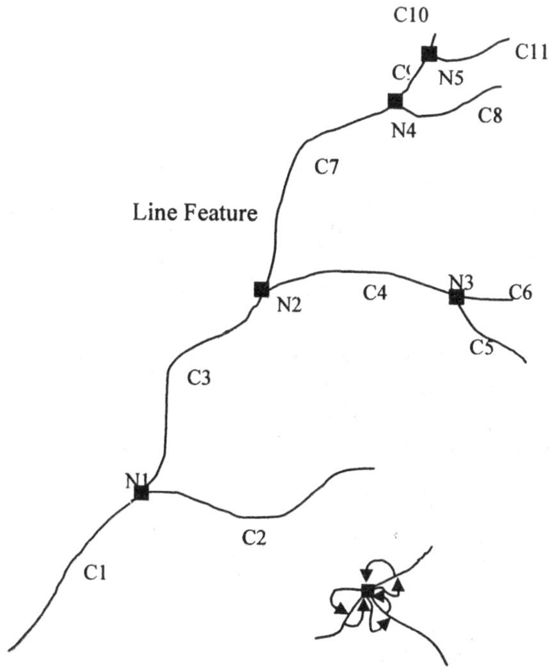

Polygon/Area Feature

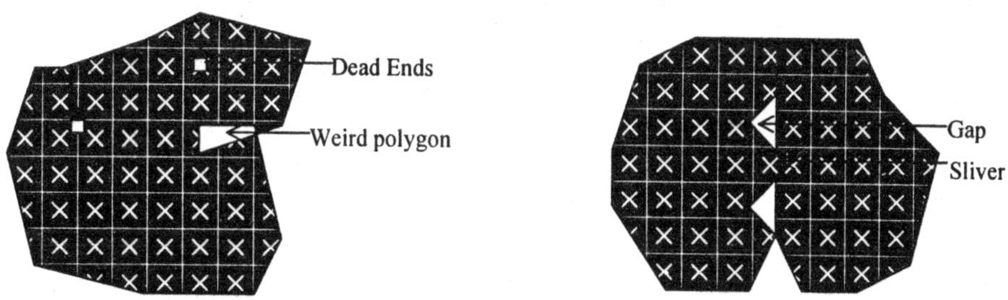

A Full topological network for polygon

A Fully Topological Polygon Network Structure Describe

Linking chain into a boundary network
Checking polygon for closure
Linking the line into polygon
Computing polygon area
Associating non-graphic attributes to the polygons.

DATA STRUCTURES

There are number of different ways to organize the data inside the information system. The choice of data structure may affect both; Data storage volume and processing efficiency. Many

GIS have specialized capabilities for storing and manipulating attribute data in addition to spatial information. Three basic data structures are—Relational, Hierarchical, Network.

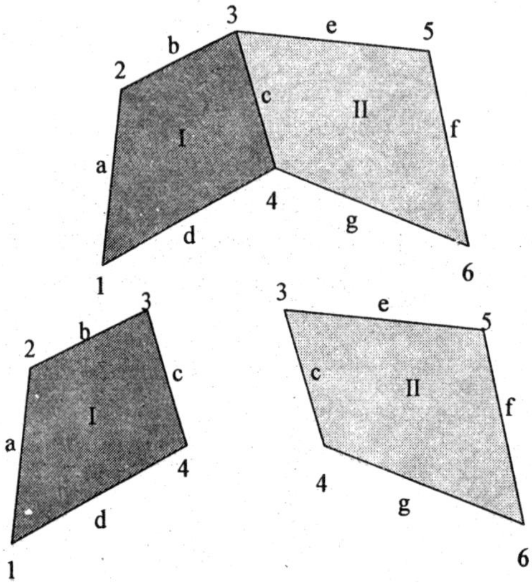

Relational Data Structure

It is the most popular model for GIS. Many GIS packages have different names, e.g. INFO in ARC/INFO, DBASE III in PC-based GIS etc. Data are stored as simple records, known as tuples, containing an ordered set of attribute values that are grouped together in two-dimensional tables.

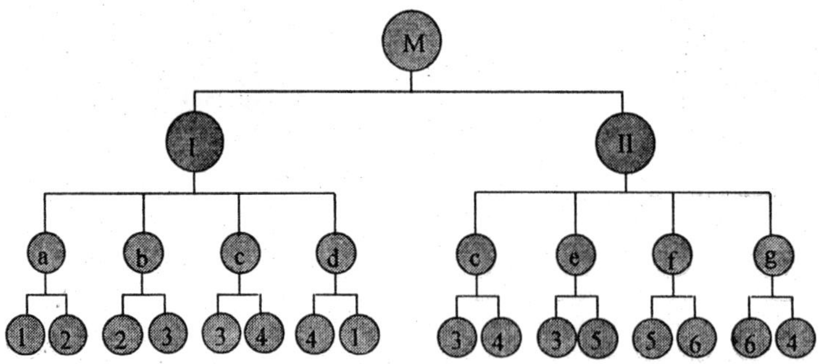

It organizes the data in terms of two-dimensional tables.
Each table is a separate file.
Each row in the table is a record.
Each record has a set of attributes.
Each column in the table is an attribute.
Different tables are related through the use of a common identifier called *KEY*.
Data are extracted by relation, which are defined by query.

Relation algebra to create new tables.

V Code	*V Name*	*Area_ha*	*Pop*	*Electric*
102	Bagru	500	2000	Y
105	Naila	1200	4090	N
205	Ghughara	900	2783	Y

Vcode	*Lit_per*	*Educ*
102	24	prm
105	33	sec
205	22	hig

Vcode	*sc_st*
102	800
105	1210
205	320

Relational data structure is time-consuming because of sequential search to find data.

Hierarchical Data Structure

Taking into consideration the data which have inherent quality of hierarchy (Tree structure), this database was introduced.

Each node can be divided into one or more additional node. Nodes have only one parent.
Stored data gets more and more detailed as one branches further out on the tree.
This data structure have one to many relationship.
Cannot go between records at the same level unless they share the same parent.
Only one link between two records.
Travel within database is restricted to path up and down.
Large index files have to be maintained and certain attribute value may have to be repeated many times leading to data redundancy which increases storage and access cost.

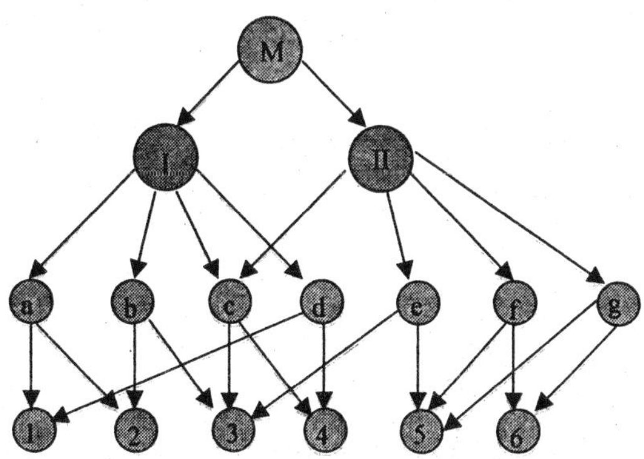

Network Data Structure

This structure is similar to hierarchical structure with the exception that in this structure a node may have more than one parent.

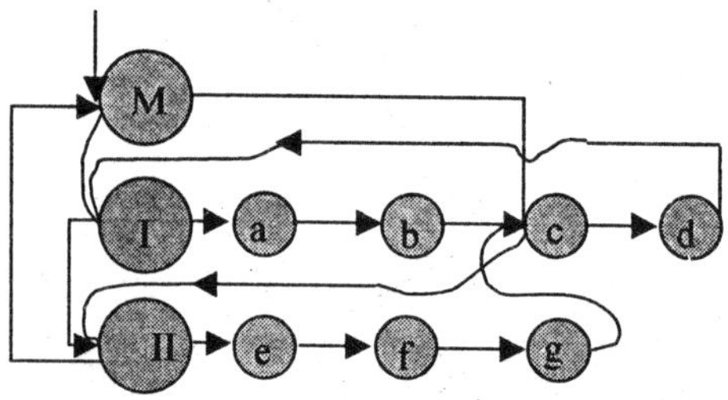

Each node can be divided into one or more additional nodes. Node can be have many parent.

Stored data get more and more detailed as one branch further out on the tree.

This data structure has one to many and many to one relationship.

Can go between records at the same level.

The database is enlarged by overhead of pointers. These pointers must be updated every time a change is made to database causing considerable overhead.

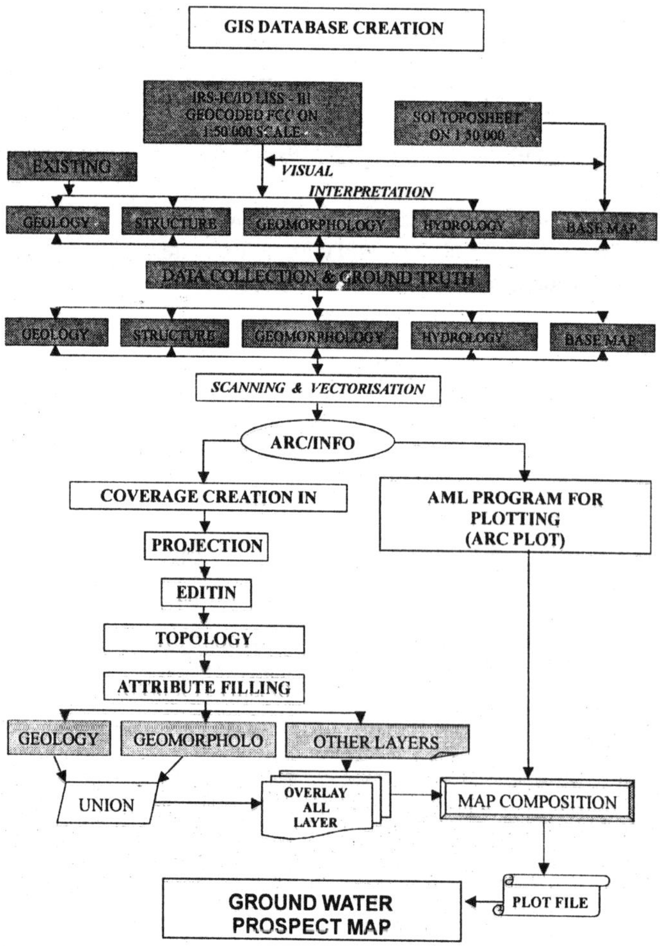

ERRORS IN GIS

Errors in GIS environment can be classified into following groups:

Error Associated with Data

Map scale—Non-availability of data at proper scale or use of data at different scales.

Density of observations—Sparsely dense dataset is less reliable.
Relevance of data—Use of surrogate data leads to errors.
Data inaccuracy—Positional, elevation, minimum mapable unit etc.
Inaccuracy of contents—Attributes is erroneously attached.

Errors Associated with Processing

Map digitization errors—due to boundary location problems on maps and errors associated with digital representation of feature.

Rasterisation errors—due to topological mismatch arising during approximation by grid.
Spatial integration errors—due to map integration resulting in spurious polygons.
Generalization errors—Due to aggregation process when features are abstracted to lower scale.
Attribute mismatch errors
Misuse of logic.

12

SPATIAL MODELLING THROUGH GEOGRAPHIC INFORMATION SYSTEM

AMI RAWAL

The fundamental language of Geography is the MAP. The map is a model of spatial phenomena. According to Demers: "It is an abstraction. It is not a miniature version of reality that is meant to show every detail of the study area."

Geographical data describe natural resources as well as cultural and human resources. The spatial distribution of such phenomena is usually shown graphically on map charts or images. Alternatively, the data may be in the form of lists of variables that are grouped by location and related to places shown on the maps. Typically the data appear as maps and images that describe such phenomena as topography, soils, vegetation, water surface bodies, land use, administrative boundaries, land ownership, population distribution, incidence of disease and so on or they are statistical data sets that can be related to such maps.

Researchers and practitioners in geography as well as in other disciplines have dealt for many years with problems relating to the analysis and manipulation of entities which exist within specific space-time framework. The most common medium for storing and displaying such coordinate based information has traditionally been the analog map document, most commonly seen in the form of a paper reproduction. The first map was apparently created before the first alphabet, so it is apparent that we have been working with these analog, storage and display devices for spatial data over an extended period of time. During this time, these devices have evolved to a high level of sophistication and today's map combines high density data storage with complex, color based displays.

WHY GIS?

Spatial data elements, recorded on maps are commonly recorded on the bases of a standard coordinate system (latitude, longitude and elevation with respect to sea level).

The identification of spatial data elements, the determination of their locations in a standard coordinate system, the measurement of their spatial attributes and the subsequent storage and portrayal of these data elements on maps is a common function. Retrieval and analysis of these map data normally involves visual inspection of the map document coupled with intuitive analysis, which is occasionally aided by simple measurement tools like scales and plan meters. It is easy to retrieve small amounts of data but the retrieval of larger numbers of map elements or attempts to determine in a quantitative fashion the complex relationships which exist between map elements is a very slow process.

Traditionally, integration of spatial data sets is carried out by transforming the two or more spatial data sets to a common map scale, creating a transparent or translucent overlay of each data set, registering these overlays so that the coordinate systems are aligned and then manually creating a composite overlay sheet that shows those locations where the various phenomena classes being studied occur in spatial position. The time involved in this process has generally been so great that it is utilized far less than one might expect. The complexity of the map display represents a significant factor in slowing down the analysis process and in contributing to error generation.

Analog map documents also display another major problem as a data storage device; they are expensive and time consuming to change when updates need to be made to the spatial data base. Updating the analog spatial database requires manual changes. No mechanism exists to ensure that changes in one spatial data element are reflected in other elements which are logically linked to it.

The technology involved in the creation of these analog devices for the joint storage and display of spatial data has reached a high level of development but it has never succeeded in overcoming some of the basic handicaps. Therefore, the advent of the digital computer as a data handling devise soon raised the question of its applicability to the storage and manipulation of spatial data. The rapidly growing use of computer for handling geographic data has been part of a more recent trend within society towards a rapidly increasing reliance on the computer as a data handling and data analysis tool.

Within the realm of geographic data handling, this trend has been driven by both 'push' and 'pull' factors. The primary push away from the limitation of manual techniques and pull towards the use of computer is manyfolds.

- First, extremely large and complex data sets can be both compactly stored and rapidly retrieved with mechanical accuracy.
- Second, many of the quantitative and analytical techniques developed in the earth sciences, urban planning and natural resource management are limited in their practical application without the capacity and very rapid data processing that computer provide to deal with the large volumes of observational data required by these techniques.
- Thirdly, at the same time, there has been a complementary pull towards using digital tools as increasing amounts of data become available directly in digital form. This growing supply of ready made 'digital' data has resulted from parallel revolution in digital data capture techniques, such as those seen in global positioning, attribute management and remote sensor images.

GEOGRAPHIC INFORMATION SYSTEMS IS

- Any manual or computer based set of procedures, used to store and manipulate geographically referenced data.

- An information technology which stores, analyses and displays both spatial and non-spatial data.
- A database system in which most of the data are spatially indexed and upon which a set of procedures are operated in order to answer queries about spatial entities in the database.
- A decision support system involving the integration of spatially referenced data in a problem solving environment.

These geographic information systems comprise many sophisticated computer software but they all contain the following major components:

(1) A data input subsystem which collects and/or processes spatial data derived from existing maps, remote sensors etc.
(2) A storage and retrieval subsystem which organizes the spatial data in a form which permits it to be quickly retrieved by the user for subsequent analysis, as well as permitting rapid and accurate updates and corrections to be made to the spatial data base.
(3) A data manipulation and analysis subsystem which performs a variety of tasks such as changing the form of the data through user defined aggregation rules or producing estimates of parameters and constrains for various space time optimization or simulation models.
(4) A data reporting subsystem which is capable of displaying all or part of the original data base as well as manipulated data and the output from spatial models in tabular or map form. The creation of these map displays involves what is called digital or computer cartography. This is an area which represents a considerable conceptual extension of traditional cartographic approach as well as a substantial change in the tools utilized in creating the cartographic displays.

SOURCES OF GEOGRAPHICAL DATA

As described in the book titled "An introduction to Geographical information Systems" by Ian Heywood, Sarah Cornelius and Steve Carver—Data are observations we make from monitoring the real world. Data are collected as facts or evidence that may be processed to give them meaning and turn them into information, although the two terms are often used interchangeably. Most of the time data are raw numbers, such as those we see listed in a table. Here what we see are numbers that have no particular meaning. To make the number useful context has to be added. For others to interpret your tables or lists of numbers they would need to know to what the data refer and which scale or unit of measurement has been used for recording the data. With these details data become information. Therefore, information is data with meaning and context added.

GIS database creation is a complex operation. It involves processes like data capture, data verification and data structuring. Raw geographical data are available in many different analogue or digital forms, such as maps, aerial photographs, satellite images or tables. A spatial data can be built in various ways. They are:

- digitize existing analogue data
- acquire data in digital form
- carry out own survey of geographical entities
- interpolate from point observations to continuous surfaces.

In all cases the data must be geometrically registered to a generally accepted and properly defined coordinate system and coded so that can be stored in the internal database structure of the GIS being used.

METHODS OF DATA INPUT

Data encoding is the method of getting data into computer. Spatial data can be obtained from many different sources in many different formats. There are number of various methods of input of data to GIS.

All data in analogue form need to be converted into digital form. Methods used for the same are:

- Keyboard entry
- Manual digitization
- Automatic digitization
- Scanning

Keyboard entry is the entry of data into a file at a computer terminal. This technique is used for attribute data that are only available on paper. Attribute data are linked to the relevant map features in the spatial database using identification codes.

Manual digitizing is the most common method of encoding spatial features from paper maps. It is an appropriate method when selections of features are required from a paper map. This method is also useful for map encoding where it is important to reflect the topology of features. When only selected features are required to be digitized, manual digitizing is preferred. For manual digitizing Digitizer is used. The digitizer is an electronic or electromagnetic table upon which a map is placed. A sensing device embedded in the table can accurately locate the data points of the map with pointing device. The pointer can be a cursor device such as the mouse, or a puck or a pen like appliance mostly corded. Positioning the cursor or pen over a point on the map and pressing a button on it, sends an electrical signal directly to the computer indicating the cursor's coordinates with respect to the digitizer's frame of reference. The principal aim of digitizer is to input quickly and accurately the coordinates of points and bounding lines. The map to be captured is fixed to the digitizer surface with tape. The digitizing process begins by digitizing at least four points that bounds the map region. Digitizing accuracy is limited by the resolution of the digitizer itself, by skill of the operator and by the accuracy of the original data.

One type of automatic digitizer is the automatic line follower. This method is used when digital version of clear, distinctive lines on map are required. The method uses line sensitive device to follow the lines on the map. The line follower produces output as (x, y) co-ordinate strings. Automatic line followers are less common due to their complexity.

Scanning is another and most commonly used method of automatic digitizing. Scanners are more advanced input devices that perform much of the work of the entity input process with little help or intervention from the user. A scanner is a piece of hardware for converting an analogue source of document into digital format. All scanners work by sampling the source document using transmitted or reflected light (Jackson and Woodsford 1991). Using a charged coupled device (CCD) scanning technology, the flatbed scanners scans the map of image one line at a time and then compiles all the lines as a composite raster image.

SPATIAL DATA MODELLING

All geographic information systems are computer representation of some aspect of the real world. GIS represents simplified view of the world as it would be impossible to represent reality in its entirety in a computer. Traditionally maps have used symbols to represent real world features. Maps reveals three basic symbol types as described by Monmonier (1991):

1. Points
2. Lines
3. Areas

These are the basic spatial entities. Each is a simple two dimensional model that can represent a feature of the real world.

Points

Points are used to represent features that are too small to be represented as area. The data stored for point feature will include geographic location and details of what the feature is. Information of co-ordinate reference can also be given along with details that are represented by point.

Lines

Lines are used to represent features that are linear in nature like road or river. Lines can also be used to represent linear features that do not exist in reality, such as administrative boundaries. A line is an ordered set of points. It is a string of co-ordinates joined together in order and usually connected with straight line. Lines may be isolated or connected as a network. Networks can be regarded as separate data type but they are actually an extension of the line types.

Areas

Areas are represented by a closed set of lines and are used to define features such as fields, administrative areas. Area entities are often referred to as polygon. As with line features, some of the polygons exist on ground and some are imaginary.

There are two fundamental methods of representing geographical entities. They are:

(a) Raster method
(b) Vector method

SPATIAL DATA MODELS

Burrough (1986) observed that the human eye is highly efficient at recognizing shapes and forms but the computer needs to be instructed exactly how spatial patterns should be handled and displayed. Computers require precise and clear instructions on how to turn data about spatial entities into graphical representations. At present there are two main approaches in which computer can handle and display spatial entities. They are the Raster and Vector approaches. The data structures that have little to do with the graphic representation of cartographic objects are simple lists, ordered sequential files and indexed file systems.

Raster Data Representation

In Raster representation, the terrain is divided into number of parcels. A parcel or a unit cell is called a grid cell. This grid cells are generally squares and they are uniform in size. In Raster data structure geographic space is divided into discrete grids so it does not provide precise locational information. Instead of representing points with their absolute locations, they are represented as a single grid cell. The assumption is that somewhere inside that grid cell, a point object can be found.

Raster structures in the form of square grid cell are pieced together to represent an entire area. Raster data structure may seem to be undesired because of the lack of absolute location information. Raster data structure has numerous advantages also. They are relatively easy to conceptualize as a method of representing space. Data from satellite can be readily incorporated into raster based GIS without any change. Overlay operations and surface operations are simple to perform with this type of data structure.

Vector Data Representation

Another method of geographic space representation is called vector. A vector spatial data model uses two dimensional Cartesian (x, y) co-ordinate to store the shape of a spatial entity. In vector system the point is the basic building block from which all spatial entities are constructed. The simplest spatial entity, the point is represented by a single (x, y) co-ordinate pair. Line and area entities are constructed by connecting a series of pints into lines and polygons. The more complex the shape of line or area feature the greater the number of points required to represent it.

As a rule, raster data structures generally demand more data storage than their vector counterpart. As spatial entities becomes more complex, the data volume requirements of different data structuring techniques becomes closer. The world is not made up of cells or co-ordinates. Therefore, any data model Raster or vector in GIS will only be an approximation of reality. However in certain circumstances, one model may give a better representation than another. The road or rail network is better represented using vector data model, whereas vegetation map can be better represented by Raster model.

REFERENCES

Ian Heywood, Sarah Cornelius and Steve Carver, 2000, *An introduction to Geographical Information System.*

Demers Michael, 1999, *Fundamentals of Geographic Information System*, Second edition.

Burrough and McDonnell, 2000, *Principles of Geographical Information Systems.*

Reddy, M., Anji, 2001, *Text Book of Remote Sensing and Geographical Information Systems*, Second edition.

13

DIGITAL CARTOGRAPHY AS THE BASIC REQUIREMENT FOR GEOGRAPHERS

SUPARN PATHAK

INTRODUCTION

After designing the database and performing series of tasks to complete the analysis, the main task is to record the analysis results and present them graphically. Usually this is done by graphical representation of data (MAP) and written description of result (INFO-REPORT). A successful map conveys several types of information such as locational information, describes the location of a particular feature on the earth's surface and their spatial relationships. A map also contains information describing the characteristics of the feature.

Design and layout must be considered when producing output. A beautiful map will help conveying the intended message with greater impact. Some of the components of good map are:

- Design the components of the map
- Using symbols effectively
- Determining the purpose of map
- Defining map parameters (size and scale)
- Designing map layout
- Preparing symbol data
- Creating a final map

MAPS

What is a Map?

A map is a geographic representation of some part of the earth's surface. Cartographic design uses symbology representing features with symbol and typography (arranging presented matter) to convey information that is both easily understood and visually appealing to the map reader.

Map body—The primary mapped area. You can display more than one image of your primary mapped area within your document. For example, you may want to portray change by showing several images with differing but related information, such as population maps of various years. Your map may also contain a locator map (a smaller-scale map used to help the reader in understanding where the main area of interest is located), an inset map (used to give more detailed information of an area within the main map that may not easily be understood), or an index map (often used to show wherein a series of maps one map exists). All are used to assist in communicating your information to others.

The primary component of map consists of **Geographic features** and **Cartographic elements**. Geographic features include area features (represented by polygons), linear features (represented by arcs) and point features [(x, y location) represented by point elements]. Cartographic elements helps to make the map easier to read and interpret, and these includes Titles (text), Neatlines, Legends, North arrow, Logo, etc.

Types of Maps

General maps	**Thematic maps**
Location positional relationship and variety of features	Structures of a distribution of single attribute/relationship.
Qualitative symbology	**Quantitative symbology**

SYMBOLS

Types of Symbols

Symbol Definition

A set of parameters defines each symbol. The parameters common to most types of symbols are colour, pattern and size.

Colour: Specifys colour in three way—by colour name, by colour model and components colour value (RGP, CMY)

Pattern: Pattern refers to the repetition and rhythm of elements in the symbol, e.g. dotted line can be composed of evenly spaced dots or two dots together followed by a gap.

Size: Size determines the width and height of text and marker or the width of line symbols or the grain of the shade patterns.

These are just some of the parameters that define symbols. Other parameters apply to specific symbol types. For example, in addition to colour, pattern and size, text symbols are defined by:

Symbol file: ARC/INFO stores symbol definitions in symbol files. Each stored symbol has a unique number referencing its own set of definition. To use a symbol, you specify the symbol set and the number of the symbol in the file. e.g. PLOTTER.LIN is for line symbols, PLOTTER.SHD for shade symbol, PLOTTER.MRK for marker symbols and PLOTTER.TXT for text symbol.

Editing symbols: Four symbol editor provide a graphic interface that allows to quickly choose and modify a current symbol or edit an existing symbol or create a new symbol and symbol sets.

CREATION OF MAP

Major steps for map creation:

1. Define the purpose and requirement of the map.
2. Determine the size and scale of the map.
3. Design the layout of the map.
4. Prepare additional data.
5. Create final map.

In **Defining the purpose and requirement of the map,** the overall aim is to define the scope of the map, establish a list of objectives and thereby develop a plan to create the map. A map is meant to communicate and it can be packed with detail or be simple in content and yet still convey its overall message. It is also important to consider the intended reader's ability to interpret the information present in the map. Thus the content of the map is determined not only by why it is being produced, but also by who will use. List of required information to display on the map.

COVERAGES/GRID/IMAGES E.G. VILLAGE, ROADS, SETTLEMENT, ETC.

Size and scale of map depends on the map's purpose and requirement:

Optimum size: Optimum size or aspect ratio for map looks better with width-to-height ratio of 4.3 (Robinson *et al.*). The kind of map and size also depends on the various output devices available (graphic screen, plotters or postscript printers) and the type of material used (paper or film) for output generation.

For all practical reasons map are 'scaled representation'. The scale of map is devised such that the important data is retained and presented in a map of particular size. The most commonly used scale are 1:250,000 (1cm = 2.5 km) 1:50,000 (1cm = 0.5 km), etc.

Once it is known what information the map will contain and the physical parameters of the maps are, **layout** of the information should be considered. Layout is the actual placement of both geographic features and cartographic elements in such a way so that it gives a balance and proportional look. Often, trial and error is the only way to assess the overall balance.

Before creating the map, one need to prepare many other files or data, in addition to the coverages. Most of this work relates to the symbols you choose to represent the various map features. The step commonly includes two tasks.

Specifying the symbol value used to symbolize each features in a coverage. In this a item in the attribute table is assigned the symbol value, and to symbolize the feature (either line, point or polygon) use this item.

Creating look-up tables and key files that define the symbols and text to appear in a symbol key legend.

Another way to use attributes to specify symbols is to assess a look-up table. A look-up table is an INFO data file that contains two items (relate item and symbol). For each unique attribute value, there is a corresponding record in the look-up table associated with a specific symbol number. One can create and update LUT table in INFO.

Key Legend: Legends are used to explain the map symbols. To do this, one has to create a key legend file that defines what to display in the legend. A key legend file is a text file created using the text editor of the operating system (e.g. vi editor), where legend symbols are specified by dot and number (.56) along with its descriptive text.

You can also define legends using items in the feature attribute table (AAT, PAT, etc.). Lookup tables (INFO data file) you create to shade the landuse polygone (LUSE.LUT) can be used to create a legend by adding an item named DESCRIPTION.

This same information can appear and be accessed in the feature attribute table of the coverage.

Once all the preparatory work is done, you can begin creating your map on the computer, using the ARCPLOT (AP) program and ARC/INFO plot system.

USING ARC/INFO TO CREATE MAP

Once you have established your map design, size and scale, it is time to translate this layout into ARC/INFO commands in ARCPLOT program to generate a map composition and a graphic file from which hardcopy plots can be generated.

There are five basic steps for creating your map using ARCPLOT.

1. Indicate the display method for the map on the graphic screen or on a graphic plotter.
2. Specify the portion of the coverage to be drawn on the map or several adjacent coverage (i.e. specify map extent).
3. Specify certain parameters about the size and layout of map to translate coverage features from computer to the display surface. This include final size of the map page, position of the geographic features and the scale. This operation is known as *map-to-page transformation.*
4. Draw the geographic features on map, such as landuse area, road and stream lines and specify which symbols to use for drawing and labelling them.
5. Add additional cartographic elements to make map easier to read and understand. These include neatlines, title, legends, scale bar, north arrow, etc.

14

MATHEMATICAL MODELLING IN GRID ANALYSIS

DIBYENDU DUTTA

It is a raster or cell based geoprocessing toolbox that is integrated with ARC/INFO. It represents space from a locational perspective and can accurately portray continuous surfaces.

Grid uses a georelational model for geographic data. It is based on a combined rester based spatial model and a relational attribute model. In Grid, the inherent power of the grid modelling structure is coupled with the capabilities of a relational database that manages all attributes associated with the cell values.

The Data Structure

Grid is based on hierarchical tile block structure. A grid is divided into uniform square units called tiles. A tile is divided into blocks. Because there are more blocks on the y-axis than on the x-axis, a block is always rectangular. A block is made up of cells, arranged in a Cartesian matrix consisting of row and columns. Cells are square.

This tile-block structure allows grid to support random access to data—rapid retrieval of information maintained from any sub-section of a grid, regardless of the size of the database. This forms the spatial indexing. Most of the cases all data for a grid resides in a single tile.

Grid-> Tiles-> Blocks-> Cells

Advantages

GRID data is accessible seamlessly.

Data compression using an adaptive run-length encoding scheme—compresses individual blocks or strips of data. It tests each group of data and then decides whether cell-by-cell or run-length code compression. Spatial indexing and compression techniques are applied separately.

Adaptive compression techniques allow for efficient use of a wider variety of tools providing more flexibility while performing spatial analysis. ARL performs well for per cell (LOCAL), per neighbourhood (FOCAL), per zone (ZONAL), and per layer (GLOBAL).

Fast processing speeds through run-length code processing without expansion. The commands that process compressed data directly include all operators, local functions and DOCELL loops.

GRID allows to store and transparently process grids at different resolutions within the same database. If two or more input grids in an expression are at different resolutions, grid automatically resamples them to the coarsest resolution unless otherwise specified.

Automatically optimizes spatial indexing and compression, making them transparent.

Data Model

1. GRID supports categorical/discrete data, i.e. categorical object has known and definable boundaries. A integer value is associated with each cell in categorical grid. All the integer grids include an INFO table called VAT. It contains at least two items, VALUE and COUNT. Any number of additional items can be added.
2. It supports also the storage and manipulation of continuous data. A continuous surface is usually represented by grid with floating point values, examples are elevation, aspect, slope etc. It does not have a VAT, because most, if not all, cell values are unique and nature of continuous data excludes other associated attributes.
3. Full supports for NODATA when inadequate information exists for a cell, the cell location can be assigned NODATA value in grid. All operators and functions process NODATA cell and return predictable values.
4. Full power of the relational database management system can be applied to VAT, e.g. ADDITEM and DROPITEM.
5. Deriving grids: Grids can be derived from other attributes in the VAT, which increase the flexibility and power of spatial analysis.
6. Inputting data into GRID: Grids can easily be produced from ARC/INFO vector coverages. Required inputs are attribute upon which to grid, cell resolution and type of gridding process.
7. An ARC/INFO lattice is a grid in GRID. Synergy is created between GRID and TIN modules. Lattice also uses tile-block spatial indexing.
8. GRID can directly input image files from external image processing and raster based packages, e.g. ERDAS files can be converted to GRID.
9. GRID also supports other generic raster files, viz. TIFF, BIL, BIP and Sun raster files.
10. Map algebra language: GRID is based on map algebra. Map algebra is a data manipulation language designed specifically for geographic cell-based system. It describes cartographic spatial analysis.
11. Through spatial language, GRID furnishes functions specifically designed to process data and information in the grid-cell format. Type of processing can be either operators or/and functions.

Operators

They complete computation on one or more input data sets. GRID supports following operators, viz.

- Arithmetic
- Boolean (TRUE or FALSE)

- Relational (<< >< =)
- Bitwise
- Combinatorial (overlay of grids while carrying attribute)
- Logical (contained within a set)
- Accumulative (sum, divide, multiply...all value within the grid)
- Assignment (output equals)

Functions Can be of 4 Types

- *Local (or per cell):* Functions compute an output grid in which the output value at each location is a function of the corresponding input value at each location. The value of a cell determines the result of a local function, e.g. sin(x), cos(x), squart(x), exp(x,2), log(x), reclassification and selection function (select all cells in a circle). For multiple inputs, operators that return min, max, majority or minority values.
- *Focal (per neighbourhood):* Functions compute an output grid in which the output value at each cell location is a function of the input cells in the specified neighbourhood of each location. The value of particular location is determined by the values of the cells around it, e.g. 3 × 3, 25 × 25, circular neighbour, wedge neighbour, annulus neighbour. The functions it can perform are return of mean, standard deviation, sum or range of values within the immediate neighbourhood.
- *Zonal (or per zone):* Functions compute an output grid in which the output value at each cell location depends on the values of all the input cells on an input value grid that share the same input value defined by an input zone grid. All cell values from the input grid that fall within the specified zone of the input zone grid is used to complete output value for each cell.
- *Global (or per grid):* Functions compute an output grid in which the output value at each cell location is potentially a function of all the cells in the input grids.
  Four types of functions:
  - Euclidean distance
  - Weighted distance
  - Surface distance
  - Others

  By applying global function to a weighted (cost) surface, the cost of moving from a destination cell to the nearest source cell can be determined. Surface functions include slope, aspect, analytic hill shading, kriging, inverse distance weighted and trend surface interpolation.

Problems it Can Solve

Best address locational problems. Two greatest advantages are: the ability to represent continuous surfaces and the ability to store points, lines, polygons and surfaces uniformly (identical representation). A continuous surface is best represented by a cell based system since the attributes of a location in the cell based system are a function of where the location is relative to a phenomena that is progressively changing as it moves across space.

Because of identical data representation data combination is easier, all data types are treated as same. But due to this some accuracy is sacrificed.

Advantages of Cell-based Data Processing System

- When the attributes of a location are influenced by the attributes of the locations that surround it.
- The attributes of a location are influenced by where its position falls within another feature.
- Cell-based systems can be used for continuous surface data modelling.
- Optimum allocation and surface determinations, e.g. determining the optimal course of a new road over a non-network surface such as soil map; drainage map creation from elevation data.

Basic Concepts in Cell Based Processing

Modelling locational problems is an inherent strength of the cell-based storage and analysis environment.

Grid: Store geographic data, but view and store surfaces differently. Grid based system divide the world into discrete uniform cells. Each cell is given a value to correspond to the feature. Location is not defined as an attribute but is inherent to storage structure. Every cell in a grid has a unique row and column.

Grid and Raster

Raster: It refers to a matrix composed of pixels or cells. Each pixel or cell stores a numeric value.

Grid: It is an implementation of generic raster data structure. Uses blocks and tiles for spatial indexing and adaptive run-length compression technique. Value associated with the cells in a grid can be discrete or continuous.

Grid Resolution: More homogeneous the area, larger cell size can be used with accuracy. With increase in size of the cell, resolution decreases.

Values: Value in a cell is an identifier that defined the class group, category or member of the cell belongs. Value can be integer or floating point. Cell locations with the same value belong to the same zone. There is usually one-to-many relationship between grid cell values (or codes) and the number of cells that are assigned the code.

Measurements: Measurement values can be broken into 4 types, viz. ratio, interval, ordinal and nominal.

- *Ratio:* Values derived relative to a fixed zero point on a linear scale, e.g. age, distance, weight etc.
- *Interval:* Time of day, pH scale etc. Values on a linear calibrated scale but not relative to a true zero point in time or space.
- *Ordinal:* It determines the position, do not establish magnitude or relative proportions, e.g. 1st, 2nd, better, stronger etc.

- *Nominal:* Values are used to identify one instance from another. They may also establish the group, class, member or category with which the object is associated. The values are qualities, with no relation to a fixed point or a linear scale.

Attribute handling within and between grids is most effective and efficient when using nominal measurements.

Grid Registration

Each location on the ground is represented by the same x, y cell address on the different grid inputs. Each location of an input grid have a cell address. Processing the grids is the same as processing matrices which is very fast in computer.

Methods of determining cell value:

- Centroid
- Predominant type
- Most important type
- Percentage breakdown

Each method is suitable for certain type of data, choice depends on the type of data to be gridded and the analysis to be performed.

Centroid

Suitable for continuous data, e.g. elevation. Each cell is assigned the value of the feature that passes through the centre of the cell.

Predominant (majority weighting)

The value of the feature that fills the majority of the cell is assigned to the location. This is good for non-continuous (discrete) data such as land cover, soils etc. When a grid is overlaid in the griddling process onto a vegetation coverage.

Most Important

Each cell is assigned the value associated with the features that have been specified as more important to the study (weightage is assigned with the categories).

Percentage Break Down

A cell is assigned several values one per feature, according to per cent each feature occupies within the cell. Difficult and costly method. Useful for statistical data.

Data compression: Two most important types,

- *Cell by cell:* For data that has a unique value for every cell. True for floating point grids or continuous surfaces, e.g. 1,1:0; 1,2:0; 1,3:4......
- *Run-length code:* When adjacent cells have same value, runs are created and data can be compressed. It stores data by row. The column range information is stored as a

from-column to a to-column along with the value associated with that cell or group of cells, e.g. 1:1-16:0, 2:1-9:0, 10-11:1, 12-16:0......

- *Quad trees:* It first divides the grid into 4 equal smaller squares when cells in the grid have different values. If all the cells within any of the 4 smaller square have the same value, the square will not be broken down any further. If the square is not homogeneous it will further be broken into 4 small squares. The process continues until each square only represents cells with the same values.

Raster Data Source

Raster digital images, Grid cell GIS data, Scanned imagery.

GRID: It describes the features and characteristics of an area and their relative positions in space.

CELL: Units of grid, square and same size which is defined by the user.

Row and Column

Cells are arranged in row and columns and produce Cartesian matrix. Each cell has unique row and column address.

VALUE: Each cell is assigned value to identify or describe the class, category or group that the cell belongs to. Within the theme which the grid describes. It can also represent magnitude, distance etc. Both integer and FP are supported.

Integer: For categorical data

FP: For continuous surface

ZONE: Two or more cells with same value, belong to same zone. Zone is consists of cells which can be connected or disconnected or both.

REGIONS: Each group of connected cells in a zone is considered as a region.

NODATA: Also known as null value, is treated differently from any other value by all operators and functions. It represents either no or insufficient information available about a particular characteristics. It can be processed in 2 ways.

- By assigning NODATA to the output cell location if NODATA value exists for the location of any of the inputs in an operator or local function, focal function or zonal function.
- Ignoring the cell and continue calculation for other valid values.
- Ignoring not possible with operators between two or with local functions. When NODATA is in a focal function, or zonal function, by default, the sum, median, majority or minority of all cells with known values can be calculated and assigned to outgrid.
- VAT: For categorical (integer grids), similar to PAT first term is value, second term count. Both are mandatory.

Data Representation

Point data: Represented by the smallest unit of a grid, a cell. Smaller the cell size, smaller the area and closer the representation of the point feature.

Linear data: Can be represented by a series of connected cells. When both the scale and the resolution are exceedingly fine, with each cell standing for a very small area, the representation is accurate.

Polygon data: Best represented by a series of connected cells that best portray their shape. Finer the cell resolution and greater the cell number, more accurate the representations.

Surface data: Represented as a matrix of evenly spaced points called lattice. In cell view, each value applies equally to the area, but in lattice view it applies to the centre point of cell. Lattice view is used for all surface calculations.

Cell resolution: It is the cell size given in map units.

Spatial Processing

1. *Location determination*: GRID stores the origin in map units. Using origin coordinates and the cell size GRID can associate the cell row-column address with any given map x, y coordinates. Cell locations are specified in terms of map coordinates.
2. *Cell by cell processing*: All operations (except zonal and global) as cell-by-cell basis. Starts from top left and proceeds towards right and then top to bottom. The result of each cell processed is written in outgrid before the second is processed. Processing occurs for all operators and local and focal functions and within DOCELL block.
3. *Multigrid calculations*: When using multiple grids, operators and local functions process all values between grids. For zonal and global functions often treat one grid as the main processing grid and use other grid as means for defining the values and parameters necessary for completing calculations.
4. *Neighbourhood processing*: Focal functions also use cell by cell processing but unlike operators and local functions, focal function consider the values of the cells within a specified neighbourhood around the processing cell. Neighbourhood can be square, circle, annulus or donut shape or wedge.
5. *Neighbourhood-notation processing*: Way to create custom neighbours and define the manner in which the neighbourhood should be processed. Following will return same result as that of FOCALSUM.

   Grid: outgrid = Ingrid (-1,1) + ingrid1 (-1,0) + ingrid1 (-1,-1) + ingrid1 (0,1) + ingrid1 (0,0) + ingrid1 (0,-1) + ingrid1 (1,1) + ingrid1 (1,0) + ingrid1 (1,-1).

Zonal Calculations

Same as neighbourhood calculations, except that zone from a input grid decides the shape of the neighbourhood, processing is zone by zone.

Calculating Distance and Direction

Centre of one cell to centre of other cell when cells are apart the distance is measured as Euclidean distance. True direction is also calculated from centre to centre of cells.

Weight-distance Calculation

Cost path calculations utilize weighted surface grids (cost surface or impedence grids). Weigthted surface assigns a value to each cell that identifies some cost to pass through the cell. Start and destination points assumes cell centre.

For horizontal and vertical node a1(cost) = (cost1 + cost2)/2
For diagonal cells cost = [(cost1 + cost2) × 1.421]/2

Area Calculation

Area of cell × number of cells in a zone. If area to be calculated for each region of a discontinuous region, then REGIONGROUP function has to be run first which will assign new identifier zone value to every continuous area within the same value.

Surface Calculations

Values associated with a location represent cell centre. Aspect and slope is calculated from a cells relationship.

15

DATA MANAGEMENT MODEL THROUGH GIS

VINOD M. BOTHALE

Analysis subsystem is the heart of the GIS; it is what the GIS is really all about. This chapter provides a simple framework for analysis that should provide an ample survey of existing capabilities of most GIS softwares. GIS can be used for counting and locating objects, making measurements of length, area, shape, of the various entities, measuring distances between the objects, identifying spatial objects, reclassifying the available data into logical groupings, carry out neighbourhood operations, studying spatial pattern and arrangement, cartographic overlays of point, line and polygon data, cartographic modelling etc. Using these kind of functionalities one can solve the problems in different application domains to get fruitful solutions.

The manipulation and analysis functions of GIS need to be used in the combination so as to arrive at the solution of a problem at hand. The functions could be categorized in various groups:

- Spatial join
- Feature extraction
- Feature merging: Reclassification, Sliver removal
- Measurement, connectivity and neighbourhood operations
- Map database merging and splitting
- Centroid determination
- Statistical analysis
- Modeling

Geographic analysis allows to study the real world process by developing and applying manipulation and analysis criteria (models). These criteria illuminate underlying trends in geographic data, making new information available. A GIS enhances this process by providing tools which can be combined in meaningful sequence to reveal new or previously unidentified relationships within or between data sets, thus increasing better understanding of real world. The results of geographic analysis can be communicated in the form of maps, reports or both.

For doing any kind of analysis for arriving at desired results, the goals and objective must be defined which will set the sequence of analysis functions to be performed on the data. Generally, following steps are involved:

- Establish objectives and analysis criteria
- Prepare data for spatial operations
- Perform spatial operations
- Perform tabular analysis
- Evaluate and interpret results
- Refine the analysis if necessary
- Produce final maps and tabular reports

The some of the typical applications which can be accomplished using manipulation and analysis functions include:

- Water resources planning
- Land resources development planning
- Waste site selection
- Dam site selection
- Irrigation water management
- Agriculture production modelling
- City planning
- Biodiversity characterization
- Habitat suitability assessment

SPATIAL JOIN

Spatial join are also called spatial overlay operations. Among the most powerful and most frequently used features of the modern geographic information system is its ability to place the cartographic representation of thematic information of a selected theme over that of another. One process could be just visually overlaying the features of one map on another and other could be actually combining the spatial and tabular information of these maps in a single map. The later process is called spatial join or overlay. This brings the information contained in different themes in a single theme for combined analysis. Any application requires more than one theme (map) data to be analyzed for arriving at the conclusion. In such cases, spatial overlay of the concerned maps is must. The overlay operation could be point-in-polygon, line-in-polygon or polygon-in-polygon type of function. Spatial join operations are nothing but the logical combinations like union, intersection etc.

FEATURE EXTRACTION

Feature extraction refers to operations like clip or erase. The clip kind of operations extracts part of the area from the theme, e.g. It can generate a landuse map or taluk from a district landuse map. Similar to clipping operation, erase type of option erases the input map at the areas specified by the "erase map".

FEATURE MERGING

Feature merging operations derive new map by merging features in the map. It is generally used to perform following kind of operations:

- Reclassification (Elimination)
- Removal of sliver or unwanted small polygons created during analysis (Dissolving)

Reclassification merges adjacent polygons or lines, which have the same value for specified item. It is used to create a simplified coverage from one, which is more complex. While the input map may contain information concerning many feature attributes, the output map contains information about the item on which reclassification is carried out. For example, landuse map at level 3 may have kharif, rabi crop land as distinct classes (under level 2 classification CROPLAND) but if one requires a simplified map then the reclassification can be done which will combine kharif and rabi Land and show them as simply "cropland".

The spatial join of two or more maps normally generate small polygons with negligible area and these areas can be merged with the adjoining relatively bigger polygons. Feature merging can also be used to do this kind of operations.

MEASUREMENT AND NEIGHBOURHOOD FUNCTIONS

A wide variety of measurement tools are needed in a GIS. With raster-based data, the precision of a measurement is limited due to raster cell size; vector systems are ultimately limited by the precision of the stored node locations. In general, the measurement include the line length (length of the road), area of polygon (Village area). Shape measurement is also necessary for some typical applications. This normally involves measurement of perimeter and area of the polygon to give the polygon complexity.

Spatial analysis involves operations that consider the spatial arrangement of information in one or more data layers. They may be broken down into two families: operations concerned with connections between location, and operations concerned with local or neighbourhood characteristics One of the classic functions of a geographic information system is to be able to identify items in proximity to each other. This function has wide variety of applications, including impact and corridor analysis and in problems of site selection.

Other common requirements could be measuring the distance between a point and the other feature (How far is village from the road), distance between polygon to polygon (which is the nearest water body for a cropland patch) of line to polygon (which is the nearest canal point from the agricultural area). There are a variety of kinds of spatial analysis that are based on networks. Applications in this general area include optimum corridor or travel selection, and hydrology and flow functions. Interesting example could be to evaluate alternative routes for emergency vehicles, based on a combination of total length of the route plus anticipated congestion on surface streets which is dependent on the time of the day. Network analysis also gives the solution to location-allocation problem to optimize demand supply scenario. Hence path finding or optimal routing, location-allocation and facility management provides solutions to the complex day to day problems.

The buffering is very common analysis function which is used to find what lies in the neighbourhood of the feature. Buffering could be point buffering, line buffering or polygon buffering. With the application of point buffering one can find out the area and attributes of features around a location in specified distance. The line buffer could be useful for road or stream related applications.

MAP DATABASE MERGING AND SPLITTING

Map database merging operation combines polygon or network features from/to or more adjacent maps into single output map. It combines both feature coordinates and attribute table and recreates the topology. Three important edge matching checks that should be performed prior to joining adjacent maps are:

- Edge coordinates should match
- Node coordinates should match
- Polygon item value should match

Map database splitting creates new maps by overlaying two sets of features. It performs series of extract or clip operations, e.g. one can generate watershed wise talukwise landuse map from a single district landuse map. The number of maps created will be equal to number of taluk in the taluk map.

CENTROID DETERMINATION

A centroid is a common way to describe the "average" location of a line or polygon. It can be defined physically as the centre of mass of a two-or-three dimensional object. There is a easy way to think about a centroid. Consider taking a polygonal area on a rigid map, cutting it out along it is boundaries and then locating the point at which this two-dimensional object will balance on a single point. This point is the centroid of the area. With the centroid as a location key, we can rapidly find objects that are relatively near a specified location. Thus one can use the set of centroid locations as a spatial index for objects in the database.

Statistical Analysis

There are variety of statistical techniques that are common in modern GIS. Whether for quality assurance, or for summarizing a dataset as a data management support or for deriving new data during analysis, several statistical procedures are commonly required. These include:

Descriptive Statistics

The mean, median, and variance of the attribute values in a data layer are often needed for continuous variables.

Histogram or Frequency Counts

The histogram of a dataset provides us with the distribution of attribute values in a region.

Histograms and frequency counts can be extremely valuable as data screening tools. Histograms are useful for examining all types of variables.

Extreme Values

Locating the maximum and minimum attribute values in a specified area is often useful. As an example we may minimum and maximum population for villages in a given area or to find out the minimum and maximum depth in a water body.

Correlations and Cross-tabulation

In correlation analysis, we wish to compare the spatial distribution of attributes in two or more data layers, usually by calculating a correlation coefficient or linear regression equation when working with interval or ratio variables. For nominal or ordinal variables, a cross-tabulation compares the attributes in two data layers by determining the joint distribution of attributes.

Some GIS provides basic statistical functions but do not have exhaustive options for statistical analysis. In such cases data from the GIS has to be exported to the statistical packages like SPSS, MINITAB etc. and import data back to GIS after statistical analysis.

16

PROJECT MANAGEMENT THROUGH GIS APPLICATIONS

B.K. BHADRA

INTRODUCTION

Geographic Information System (GIS) form the backbone of scientific management of natural resources. Developments in computer technology have modernized the management of geographic information and resulted in a boost in GIS applications. A variety of geographic data and tabular information generated by the different agencies on different projections, scales and accuracy can be input into a single GIS database. Using GIS different forms of data (vector, raster etc.) can be brought to a common format and analyzed for meaningful decisions. The output from GIS system can be generated on desired scales. A variety of input and output media are possible to use on a GIS system.

GIS systems support logical querying of information stored in a GIS database, e.g. we can find out distance between any two points/places of interest, area of a polygon; attach list of information attributes or charts, diagram, pictures, etc. with a given feature in a GIS base; find out shortest route for reaching a place (while imposing many of the pre-conditions on the route search) or analyze any other network related problems. Further, we can also find out feature sets qualifying certain conditions, e.g. smaller than, bigger than, equal to, less than, smallest, biggest, within a radius of given km, etc. Buffer creation along a feature is another very important utility added with the GIS.

GIS SYSTEMS

Basic functions of GIS include capability to store voluminous data on different aspects of a theme/application, retrieval, integrate information stored in different layers, process, analyze and output analyzed information in desired scales/form in an efficient manner. Additionally, a GIS has facility for modelling, statistical analysis of stored information, map making/designing and limited scale image analysis facility.

A number of standard GIS systems and packages are available in the market. ARC/INFO, ARC/VIEW, GRASS, PAMAP, ILWIS, IDRISI, GIPSY, etc. are some of the commonly used GIS software in the country.

GIS APPLICATIONS

The awareness about advantages of GIS is increasing day by day among the decision-makers and organizations. A number of GIS studies in different thematic areas have been carried out in different application areas including geology. A number of GIS studies have been initiated by different organizations in the country on the project (experimental/operational) mode. The areas covered under various GIS application studies include Urban Information System (e.g. GIS for NCR and Jodhpur City Information System), Wasteland Information System, Agriculture and Soils Information System, Geological Information System (Project Vasundhara), Forest Information System, Water Resources Information System, Natural Resources Information System at district level (under NRIS project of DOS), NRDMS project of the DST, etc.

GIS APPLICATION AREAS

1. Agriculture

Crop/Non Crop
Crop Types
Crop Acreage
Crop yield
Horticulture
Plantations

2. Soil

Soil Types—Texture, Colour etc.
Soil Suitability
Soil Moisture
Soil Association

3. Forestry

Forest Type
Extent
Density
Clearings
Encroachments
Timber Volume
Forest Fire Areas
Damage Assessment

4. Environment

Application Areas

Environmental Impact.
Mining
River Valley Projects
Urbanisation
Hazardous Waste
Thermal Power Projects
Chemical Industries
Change Detection
Coastal Erosion
Oil Slick
Forest Degradation

5. Water Resources

Type/Extent Water Bodies
Reservoir Volume Estimation
Ground Water Prospecting
Snow Covered Areas
Snow Melt Runoff
Reservoir Sedimentation
Water Logged Areas
Leakage Along Canals
Drainage Network
Dry *versus* Wet Sources

6. Urban Studies

Urban Infrastructure
Sprawl/Growth
Facilities Planning/Mapping
City Planning
Urban Landuse
Alignment for New Roads

7. Integrated Studies

Land Suitability Analysis
District Planning
Watershed Prioritisation
GIS Studies

8. Marine Studies

Fisheries

Ocean Wind
Ocean Waves
Sea Ice
Ship Routing
Geoid

9. Coastal Studies

Land Use
Erosion
Wetland
Bathymetry
Aqua-culture

10. Geological Applications

Geological Mapping
Geomorphological Mapping
Structural Mapping
Mineral Exploration
Oil Exploration
Engineering Geology
Ground Water Studies
Geo-Environmental Studies
Cyclones
Floods
Drought
Forest Fires
Crop Pest and Disease
Volcanoes
Earthquakes
Landslides
Coal Fires

17

MODEL OF DATA MANAGEMENT IN GIS

RAKESH PALIWAL

The theme covers introduction to the Arc View and its use to create, edit, display, query, and analyze spatial and non-spatial data and also to create presentation-quality maps. The following are the goals to deal with spatial database:

Goals

- Creation and editing of spatial database
- Displaying of spatial database in a view
- Feature identification
- Querying a database (spatial and non-spatial)
- Creation maps for presentation

Arc views and themes: Creating views to display geographic data (themes); creating themes from x, y coordinates; manipulating the symbology and classification scheme for each theme; and labelling themes.

Arc View tables: Creating an Arc View table from a variety of tabular data sources, selecting from a table, joining and linking multiple tables, modifying the structure of a table, editing values in a table, and creating a chart for presenting tabular data.

Creating and editing themes: Using Arc View to create shape files, adding and editing shape file features, updating attribute information, and geocoding street addresses

Spatial query and analysis: Selecting features based on relationships between multiple themes and merging theme features.

Arc View layouts: Creating a final map for presentation and printing and combining views, tables, charts, images, north arrows, logos, and scale bars to create a final map.

ARCVIEW BASED PROJECT PREPARATION

This section provides information on project preparation in an ArcView environment. Table 17.1 provides the menu/button and their utility.

Table 17.1

Menu/Button	*Utility*
View	Open new view, Open Existing view, Print view
Tables	Create New Table, Open Table, Add table (e.g. *.DBF, INFO)
Layouts	Open new layout, Open existing layout, Print layout
File: New Project	Open new project
Open	Open existing project
Project	Save existing project
Save Project	Save project to new project name
Save as Project	
Project: Properties	To give creator's name and to change 'Selection Colour'
Rename	Rename view, table, layout etc.
Delete	Delete view, table, layout etc.

View Module

The view module enable viewing themes with desired symbology, spatial analysis, buffering etc. The basic features of this module are in Table 17.2.

Table 17.2

Menu/Button	*Utility*
Add Theme	To add new theme
Save Theme	To save the project with themes
Zoom In	To zoom in
Zoom Out	To Zoom Out
Zoom to full extent	To Zoom to full extent
Zoom to active theme	To Zoom to active theme
Zoom to previous extent	To Zoom to previous extent
Zoom to selected	To Zoom to selected
Theme properties	To see the theme path, to change theme name in legend
Query builder	To generate query
Clear Selected Features	To clear the selected features
Identity	To identify data related to particular point, line, and poly
Pointer	To bring mouse pointer to Normal mode
Select Feature	Select feature to zoom or view database of the selected feature
Measure	Tool to measure the length on the theme
Text	To assign title or any text to the theme
Draw Point, Line, Box	To draw point, line and box etc.
View Properties	To apply the theme properties, i.e. Map unit: Meters and Distance unit: Mile/meters/km etc.
Graphics: Attach Graphics	To attach the graphics, e.g. title, text, points, box etc.
Theme: Auto Label	To add the label on the theme e.g. Village Name from Village—Name field
Remove-label	To remove the label
Edit: Copy themes	To copy theme
Delete themes	To delete theme
Paste	To paste theme in same view or in another view and also to paste graphics, i.e. points, text, line, box etc.

Table Module

This module enable non-spatial data updation, joining of two tables, inserting new fields and perform arithmetic operations and statistics generation. Table 17.3 gives an overview of this module.

Table 17.3

Menu/Button	*Utility*
Add field	To Add field
Add Record	To Add Record
Delete Field	To Delete Field
Delete Records	To Delete Records
Start Editing/Editing	To update the database
Save Editing	To save updation
Stop Editing	To stop update or lock the table
Query/Query Builder	To generate simple or complex query
Summarize	To generate summary of the field and also as spatial database, i.e. new theme
Calculate	To calculate the field or record value
Statistics	To generate Statistics, e.g. length of the NH, MDR etc.
Sort: Ascending	To sort data of field in ascending order
Descending	To sort data if field in descending order
File: Export	To export the whole database/selected database to *.DBF file format to use in MS Excel etc.

Layout Module

Layout module enable map composition with title, scale, legend, etc. The output generated in this can be exported to other formats (jpg, tiff) or for generating prints. Table 17.4 shows the button/menu of this module.

Table 17.4

Menu/Button	*Utility*
Text	To add text or title the map
Legend Frame	To assign legend of the theme
Scale Frame	To add scale
Hide Grid	To hide the grid
Export	To export Map in JPG, TIFF or WMF format to use in MS Word or Website

SPATIAL DATABASE FORM AND CONVERSION TO ARC VIEW GIS FORMAT

The outputs of the study are essentially provided in two forms, viz. (a) Direct ARC INFO GIS coverage as layers and themes and (2) *.e00. The *.e00 are exported coverages for each of the layers. These files are used to import covers in ARCVIEW GIS software. The import facility that is available in ARCVIEW enable to convert them to use in this software. This is shown as

START ⟶ PROGRAMS ⟶ ESRI ⟶ IMPORT 7.1

The imported cover or direct cover can be accessed in ARCVIEW using ***Add Theme*** utility and then can be converted in shape file using ***Convert To Shapefile*** utility.

The shape format is the format of ARCVIEW GIS software which consist of the following files:

1. SHP : Shape file
2. SHX : Shape index file
3. DBF : Database file

SPATIAL DATABASE ANALYSIS

Display of Maps

Access and Preparation to View Spatial Database

All the themes generated can be accessed in Arc View's '***VIEW***' utility after import of *.e00 files. Each of the theme can be arranged and viewed or overlaid as per requirement. The **Legend Editor,** viz. Pen Palette, Colour Palette, Marker Palette, Fill Palette, Font Palette and Palette Manager etc. can be used to modify colors, size, and pattern of each of the feature displayed for better representation.

Feature Identification and Query Generation form the Spatial Database

Feature identification can be done using ARC VIEW through various inbuilt and customized queries. An appropriate query on each theme can be generated as the spatial database is designed for various simple and complex queries using ***Query Builder*** utility. The desired query can be saved in the form of a new *shape file.*

Use of Buffering Utility

The buffer of point (tehsil headquarters), line (National Highway/State Highway) or poly features could be done using the ***Create Buffers*** utility of the ARCVIEW software.

Map Generation

Map of various themes can be generated using '***LAYOUT***' utility. Title of the map, legends and scale can be added using ***Text, Legend Frame,*** and ***Scale Bar Frame*** utility. The thematic outputs can be printed or exported in JPG or TIFF format for further use in reports.

Spatial Database Updation

The theme (point, line and poly) could be updated using ***Start Editing*** utility of ***VIEW*** module.

Non-spatial Database Updation

The non-spatial attribute information could be updated in ***TABLE*** module using ***Start Editing, Add Filed, Delete Filed, Edit*** etc. utility.

Statistics Generation

The data statistics could be generated using ***Statistics*** and ***Summarize*** utility in TABLE Module.

DATA LINKING/JOINING

The external database in the form of *.DBF or INFO format could be linked to the spatial database using ***Join*** utility on the basis of Unique Identification Number.

18

MODEL OF THREE-DIMENSIONAL VIEW IN GIS

MILAP PUNIA

INTRODUCTION

Digital computer models are used in landscape design and planning and other disciplines for visualisation of proposals, evaluation of alternatives, and simulation of impacts. By digital models, we mean those intangible ones inside computer memory, which take form are appearance only when rendered, either to a computer screen, an immersive display environment, or computer generated physical model. These physical artefacts are what are most often seen and evaluated, but it is their source—the 'internal representation', or data model, data structure and associated software—which is critical in modelling enterprise, and is the real subject of our talk and its research agenda.

By looking at digital landscape model, viewer's response or experiences/preferences which are quite subjective in nature to the representation of a landscape in some way equivalent to same viewer's response to landscape *in situ*? Can be evaluated, that were otherwise very difficult to model. Or if so, or if not, what aspects of the representation are important (realism, colour, navigability, etc.)? This represents a vital and significant area of research into visualisation.

Visual questions are not the only ones that call for vital landscape models. GIS system creates maps of landscape preferences, suitability and visibility; CAD systems make rendering of road alignments geometry or estimates of cut and fill volumes; landscape ecologists use differential equations and particle systems to describe dynamic process like vegetation succession. For these, vision and visualisation—and even the details of human perception—are important.

Digital landscape models are made for purpose of 'visual inference', or for simulating and understanding behaviour or other invisible aspects of the landscape, require abstraction and simplification. Yet for many visual purposes, 'realistic' depictions are desirable. The conflicts between these two demands are substantial for landscape modelers. For the basic landscape elements—terrain, vegetation, and water—some standard techniques for convincing static visual representation have been developed, but many complicating questions and obstacles remain. In addition, landscapes are essentially dynamic, and digital techniques for representation of these dynamics are still in their infancy. Surveying these techniques, complications, and possibilities can lead us to some generalised about the promises and problems of landscape modelling.

Generally all, GIS software's gives their output in 2D and 2 and half D, but some of them have this capability to model the digital elevation data sets in 3D and offer enhanced rendering capability to look at topography from various dimensions and infer some results.

Landscape Modelling

Landscape means the natural world, in which we live, garden, work, and build, including both natural systems such as plants and weather, and also built systems, such as roads and cities.

Of course, in the real world that we live in, the landscape includes structures of all kinds, including buildings and bridges and cars, and a wide variety of animals whose activities are vital to the function and look of the landscape. An important form of dynamics in landscapes is their dynamic interaction with observers, or participants, or computer-users. Using modern technology including Java, JMF and VRML, "active worlds" can be modelled digitally.

Landscape Elements

The essential elements that most landscape models require are:

Landform (Land)
Vegetation (Forests, trees etc.)
Water
Structures (Buildings and sites)
Animals (including people and wildlife)
Atmosphere (sun, clouds etc.).

Each of these has it own techniques, and changes, for modelling and visualization.

Landform: Landform, or terrain, is the basis of any landscape, forming the (roughly) horizontal foundation upon which all else is arrayed. Landform is made up of various underlying geological substrates, with diverse surface coverings: exposed rocks, sand, gravel, synthetic paving, grass, leaf litter, and so on.

Vegetation: Trees, and assorted other vegetation including shrubs and grass, are perhaps the most visible elements of the usual landscape. They surround us at eye level with their trunks; they embrace us overhead with their canopies, their textures are underfoot. Leaves frame our views, grasses carpet the foreground. Foliage provides texture and colour in the landscape—infinite varieties of green and brown, then, in temperate climates vibrant yellows, oranges, and reds. Trees with their verticality and shadows accentuate the terrain they stand upon; groundcovers, like grass, provide the texture and grain of the landform.

Water: Water, even when unseen in the landscape, is everywhere, and all-important. Water is never alone in the landscape; it is held and shaped by landform, which is revealed at the edges, banks and beaches.

Structures: Structures are essential elements of landscape. It includes building and architecture and should be modelled by landscape modellers.

Animals: Animals, including people, are harder to model, but are essential elements and should not be omitted from landscape models and visualizations.

Atmosphere: Landscape modelling and visualization requires attention, beyond the constituent elements (landform, plants and water) to the two essential ingredients that shape how the landscape looks and feels: the atmosphere, and the light that passes through it. Sunlight, moonlight, or streetlights (or the lack of them) have everything to do with how we see the landscape and its elements.

Fig. 18.1: View of Landscape Over Luni basin

Although many advances have been made in being able to model landscapes and landscape elements, challenges still remain. The landscape will always be bigger, more complex, more encompassing, than any computer representation. And so techniques of abstraction, simplification and representation will always be essential.

Landscape Visualization

Viewing data in three dimensions gives you new perspectives. 3D viewing can provide insights (effective perception) that would not be readily apparent from a planimetric map of the same data. For example, instead of inferring the presence of a valley from the configuration contour lines or elevation datasets, we can actually see the valley and perceive the difference in height between the valley floor and a ridge.

Visualization has been an integral part of environmental simulation to facilitate comprehension of complex environmental processes. Traditionally, data analysis and visualization are often considered and performed as post-processing steps after a simulation has been run. Despite the advantage of these approaches, any user interaction with the model computation is often not allowed.

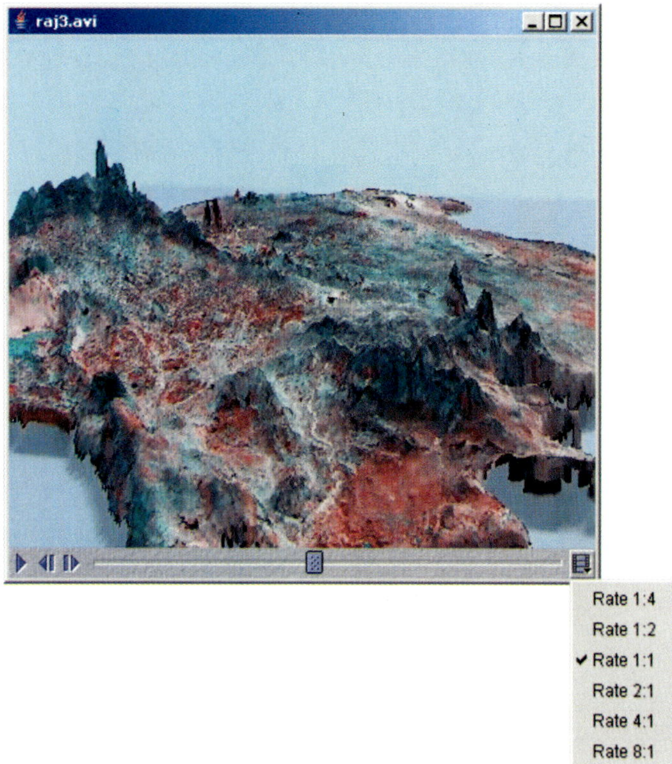

The 3D visualization of the model is accomplished by Java/VRML interaction through the External Authoring Interface (EAI). The interaction through process steering enables users to experiment with model computation and observe the model behaviour through dynamic 3D graphics, thereby enhancing investigation of the dynamic environmental process. The performance of the Java/VRML approach has also been examined through the comparison with the Java 3D and the conventional 2D approaches.

VR is a way for humans to visualize, manipulate and interact with computers and complex data. Through the Virtual Reality Modelling Language (VRML), an ISO standard for describing interactive 3D objects and worlds, a virtual environment can be modelled and constructed to engage user senses and provide more interactivity. Visualization plays critical roles in the environmental modeling process.

There are basically two ways of creating dynamic 3D content in VRML. One is animating a series of VRML files that have been generated in advance through the events routing (this method is also called post-processing animation), and the other is interacting with a VRML scene with Java.

VRML a Tool for 3D Modelling

VRML (**V**irtual **R**eality **M**odelling **L**anguage) is a hierarchical scene description language that defines the geometry and behaviour of a 3D scene or "world" and the way in which it is navigated by the user. VRML world files have the file extension .wrl (or .wrz for gzip compressed files) and require either a stand-alone application or web browser plug-in to be viewed.

Fig. 18.3: View for Hadoti, Marwar and Vagad Region

VRML is the only standardised (ISO/IEC 14772) 3D format suitable for Web delivery (due to its generally small file size). It is versatile, compact, extensible and constantly evolving-theoretically there is no limit to what can be achieved with VRML given sufficient computing power.

Due to rapid advances in computing speed and the availability of comprehensive production tools web 3D technologies are becoming more appealing and accessible to the wider market. As they say "today's supercomputer is tomorrow's personal computer"—we now have enough power on the average desktop to become fully immersed in these interactive 3D environments.

VRML is a graphical system that creates views of 3-dimensional images that change as the viewer's angle and position changes and light sources change. It can be used to create an environment or world that appears realistic. With VRML and certain software tools, you can create and view distributed, interactive 3D worlds that are rich with text, images, animation, sound, music and video.

VRML is not the only Web3D or Virtual Reality technology available. Now that the Web has matured and computers are sufficiently powerful, many companies are developing competing technologies

3D Visualization with VRML and X3D Over the past years, VRML has been the main standard for representing 3D content on the WEB. It provides a versatile platform for a variety of applications that uses 3D as a central metaphor or interface. One of the strengths of VRML is its tight integration with a variety of other WEB technologies and its ease of incorporating the benefits of those technologies, from graphic, audio, and video formats to scripting languages and network protocols. Another powerful feature of VRML is its easy extensibility and ability to add new node types and capabilities to the base language.

DISPLAY 3D IN ARCSCENE SOFTWARE

2D vector and raster data can be viewed in 3D if a layer is created and set its 3D viewing properties. We can assign z-values to features based on an attribute, a constant, or an expression. You can also specify a surface from which to derive the z-values of the features. As in above Figures 18.1 and 18.3, Z values are generated from SRTM layer of 90 metre grid resolution. Landsat 7 ETM plus mosaic of Rajasthan state is draped over the height values derived from SRTM data for enhancing the landscape elements.

3D Analyst adds three tabs to the Layer Properties dialog; these tabs allow us to modify how a layer is rendered in 3D. The tabs are called:

- Base Heights
- Extrusion
- Rendering

Most of the time, we want to create a layer that can be previewed in 3D, so we will set the height of the features in the layer using the Base Heights tab. Sometimes we may want to extrude features up or down from the base height—for example, if we want to create lines that show the depths of well point features. We can set how features are extruded on the Extrusion tab. If we want to shade features in our layer based on a light source, we can set that property of a layer from the Rendering tab. This tab also lets us fine-tune how our data is displayed while navigating.

Finally, these rendered elevation data sets along with some textured draped satellite image can be saved as dynamic animation in .avi, 2D and .wrl formet. But incorporation of other parameters that influence the perception and subjective experiences of landscape and its environment in dynamic sense is not so easy.

SOME CHALLENGES AND OBSTRUCTIONS

Modelling digital landscape is very complex phenomena and it cannot be accomplished fully without incorporating following:

1. Physical laws and systems
2. Sheer size/magnitude
3. Level of detail and resolution
4. Complexity and interaction
5. Dynamics
6. Objects *vs* fields
7. Level of abstraction
8. Human perception
9. Computer science and algorithms

Right know our main focus is on dynamics and human perception. So, by doing this at this stage can facilitate the user to understand the 3D rendering in effective way.

A special form of complexity in the landscape is that its dynamic: changing over time, and at various time scales. How to capture and represent these dynamics is a pressing problem? Dynamics in the landscape can be understood as three major types:

1. Movement *through* the landscape
2. Movement *of* the landscape
3. Interaction *with* the landscape

Interaction *with* the landscape is barely addressed in any digital model. We are still far from having haptic interface and tactile feedback mechanism available for most modellers, and we do not have good theoretical or empirical basis for determining the kinds of interactions we need to model. In this regard, virtual landscape offers a quite different range of kinds of interactions to model and explore, i.e. VRML, JAVA3D. Similarly, all these models and representations are made for, and by, humans with complex and specialized—and still only barely understood—perceptual systems. What difference does colour, or line weight, or screen resolution, make to a human evaluator? Under what circumstances? What role does representational conventions—as opposed to content—play in visual preference assessment?

19

MATHEMATICAL MODEL IN SLOPE ANALYSIS

MILAP PUNIA

INTRODUCTION

The design and management of water resources and other protected areas required for the researchers and managers to have access to a broad base of physiographic and geomorphic information. Derivation of landscape features and physiographic attributes, e.g. slope, drainage, watershed boundaries, based upon the satellite photogrammetric generated DEM, is a primary component of morphometric analysis (Zomer *et al.*, 2002).

Until relatively recent times, the field surveying techniques were the only means available for obtaining topographic data. The data was derived by measurements on the physical terrain surface. The basic principals of surveying have, therefore, remained practically unchanged for decades, although the changing technology has been reflected in improvements in surveying instrumentation. The field methods are, however, time consuming, subject to weather conditions and are very costly. The advent of high-resolution satellite data and development of Photogrammetric principals brought about a major change. Three-dimensional models for measurements have now replaced the physical terrain surface. The Photogrammetric techniques and advanced computers have made the digital elevation datasets available at different resolutions to researchers and resource managers.

Digital Elevation models can be used to derive a wealth of information about morphology of topography (U.S.G.S, 1987). The algorithms tradionally included in most raster processing system use neighborhood operations to calculate slope, aspect, and shaded relief (Klingebiel *et al.*, 1998) and point of inflection (Peuker and Douglas, 1975). While watershed and overland flow paths are closely related to slope, aspect, and inflection information, they also present non-neighbourhood problems such as determining direction of flow in the interior of a large flat area. To overcome these limitations, the River Tools Software provide an analyst with the ability to extract from DEM information on morphologic features and properties, specifically topographic depression and flow direction, that may be further processed to delineate application specific watersheds and overland flow paths.

These tools are relatively computer-intensive but require little analyst intervention, thus saving analyst time spent in manual interpretation, digitizing and are cost effective. In addition the results produced have the advantage of precise registration with the DEM. While the algorithms are essentially raster based, the products (watershed polygons, drainage line network, and tabular attribute information defining watershed linkages) can readily be converted to vector form (Arther and Spencer, 1978).

In this study an attempt has been made to generate a digital elevation model from IRS-1C satellite stereo pair and further processing for extracting morphometric parameters of a test site, i.e. small watershed on the northern-western side of Datia Township in Madhya Pradesh.

AIMS AND OBJECTIVE

As part of research effort aimed at delineating methodologies for geophysical/morphometric analysis and mapping in remote and/or rugged terrain, in this study morphometric attributes are extracted from DEM that is generated from IRS-1C Stereo pair.

1. Generation of DEM from IRS-1C Pan stereo pair of 5.8 m resolution.
2. Extraction of morphometric parameters like drainage pattern, watershed boundary and e.t.c. from DEM generated through IRS 1-C Pan stereo pair.

STUDY AREA

The study area is north-western part of M.P in Datia District corresponding to 54 K/6 of SOI topographic mapsheet. The relief of area is about 200 mts. The topography is like of rugged hills all around and covered with forest towards the west of Datia town.

DATA USED

SOI Topographic Map

These are the topographic maps available from Survey of India at 20m contour intervals at the scale of 1:50,000. These toposheets are giving various information to us like water resources, roads, railways along with various administrative (district, state) boundaries, forest boundaries, etc.

IRS-1C PAN Stereopairs

These are the satellite imageries available from National Remote Sensing Agency (NRSA), Hyderabad. The first stereopair is of 29 October,1998 and 11 November,1998, while second stereopair is of 3 May,1999 and 9 May,1999 respectively of the study area. The path and row numbers are 97 and 53 respectively. The spatial resolution of PAN data is 5.8m at nadir.

METHODOLOGY

Methodology is shown down below using a flowchart:

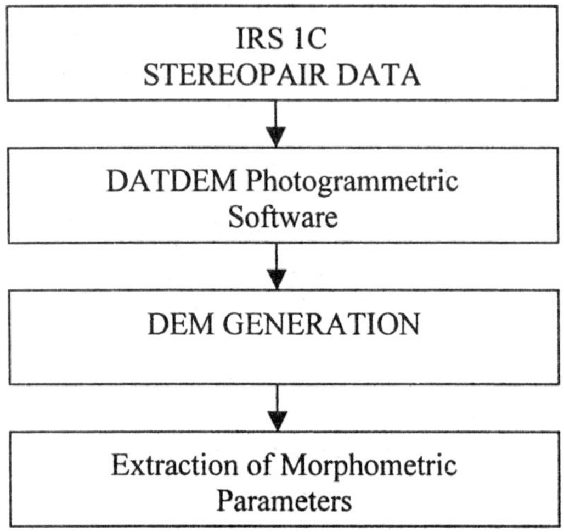

DEM Generation from IRS-1C Pan Stereo Pair

DEM from IRS-1C Pan stereo pair was generated using digital photogrammetry Software, DATDEM developed at SAC/ISRO on Unix platform. The level of accuracy is dependent upon factors like accurate georegistration of complete set of ground control points & B/H ratio. Two available stereopairs, corresponding to 54k/6 mapsheet, were processed for the study. The overlap area of stereopair is shown below in Fig. 19.1.

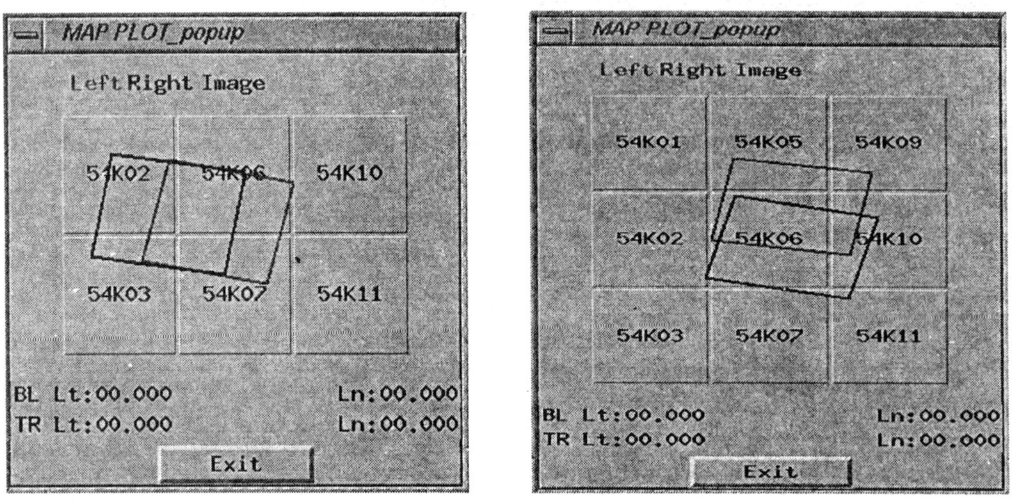

097_053_D8D7 097_053_D9D8

Fig. 19.1: Interactive map for 54 K/6 with IRS-1C Pan Stereo Pair

The Base to Height (B/H) ratio is the ratio of distance between two sub-satellites points (viewing objects) and mean satellite altitude. Better base to height ratio will provide better perspective view for stereo pair. Preferred B/H ratio should be minimum 0.5. In our case it is 0.6338 for one pair and for other it is 0.7298, which is satisfactory. The details of the stereopairs are as follows (Table 19.1).

Table 19.1: Details of the Stereopairs

Path–Row	*PAN Sub scene*	*Date of Acquisition*	*Incidence Angle*	*B/H*
97—53	D9	11 November, 1998	20.445900	0.6338
97—53	D8	29 October, 1998	19.650600	
97—53	D8	3 May, 1999	16.3668	0.7298
97—53	D7	9 May, 1999	18.800501	

The extraction of DEM from a stereopair is a multistep process. Firstly, the IRS-1C PAN-stereopair data is loaded. These stereopairs upon processing in the DATDEM software generates the DEMs. These DEMs were generated from independent set of GCPs using DATDEM. The ASCII files from DATDEM are imported into River tool software for further extraction of various geophysical/morphometric parameters.

Patching DEMs of two Stereo Pairs

We have two DEMs of upper and lower portion of NW and SW of toposheet No. 54 K/6. So for getting drainages which are common to both areas, first their ASCII files were appended to get a seamless patching of both DEMs in DATDEM software and then secondly, their irregular ASCII XYZ files were imported in River tools software and converted them into .rtg files for patching.

Extraction of Morphometric Parameters

Digital elevation dataset generated from IRS-1C satellite stereo pair is imported into River Tools software as Irregular XYZ ASCII files and placed into different folder. Now, dataset is preprocessed for morphological analysis. For this analysis, data have to go through few conditioning phases:

1. Filling depressions in a DEM,
2. Building flow directions, and
3. Flow accumulation in datasets.

All these phases are based on neighbourhood operations where calculations and decisions are made for a cell based on values in the eight cells that are spatially adjacent in the raster (Jenson and Domingue, 1998). For instance, a cell that is equal in elevation to all of its neighbour's elevations, meets the criteria to be classified as a member of a flat area. So depressions areas surrounded by the higher elevations, in the DEM data are nemesis of determining hydrologic flow directions because the depressions must fill before the flow can continue. Some depressions are data errors introduced in the DEM extraction process, while others represent real topographic features. Here the depressions are filled by increasing the values of cells in each depression by the cell with the lowest value on the depression boundary. Then they become flat areas across which water can be routed. It is assumed that these and other flat areas may be quit large and have more than one outflow point.

Further analysis was done for one watershed. Specifying the basin outlet point did identification of watershed. This was done in shaded relief grid for the DEM independently (Figure 19.2.1). Approximately outlet is masked on DEM. After extracting drainage network all relevant morphometric parameters were extracted for the specified watershed (Figure 19.2.2).

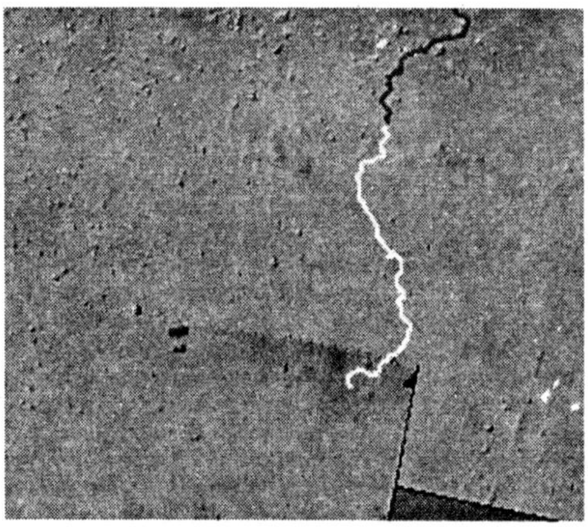

Fig. 19.2.1: Stereo Pair

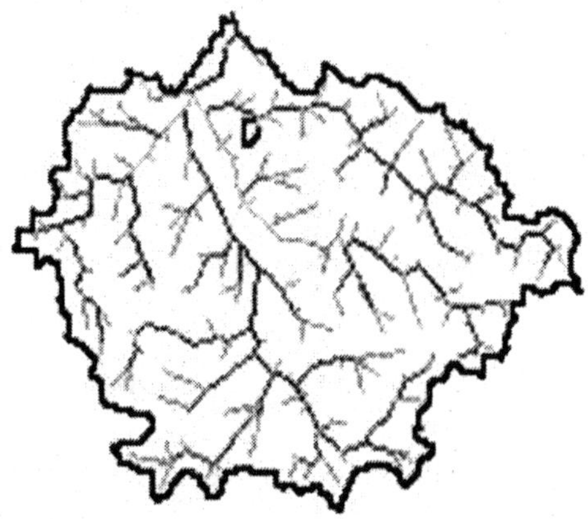

Fig. 19.2.2: Extracted Watershed

ANALYSIS

A series of map products, and morphological parameters are produced based upon the DEM using River tool software. These results are interpreted in terms of the usefulness of the DEM for the purpose of morphological analysis.

Morphometric analysis of drainage network is done using regression plots between various attributes like Strahler's order, Stream numbers, Drainage density, Slope, Length of stream, Basin area, Bifurcation ratio etc. Few values are given in Table 19.2 for interpretation.

Table 19.2: Values of Various Parameters for a Selected Drainage

Parameters	*Stereo Pair*
Outlet elevation:	220.552 (m)
Basin area:	44.531250 (km^2)
Basin relief:	0.062000000 (km)
Strahler order:	5
Network magnitude:	201
Network diameter:	37

(*Contd.*)

(*Contd.*)

Parameters	*Stereo Pair*
Longest length:	12.454951 (km)
Total channel length:	135.45268 (km)
Drainage density:	3.0417445 (km^{-1})
Number of streams 1 order	201
Number of streams 2 order	41
Number of streams 3 order	7
Number of streams 4 order	2
Number of streams 5 order	1

Number of Streams *vs* Order

The *Horton's law* of stream numbers states that there exists a geometric relationship between the number of streams of given order and corresponding stream order. A plot of the logrithm of N (no. of strems) versus order gives approx. a straight line with negative slope.The magnitude for IRS-1C satellite stereo-pair is 3.83. This states that the number of stream segments of each order form an inverse geometric sequence with order number. So, it is observed that number of streams is decreasing as the order is increasing.

Length of Streams *vs* Order

The *Horton's law* of stream length states that there exists a geometric relation between the average length of streams of given order and the corresponding order.

The order *versus* stream length relationship is observed with positive slope with straight line fit (Figure 19.3).The higher the stream order the longer the mean stream length (a positive relationship).

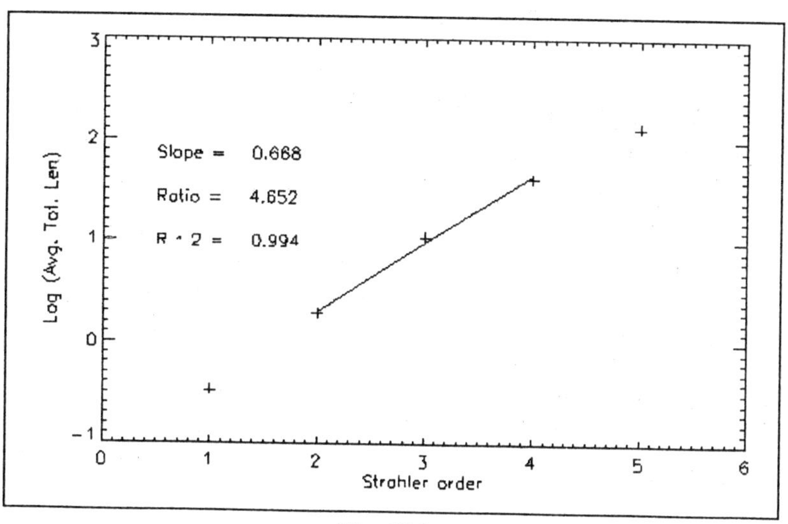

Fig. 19.3

Drainage Density *vs* Order

In case of curve between drainge density and Strahler's order, the distribution of number of first order, seconed order and higher orders streams can be seen (Figure 19.4). In stereo, we are getting upto fifth order. This is clear as the resolution is better in the same order. The average value of drainage density is decreasing as stream order is increasing. The measurement of drainage density provides a hydrologist or geomorphist with a useful numerical measure of landscape dissection and runoff potential (Rafael, 1990).

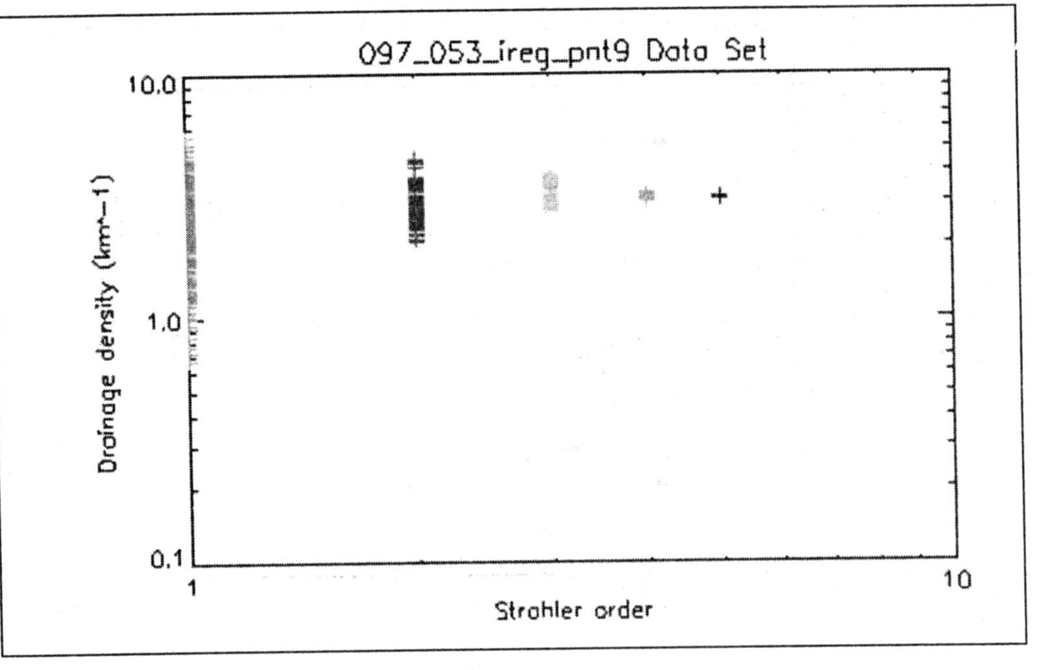

Fig. 19.4

Bifurcation Ratio

Not only are the numbers and lengths of particular stream orders important, but their ratios are quite instructive as well. The relationship between the number of streams in successive stream orders is called the **bifurcation ratio.** The utility of the bifurcation ratio lies in its ability to succinctly express the organization of a drainage basin and allow statistical tests. Their drainage patterns differ with stereo possessing a 1st/2nd bifurcation ratio of 4.9 and a 2nd/3rd ratio of 5.85. By this, we can accurately predict which watershed has the flashiest hydrograph at its mouth (Figure 19.2.1). So the maximum runoff is obtained in stereo.

Slope Maps

The slope values in stereo vary from 0 to 3.2. The values in stereo are reaching high figures, which is actually duc to spikes present in it. These spikes are easily recognized in the surface plot (Figure 19.5).

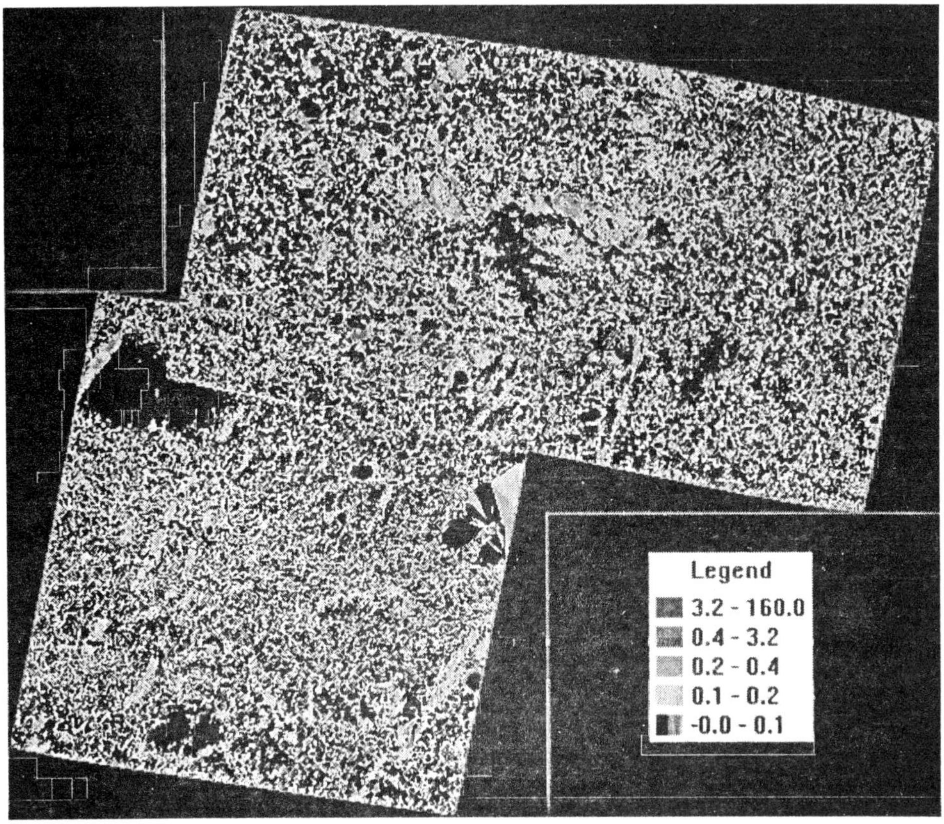

Fig. 19.5

CONCLUSION

Overall, the results of this study show that high resolution remote sensing stereo pair can provide essential and better morphometric information and a broad basis for further analysis of remote rugged or mountainous areas, especially in the absence of other data sources. Advantage ascribed to this methodology include cost effectiveness, timeliness, and equal access to high spatial resolution without any dependency. With ongoing improvement in the sensor technology and ground pixel resolution DEM extraction from stereo imagery will be becoming an increasingly useful tools for researchers and natural resource managers.

REFERENCES

Bras, L. Rafael, 1990. *Fluvial geomorphology*, Addison-Wesley Publishing Company, New York. pp. 567-604.

Gregory, K.J., and Walling, D.E., 1976. *Drainage basin form and process—A geomorphologic approach*, Edward Arnold Publisher, London, pp. 37-92.

Jenson, S.K. and Domingue, J.O., 1998. Extracting topographic structure from digital elevation data for geographic information system analysis: *Photogrammetric Engineering and Remote Sensing*, Vol. 54, No.11, pp. 1593—1600.

Noma, A. Arther and Spencer, S. Norman,1978. Development of Digital terrain data base system. Proceedings of the DTM Symposium, St.Louis-Missouri, American Society of Photogrammetry, pp. 493-505.

Zomer, R., Ustin, S., and Ives, J., 2002. Using satellite remote sensing for DEM extraction in complex mountainous terrain: Landscape analysis of the Makalu Barun National Park of eastern Nepal, *IJRS*, Vol.25, No.1, pp.125-143.

20

GENERATION OF DIGITAL ELEVATION MODEL

MILAP PUNIA

INTRODUCTION

The term Digital Elevation Model (DEM) has several different meanings:

- DEM is a generic term for digital topographic, in all its various forms. It is called a "model" because computers can use such data to model and automatically analyze the earth's topography in 3-dimensions, minimizing the need for labour-intensive human interpretation. Unless specifically referenced as a Digital Surface Model (DSM), the generic DEM normally implies elevations of the terrain (bare earth z-values) void of vegetation and man-made features. This bare-earth DEM is generally synonymous with a Digital Terrain Model (DTM). DEM elevations of lakes and rivers normally imply the water surface.
- According to US Geological Survey (USGS), a DEM is the digital cartographic representation of the elevation of the terrain at regularly spaced intervals in x and y directions, using z-values referenced to a common vertical datum. There are several types of standard USGS DEMs archived in the National Digital Cartographic Data Base (NDCDB) with uniform Universal Transverse Mercator (UTM) grid spacing (e.g., 10 × 10 meters or 30 × 30 meters) or arc second grid spacing in geographic coordinates (e.g., one elevation point every 1-arc-second in latitude and longitude).
- As used by others in the US and elsewhere, a DEM has bare earth z-values at regularly spaced intervals in x (Eastings) and y (Northings); however, grid spacing, datum, coordinate systems, data formats, and other characteristics may vary widely.

Related Definitions

Other related terms are defined as follows:

- A Digital Terrain Model (DTM) also has multiple definitions. In some countries, DTMs are synonymous with DEMs, representing the bare earth terrain with uniformly spaced z-values. As used herein, DTMs may be similar to DEMs, but they may also incorporate the elevation of significant topographic features on the land, plus mass points and breaklines that are irregularly spaced so as to better characterize the true shape of the bare earth terrain. The net result of DTMs is that the distinctive terrain features are more clearly defined and precisely located, and contours generated from DTMs more closely approximate the real shape of the terrain. Such DTMs are normally more expensive and time consuming to produce than uniformly spaced DEMs because breaklines are ill suited for automated collection, but DTMs may be technically superior to standard DEMs for many applications.

- A Digital Surface Model (DSM) is similar to a DEM or DTM, except that a DSM depicts the elevations of the top (reflective) surfaces of buildings, trees, towers, and other features elevated above the bare earth. DSMs are especially relevant for telecommunications management, forest management, air safety, and 3-D simulation and fly-through.

Acquisition of DTM data

The quality of the DTM and its derivatives generally depends more on the quality of the data input than on the type of data structure and interpolation algorithms. The following sources for digital elevation data exist.

Ground Surveys

Such data sets are often restricted to large scale topographic mapping, such as contaminated sites, planned reservoirs, and groundwater surfaces. The quality of the data is normally good.

Photogrammetric Methods based on Stereo Aerial Photography

The relation between stereo parallax and object elevation in the scene for orthogonal and central projection imagery is a clearly defined to which photogrammetry offers the solution. The principle is shown in Figure 20.1a where, for reasons of geometry, the height of a ground object (h_p) depends on the base B (the true distance between the two images), the flying height (H), the focal length (f) and the total parallax, as measured from the nadir positions ($p_l + p_r$). For oblique viewing, a correction has to be made (Figure 20.1b).

Automated DEM Generation from Satellite Stereo Imagery

The IRC-1C/1D and SPOT satellite has a steerable mirror so that nadir and off-nadir imagery is produced. The different look angles at orbital passes and the central projection creates stereo, parallax, (see Figure 20.1c), here the distance between the top and bottom of a vertical object on the earth's surface is different on the two images. In the case of SPOT imagery the parallax depends on the position of the object in the image because of the curvature of the earth and the oblique viewing. The differential image parallax due to elevation is illustrated in Figure 20.1d (symbols as in Figure 20.1a, b). The height measurement is based on the disparity in the image, or the parallax shift dP at reference height.

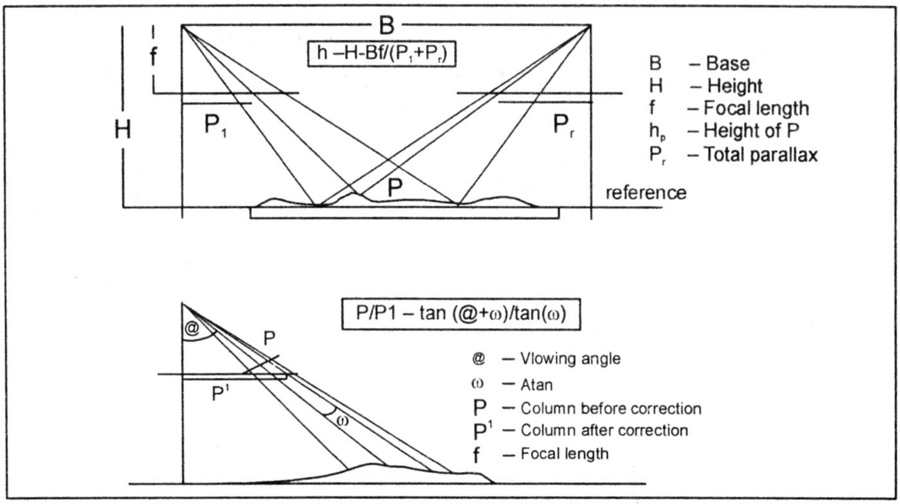

Fig. 1: a, b. Image parallax compared with relative elevation in epi-polar geometry and central elevation in line projection. Fig. 'a' shows oblique viewing angles. Fig. 'b' shows vertical projection; correction for panoramic distortion.

$$h = \frac{h.Dp}{(b + Dp)}, \text{because } dp = \frac{H.dp}{f}$$

$$h^1 = \frac{f.dP}{(b + dP)}; h^1 = \frac{f.h}{H} \text{ and } b = \frac{f.B}{H}$$

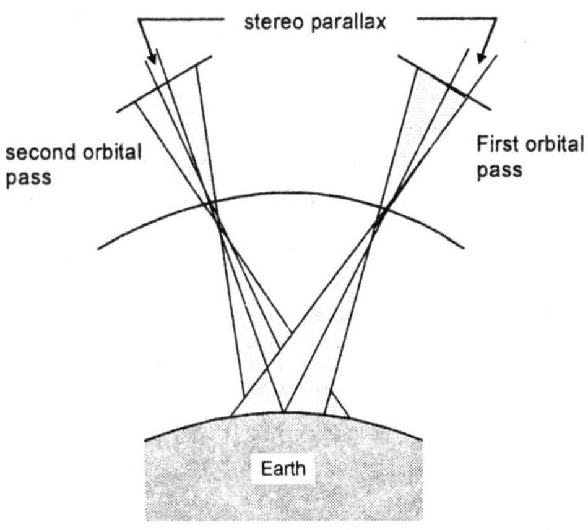

Fig. 20.1.1c: Parallax shift between corresponding left-hand and right-hand imaged height objects.

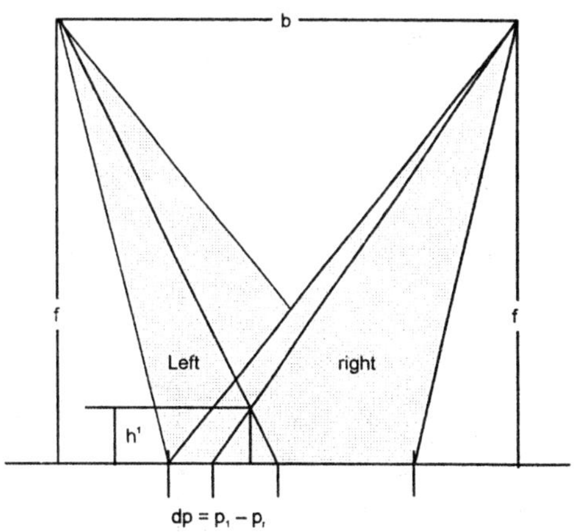

Fig. 20.1.1d: Differential Image parallax or disparity related to relative elevation, relative to a reference elevation. For disparity measured in the scene at reference height:
h = H.dP/(B + dP), because dP = H.dp/f.

Both images can be placed in a photogrammetric instrument. After relative and absolute orientation of the stereomodel using ground control points, the height information can be extracted by visual means, as with stereo aerial photographs.

In order to determine the parallax shift automatically, corresponding pixels of the same ground location on both images must also be found by automation. This is to allow the calculation of the height from the shift in position, i.e. the parallax. The process is known as 'stereo matching'.

Laser

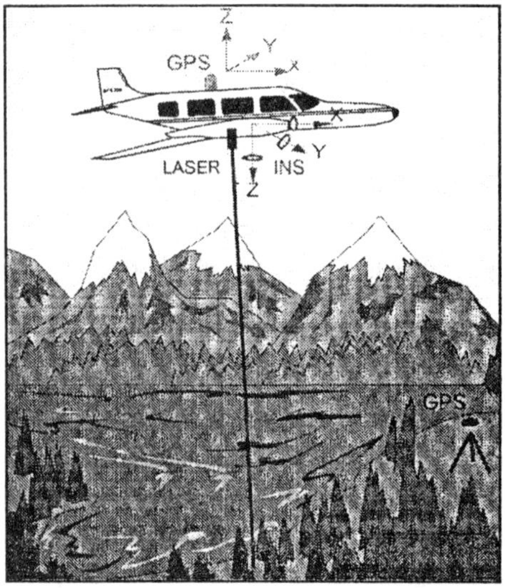

- GPS
  - Position of the aircraft
- INS
  - Attitude of the aircraft
- LASER
  - Range aircraft-ground

VIDEO CAMERA
  - Visual control

Fig. 20.2

Laser scanning can produce high accuracy and capability to penetrate tree foliage, and it is an attractive where photogrammetric data collection is difficult or expensive. There are various applications and advantages, namely determining forest areas, coastal monitoring, urban surface models, offers near real-time production, open pit mining, waste dumping, and special hydrological and geological applications.

Recently, NRSA, Hyderabad has procured ALTM package for LASER scanning data over Indian region, right know evaluation of its products is going on.

Radar Interferommetry

SAR is an active instrument, producing images under all weather conditions. Satellite SAR covers large areas with high location accuracy, e.g., ERS SAR scene: 100 km by 100 km and its absolute localisation accuracy < 50m. Radargrammetry produces a DTM from a stereo pair of radar images. INSAR is based on the combination of two SAR images from slightly different positions/orbits, exploiting the phase difference; high accuracy possible in open terrain.

- System Airborne and space borne
- **SAR** (Synthetic Aperture Radar)
- **INSAR**
- RS Internet services
- Radar processing software

Fig. 20.3: Mt. Etna

Digitizing Maps

Digitizing of maps is:

- economic, especially if contour separates available
- offers automation
- limited quality

Digitization can be manual or semi-automatic.

1. Digitizing tablets and SW (manual line following)
2. Line scanners (semi-automatic line following)

Accuracy depends upon map quality, contour interval, operator/system performance and best expected accuracy is 1/3 of the contour interval.

PRODUCTS THAT CAN BE DERIVED FROM A DEM

Various derived products can be obtained from DEMs whether the latter are in the form of altitude matrices, sets of irregular point data or triangulated networks. Products derived from DEMs are:

Orthophotographs
Block diagrams, profiles, and horizons
Volume estimation by numerical integration
Contour maps
Line of sight maps
Maps of slope, convexity, concavity, and aspect
Shaded relief maps
Drainage network and drainage basin delineation
Viewshade
Inter-visibility products
Display and visualisation for better depth preception

A CASE STUDY OF GENERATING DEM FROM IRS-1C SATELLITE STEREO PAIR

DEM from IRS-1C Pan stereo pair was generated using digital photogrammetry Software. The level of accuracy is dependent upon factors like accurate georegistration of complete set of ground control points and B/H ratio. Two available stereopairs, corresponding to 54k/6 mapsheet, were processed for the study. The overlap area of stereopair is shown below in Figure.This study area comes in Madhya Pradesh and it is west of Datia township, where relief value is about 320 metres.

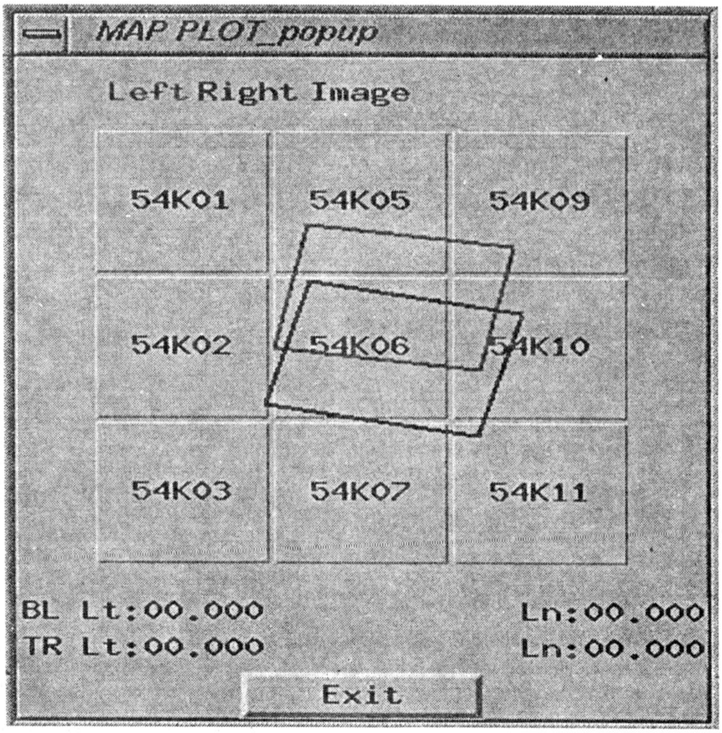

The Base to Height (B/H) ratio is the ratio of distance between two sub-satellites points (viewing objects) and mean satellite altitude. Better base to height ratio will provide better perspective view for stereo pair. Preferred B/H ratio should be minimum 0.5. In our case it is 0.6338 for one pair and for other it is 0.7298, which is satisfactory.

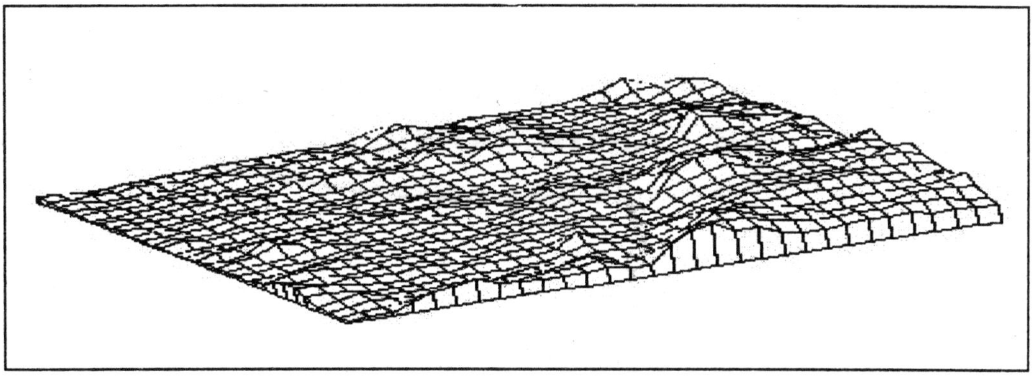

Fig. 20.5: DEM for West of Datia Township

The resultant DEM is as given above. The satellite stereo pair was of overlapping area. These stereo pair is processed in photogrammetric software where their conjugate points were given geographic location and height values with the help of the Survey of India topographical maps. After that inbuilt algorithm in the software matches their surrounding locations and computes the values by using epipolar lines principal of parallax in stereo pairs and finally, computes the value for whole image and generates the digital elevation model.

21

COORDINATION SYSTEM TRANSFORMATIONS OF THE CARTOGRAPHIC DATA

M.P. Punia

MAP PROJECTION

Map projections allow areas on the surface of the Earth (a spheroid) to be represented on a map (flat surface). A projection more precisely equates locations on a map with their true locations on the Earth.

However, expressing a three-dimensional surface in two dimensions involves distortion of some parameter, either shape, area, distance, or direction, and various projections cause different kinds of distortion. The individual characteristics of the projections make each useful for some applications and not useful for others.

Following are definitions and discussion of projection parameters required by various projections.

Semi-Major Axis of Ellipsoid: This defines the size of the Earth by the radius at its widest part. The value measured by Clarke in 1866, 6,378,206 metres.

Semi-Minor Axis of Ellipsoid: This is the earth's radius at its narrowest part. Clarke's value of 6,356,584 metres is the default.

A datum is a set of parameters defining a coordinate system, and a set of control points whose geometric relationships are known, either through measurement or calculation (Dewhurst 1990). A datum is defined by a spheroid, which approximates the shape of the Earth, and the spheroid's position relative to the center of the Earth.

There are many spheroids representing the shape of the Earth, and many more data based upon them. For a thorough discussion of spheroids, refer to the section titled 'The shape of the Earth' in 'Map projection concepts'.

A horizontal datum provides a frame of reference for measuring locations on the surface of the Earth. It defines the origin and orientation of latitude and longitude lines. A local datum aligns its spheroid to closely fit the Earth's surface in a particular area and its "origin point" is located on the surface of the Earth. The coordinates of the "origin point" are fixed and all other points are calculated from this control point. The coordinate system origin of a local datum is not at the centre of the Earth. NAD27 and the European Datum of 1950 are local data.

In the last fifteen years, satellite data has provided geodesists with new measurements to define the best Earth-fitting ellipsoid, which relates coordinates to the Earth's centre of mass. An Earth-centred, or geocentric, datum does not have an initial point of origin like a local datum. The Earth's centre of mass is, in a sense, the origin. The most recently developed and widely used datum is the World Geodetic System of 1984 (WGS84). It serves as the framework for supporting locational measurement worldwide. GPS measurements are based upon the WGS84 datum.

SPHERES AND SPHEROIDS

Although the Earth is best represented by a spheroid, the Earth is sometimes treated as a sphere to make mathematical calculations easier. The default sphere in ARC/INFO has a radius of 6,370,997 metres. The assumption that the Earth is a sphere is possible for small-scale maps, those less than 1:5,000,000. At this scale, the difference between a sphere and a spheroid is not detectable on a map. However, to maintain accuracy for larger-scale maps (scales of 1:1,000,000 or larger), the Earth must be treated as a spheroid.

A sphere is based on a circle, while a spheroid (or ellipsoid) is based on an ellipse. The shape of an ellipse is defined by two different radii. The longer axis is called the major axis and the shorter axis is known as the minor axis. The two radii are called the semi-major and the semi-minor axes respectively.

The difference in magnitude between the two axes of an ellipse can be expressed as a fraction or a decimal. This value defines the degree of ellipticity or flattening.

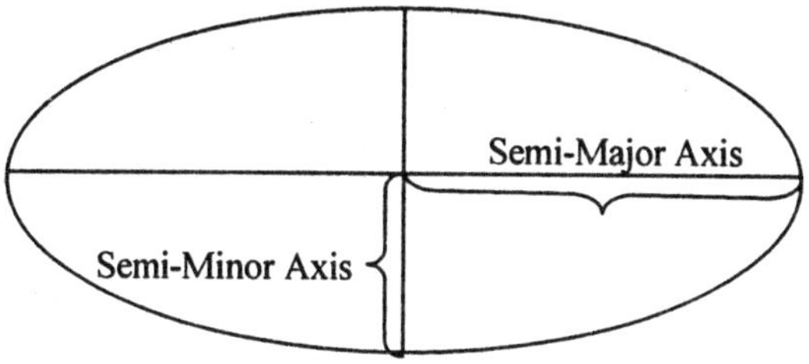

Values of ellipticity range between 0 and 1. An ellipticity of 0 means the two axes are equal, resulting in a circle. An ellipticity of 1 means the figure has only one axis, and appears as a straight line, whose length equals the length of the major axis. In general, large elliptical values describe narrow ellipses, and small elliptical values represent almost circular ellipses.

Just as rotating a circle about an axis defined by its diameter creates a sphere, rotating an ellipse about either its major or minor axis produces an ellipsoid. An ellipsoid which approximates a sphere is called a spheroid. An ellipsoid which approximates the shape of the Earth is formed by rotation around the minor, or polar, axis.

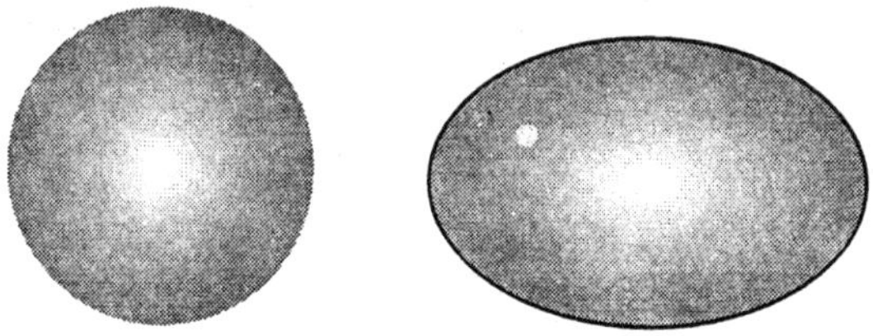

The ellipticity of a sphere is 0.0, whereas the ellipticity of the Earth is approximately 0.003353. Flattening is observed at the poles, and bulging occurs at the Equator. Therefore, the semi-major axis describes the equatorial radius, and the semi-minor axis represents the polar radius.

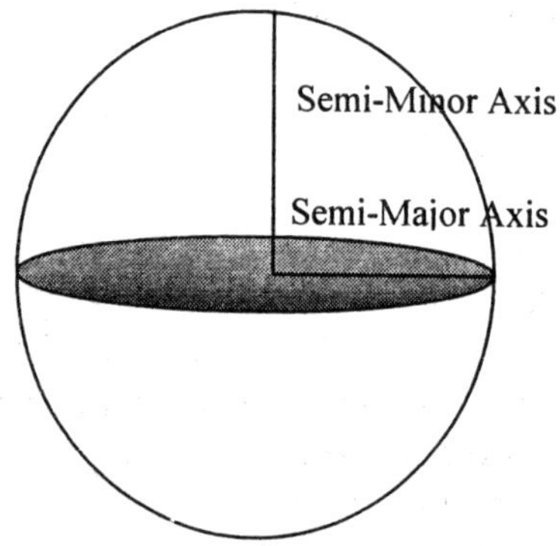

Defining Different Spheroids for Accurate Mapping

The Earth has been surveyed many times to understand better its surface features and their peculiar irregularities. The surveys have defined many spheroids to represent the Earth. Generally, a spheroid is chosen to fit one country or a particular area. The semi-major and semi-minor axes that best fit one geographical region are not necessarily the same ones that fit another region. Until recently, values determined by Clarke in 1866 described the spheroid commonly used with the reference datum for North America; this is often referred to as the North American Datum 1927 (NAD27). The semi-major axis of the Clarke 1866 spheroid is 6,378,206.4 metres, and the semi-minor axis was calculated as 6,356,583.8 metres.

Because of gravitational variations and variations in surface features, the Earth is neither a perfect sphere nor a perfect spheroid. Satellite technology has revealed several elliptical deviations; for example, the South Pole is closer to the Equator than the North Pole. It should be noted that satellite-determined spheroids are starting to replace the older ground-measured spheroids for reference calculations. A factor that must be taken into account before changing spheroids of reference is that it will change all previously measured values. Because of the complexity of changing spheroids, ground-measured spheroids are currently still in use and are still valid.

22

ARC/INFO—A TYPICAL GIS SYSTEM

M.P. Punia

ARC/INFO

ARC/INFO is a geographic information system used to automate, manipulate, analyse and display geographic data in digital form. It is developed and supported by "Environmental Systems Research Institute" (ESRI), California.

It is characterized by:

Its Data Model—supports various data models like Coverage, Grid, Tin, Lattice, Images etc.

GIS functions – Data input, Analysis, Data manipulation and Display.

Its Modular Design – Logical grouping of functions.

Ability to integrate many types of data—Map data, ASCII text files, DEMs, Classified images, RS data, AutoCAD files, other GIS package files etc.

Utility for developing application specific user interfaces.

Macro language AML – for programming.

Open architecture, which allows for the integration of numerous RDBMS.

Ability to operate on many types of computer with variety of graphics hardware.

Support for various Input and Output peripherals

ARC/INFO organizes geographic data using a relational and topological model. This facilitates efficient handling of: 1. Locational data describing locations and 2. Attribute data describing the characteristic of the feature.

The cartographic x, y coordinate data are used to identify geographic location and topology data are used to identify arc, node and polygon relationship. The advantage of topology structure is that it significantly reduces the amount of required storage space. Creation of topology file is fully automatic in ARC/INFO.

For descriptive data, a fully relational database management system lets one easily create and manage tables of statistical data including real and integer numbers, date, text and references to additional related tables of interpretive data. These data tables are linked to the map data.

FEW TERMINOLOGIES IN ARC/INFO

WORKSPACE – It is user's work area. A directory containing geographic data sets and INFO directory for feature attribute table.

COVERAGE – Map features are logically organized into set of layers or themes of information. Each layer in ARC/INFO is called Coverage. A coverage consists of topologically linked geographic features and their associated descriptive data stored.

A Coverage stores primary features—Arc, Node, Polygon and Label point, and secondary features like tics, links, map extent, annotation etc.

TICS – Points that represent location on the earth's surface for which real-world coordinates are known. Coverage uses a common set of tics so that layers can register together.

BND – The coverage or grid boundary file, which contains the minimum bounding rectangle.

LABEL – Link between a polygon and its associated attribute.

FEATURE – Type of feature represented in the coverage, viz. arc, node, polygon, label, tic.

FAT – Feature attribute table. A table used to store attribute information for specific feature class.

Coverage.PAT – for polygon and points.

Coverage.AAT – for Arcs.

Coverage.NAT – for nodes.

Coverage.TAT – for text.

NODE – Beginning and ending location of an arc.

VERTEX – A point within an arc that provides its shape.

WEED TOLERANCE – The minimum allowable distance between any two vertices along the arc.

FUZZY TOLERANCE – Smallest distance between all arc coordinates.

DANGLE LENGTH – Minimum length allowed for dangling arcs during topology creation.

SNAP DISTANCE – Within this distance the feature can be moved to coincide exactly with coordinates of another feature.

ENVIRONMENT – A set of parameters defining various display, editing and data manipulation condition that remain active during a session.

ITEM – In an attribute table a field of information displayed as a column.

PAN – To move the viewing window up, down and sideways.

ZOOM – To enlarge a region and show greater detail.

SYMBOL – A graphic pattern used to represent a feature. Line symbol represents an arc feature, Marker symbol represents a point feature and Shade symbol represents a polygon feature.

INFO – A tabular DBMS used by ARC/INFO to store, manipulate feature attributes and related tables.

UNIVERSE POLYGON – The first record in a .PAT file. It represents an area beyond the outer boundary of the coverage and always has user id as O. Its area equals the-ve sum of all polygons in the coverage.

RMS – Root mean square error. Deviation between tics location in the input coverage and those in the output coverage.

GUI – Graphical user interface. A graphical method of controlling how a user interacts with a computer to perform various tasks. User performs desired tasks by choosing options presented in the form of icons and lists.

23

ATTACHMENT OF ATTRIBUTE DATA TO EXISTING CARTOGRAPHIC INFORMATION

M.P. PUNIA

ARC/INFO MODULES

ARC is the main program environment in ARC/INFO. It contains commands which start other subsystems and also has extensive capabilities of:

Data Conversion

Map coverage digitizing/editing, error discovery.

Workspace and file management operations to list, rename, copy, delete and describe map data.

Coordinate projection and transformation (shift, rotation, scale) functions.

Management and manipulation of feature attributes.

For analytical operations, etc.

ARCEDIT MODULE

ARCEDIT is a unique graphic and database editor. It combines the capabilities of CAD functions with the power of geographic database. It combines all facilities for digitizing map coverage with a comprehensive set of editing commands.

ARCEDIT takes advantage of feature based editing. One can move, copy, add, delete, reshape and update point, line and polygon. It enables:

Editing feature attributes.
Each individual vertex within a line can be moved, added or deleted.
Lines can be reshaped and splined.
Parallel lines can be generated.
Usage of other data type as background display.
Construction of topology—creates feature attribute table.
Diagnosis and correction of digitizing errors—node and label errors.
Map sheets edge matching.

Annotation text can be scaled, rotated, spaced proportionally or follow a line feature.

Tabular attributes can be created and updated for any feature.

ARCPLOT MODULE

ARCPLOT is the interactive cartographic and mapping subsystems of ARC/INFO.

A map contains information describing the characteristics of the feature. A quality map helps in conveying the intended message/information with greater impact. Some of the components of a good map composition are proper scale and resolution, effective symbolization, map layout and generation of tabular reports etc.

ARCPLOT provides all the facilities to compose maps interactively on the graphics screen. It can perform all ARC/INFO mapping tasks to produce great variety of map graphics from simplescreen displays to high quality cartographic plots.

Features from any number of map coverages can be selected and drawn with different symbols.

Has an extensive symbol library and facilities for customizing symbols.

Preparation of legends.

Using map composer, one can easily create maps with all the basic map attributes.

Using ARCPLOT one can also query database

- Where do all features exist that meets certain set of conditions?
- What are all the conditions that exist in that area?
- What are the characteristics of a specific feature?

INFO

It is a complete relational database manager for the tabular data associated with the map features. It helps in organizing information and provides query and report tools to understand analyze information.

ARC/INFO keeps track of and update map feature attribute tables and maintain the relationship between each map feature and corresponding tabular record.

Features

Manipulation and updation of each feature's attribute by performing logical and arithmetic operations on the rows and columns of the table.

Data types include integers, decimal numbers, characters and dates.

Provide facilities for data definition, create/redefine items, data entry, update, sort and perform query.

Joins the data files to feature attribute table.

NETWORK MODULE

Network support the infrastructure modelling of the modern world. It is the system of interconnected linear features through which resources are transported or communication is achieved. The ARC/INFO NETWORK module facilitates the modelling of spatial networks.

It Provides Tools to

Find the shortest or minimum impedance path through a network (PATH). Provides a tool for vehicle routing.

Most efficient path to a series of locations (TOURING).

Assigns a portion of network to a location based on certain predetermined criteria. (Supply/ Demand).

Optimum Facility Siting

Whether the location is connected to other (TRACING).

Model the accessibility of location and interaction between location based on cost of travel and communication (Spatial Interaction).

The ARC/INFO data structure combined with data management and display function of NETWORK provides these facilities.

Applications such as marketing research, location of service centres, assigning students to a school, emergency vehicle dispatching are few of the many possibilities.

GRID MODULES

GRID is a raster or cell based geo-processing toolbox integrated with ARC/INFO. A cell based structure divides the area into individual locations.

Attributes of a location are influenced by the attributes of a location surrounding it. GRID is best suitable to address locational problems—where the locations and its surroundings are of importance.

A GRID is similar to ARC/INFO coverage. Data belonging to different themes are stored as separate GRIDs. GRID comprises of cells arranged in a matrix form. An integer value is associated with each cell. Each GRID has an associated Value Attribute Table (VAT) stored in INFO. Full power of relational database management system, as in case of INFO, can be applied to VAT.

Because Cells are uniform and representation of line, point and polygon are identical, different data types can be mixed with no prior preparation.

Features

GRID uses the graphics environment of ARCPLOT and ARCEDIT for all visual display.

Every query capability provided by ARCPLOT can be used on GRID.

GRID can be easily produced from ARC/INFO vector coverage.

GRID can directly input image files from Image Processing Packages. It supports several image file formats like TIFF, BIL, BIP, Sun Raster files, ERDAS files, etc.

GRID supports various operators like—Arithmetic, relational, Boolean, logical etc.

Greater analytical capabilities—Per Cell, Neighbourhood, Zonal, Per layer etc.

GRID compresses data and optimizes the database size.

Once the GRIDs are registered, the computation is very simple. No geometric detection, topology building and error checking is essential.

TIN MODULE

A set of software programs used to store, manage and analyze three-dimensional surfaces for ARC/INFO. TIN allows efficient generation of various surface models for analysis and display of terrain. TIN software package supports two data models for representing surfaces—Tin and Lattice.

Tin-Set of adjacent, non-overlapping triangles used to represent the facet of the surface. It is created from a set of irregularly spaced points having x, y coordinates and z value such as elevation. TIN data sources are ARC/INFO coverages (point, line, polygon), Contour maps, Stereo plotter data, ASCII files, etc.

Lattice—A lattice represents a surface using an array of equally spaced points and having a constituent sampling distance in x, y directions. Lattice data sources are Tins, DEMs, Point coverage having random points, Images, GRIDs, etc.

Features

Fully integrated with rest of ARC/INFO.

Provides data conversion, modelling, analysis and display capabilities for terrain and surfaces.

Contour stored as a series of arcs and points can be loaded.

Modelling capabilities include calculation of Slope, Aspect, Volume, Surface length, Profiling, Determination of stream network, Generation of contours, Visibility analysis.

Display capabilities include previewing of profiles or 3-dimensional surfaces displayed at specified view angle and orientation.

Surface drapes—Surface views can be draped with Mesh of lines, Coverage, Images, Contours etc.

ARC VIEW MODULE

Arc View is a software tool that creates an environment to display and query the contents of Spatial database. Arc View has tools to create, manage, draw, query, classify and symbolize views of geographic data. It is an icon based software package.

Arc View simplifies the display and query of geographic data by managing it in a file called View. A View can be thought of as a collection of themes. One can display all or some of the themes and can even control the order in which they are drawn.

Arc View is used to

Display the current status of your database.
Draw spatial data with desired colors and patterns.
Examine images and display Remote Sensing data.
Integrate vector and raster data.
Perform spatial and logical queries.
Browse the tables of database.

Few Components

Display window – Canvas on which views are drawn.

Legend editor – One can symbolize the features based on the unique value or based on an attribute.

Image file editor – Displays an image with a single band or composite of three bands.

Query builder – Ways to ask questions about spatial database based on logical expressions.

Table—Display a table of feature attributes associated with the theme. Logical query and few statistical calculations can also be performed.

Tools – Identify a feature, measure area, zoom, pan, select etc.

ARC TOOLS

General-purpose menu-based interface to ARC/INFO to perform common Geo-processing task with a better speed and ease. Arc Tools was developed using an object-oriented programming, which simplifies application development. Arc Tools supports three primary classes of objects – Collection of themes, Group of map objects, A set of GRID operations.

Arc Tools are functionally grouped into four tool sets—Map tools, Edit tools, Grid tools and Command tools.

Map Tools – Operates within ARCPLOT. Provides an environment for data display.

Query, map layouts creation and output file generation in standard graphic and plotter format.

Edit Tools – Operates within ARCEDIT. It has facilities to generate geographic features as well as tabular data.

Grid Tools – Operates within GRID. Provides sophisticated environment for Performing spatial analysis and data display. Allows editing of grids, model definition and analysis of results.

Command Tools – Operate within ARC. Extensive tool set for batch-oriented operations such as Overlay and Buffer analysis, data conversion, layer creation.

ARC STORM MODULE

Arc Storm (ARC Storage Manager) is a geographic data storage facility to manage ARC/INFO databases intended for public use (Multi-users). It provides a single point access to the database for multiple users without having to know where data physically resides.

ArcStorm manages the edits of multiple users on geographic data at feature and attribute level that reside in RDBMS like ORACLEI INGRES, SYBASE etc. All modifications are recorded as edit history.

Arc Express Module

Arc Express allows rapid display and navigation of very large data sets on X-window based workstations. It works in both ARCEDIT and ARCPLOT to increase drawing speed for interactive display, edit and query operations. Operations such as pan, zoom and redraw are performed within seconds. This allows support for many applications in which real-time display of database is critical.

ARC MACRO LANGUAGE

ARC Macro language provides full programming capabilities and a set of tools to tailor the user interface for an application. AML can be used throughout the ARC system to perform variety of tasks. AML allows to automate frequently performed actions, customize application, and develop menu-driven interfaces to meet the needs of end users. AML programs organize a sequence of ARC/INFO commands to automate the task. AML is similar in principal and functionality to command languages in its ability to execute actions like branching, variable manipulation and argument transfer etc. AML also provides tools to establish a screen layout that presents information in a logical and visually pleasing manner.

24

GEOGRAPHICAL INFORMATION SYSTEM AS A TOOL FOR GEOGRAPHICAL STUDY

RAJEEV JAIN

Rapidly declining computer hardware costs have made GIS affordable to an increasingly wider audience. More importantly, we have come to realize that geography (and the data describing it) is part of our everyday world; almost every decision we make is constrained, influenced or dictated by some fact of geography. For example:

- We send fire trucks to fires by fastest available routes.
- Central Government often awards grants to local governments based on population.
- We study disease by identifying areas of prevalence and rate of spread.

This demand for geographic information parallels the need for GIS, explaining its rapidly growing popularity.

GIS has become accessible to planners, policy makers and decision makers. It allows entry of many types of data in a single spatial framework. GIS offers the capability of collection, compilation, storage retrieval, analyzing, manipulation, displaying and integrating environmental, economic and social data in a single system.

It facilitates the followings:

- Overlays of data for purpose of comparison
- Updating of information to illustrate changes over time
- Changes of scale for micro-analysis
- Derivation of non-available data through manipulation of known factors
- Integration of physical and social science data sets
- Incorporation of remotely sensed data such as Satellite Imagery for continuous environmental monitoring
- Modelling of physical, economic and social processes for the purpose of simulation and prediction.
- What is GIS?
- Questions a GIS can answer

- Some applications of GIS
- Components of GIS
- Using a GIS

Definition of Geographic Information System (GIS)

Geographic Information System (GIS) "is computerised information storage, processing and retrieval system that have hardware and software specially designed to cope with geographically referenced spatial data and corresponding attribute information (tables, charts and statistics)".

Geographic Information System (GIS)—is an organized collection of computer hardware, software, geographic data and personnel designed to efficiently capture, store, update, manipulate, analyze and display all forms of geographically referenced information.

A computer system capable of holding and using data describing places on the earth's surface.

WHY USE A GIS?

- GIS is a powerful tool for handling spatial data.
- Digital format data is compact and large quantities of data can be maintained and retrieved at a greater speed and lower cost.
- Ability to manipulate and integrate different types of data in a single analysis.
- Ability to perform complex spatial analysis rapidly providing both qualitative and quantitative advantage.
- Planning scenarios, decision models and interactive process are normal functions of GIS.

Remote Sensing—Refers to the science and art of obtaining useful information about an object or phenomenon without actually coming into contact with the target.

Geographical Information Systems (GIS) are hardware and software systems which are designed to capture, store, manipulate and display spatial information and related attributes.

Remote Sensing and GIS in Development Planning

Remote Sensing is an emerging technology which contemplates collection of spatial spectral and temporal data of earth surface using air and space craft. Remote Sensing technique is powerful tool for evaluation of Natural Resources, both renewable and non-renewable with speed, accuracy and economy. In conjunction with ground survey, it offers an efficient time saving and cost effective method for assessment of Natural Resources on repetitive basis for planning the development activities of state.

The technology comprises application of Satellite data, Air borne data and Ground truth data. The data products from satellite are in Black and White imagery, false colour composites and computer compatible tapes, aerial photographs are of various scale can be analysed visually and digitally.

These information on Natural Resources help not only in assessing their latest status but also in planning their development.

Under Geographical Information System (GIS) environment, various information of natural resources in layer forms are integrated and when coupled with collateral data provide unique data base for district level planning taking care of natural resources development as well as infrastructure strengthening.

Both remote sensing and geographical information systems technology are involved in the analysis of phenomena which have geographic or spatial significance.

Both technologies have evolved rapidly, their early evolution was along separate paths. These paths are now converging rapidly since GIS technology is ideally suited to the analysis of spatial phenomena, and remote sensing is the most common source of spatially continuous data. Together these technologies offer the most promising and practical means to monitor and manage our use of the earth's natural and cultural resources.

REMOTE SENSING

Definition

Remote sensing is an art or science by which information about an object can be gathered from a distance without being in physical contact with it.

Two aspects of Remote Sensing:

(i) Air borne
(ii) Space borne

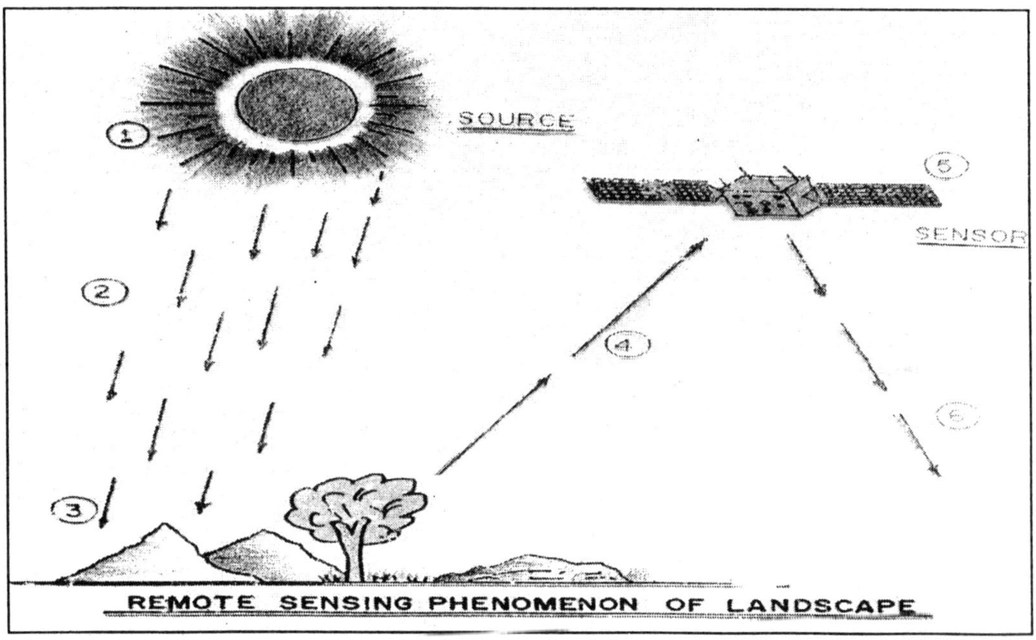

Based on sensor platform, remote sensing is classified into two categories:

Air Borne

In the form of aerial photographs which depict stereoscopic view of land scape, which is highly useful for micro level planning. Aerial Photographs are available in scale 1:10,000 to 1:60,000.

Space Borne

In the form of Satellite data which provides synoptic view of large area, repetitive coverage and data available in multi spectral bands.

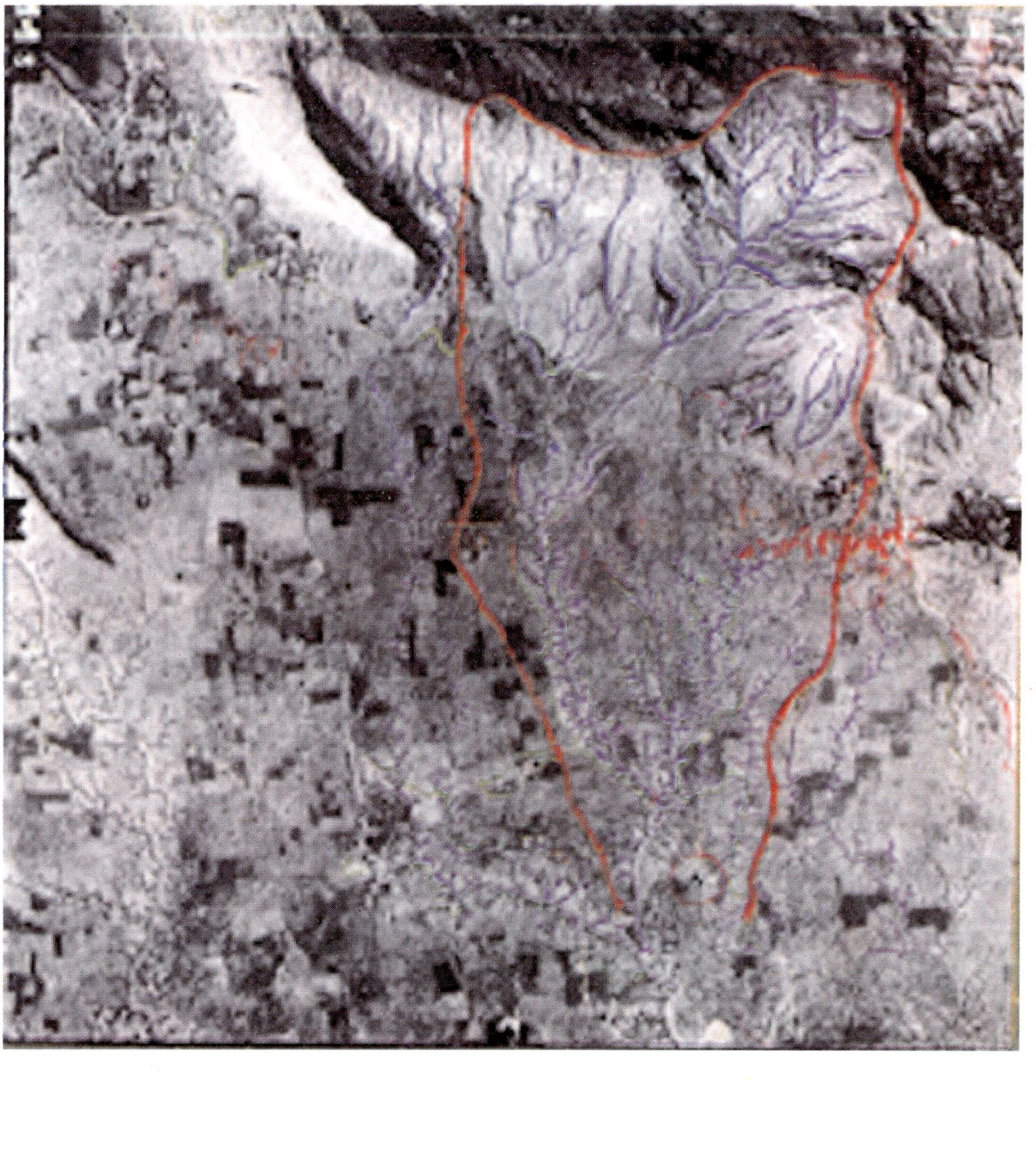

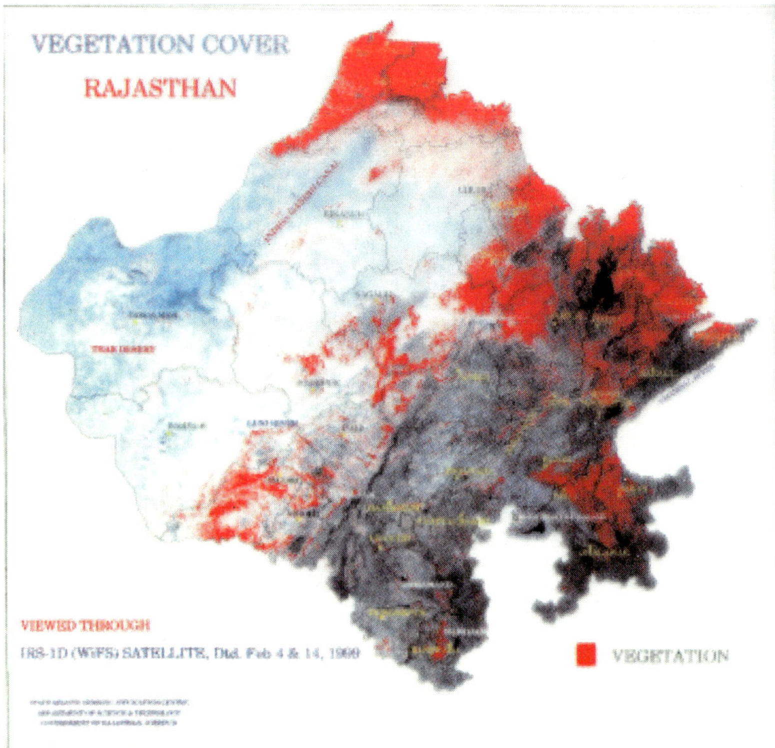
VEGETATION COVER
RAJASTHAN
VIEWED THROUGH
IRS-1D (WiFS) SATELLITE, Dtd. Feb 4 & 14, 1999
VEGETATION

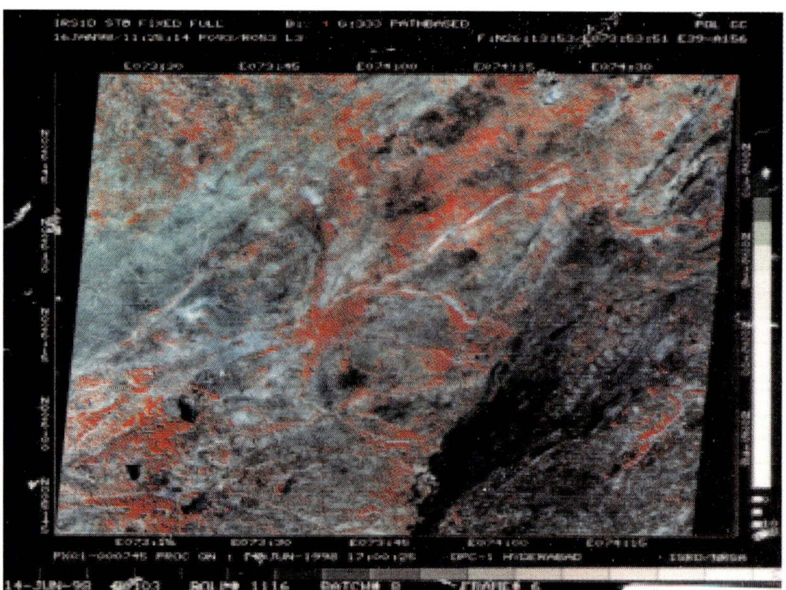

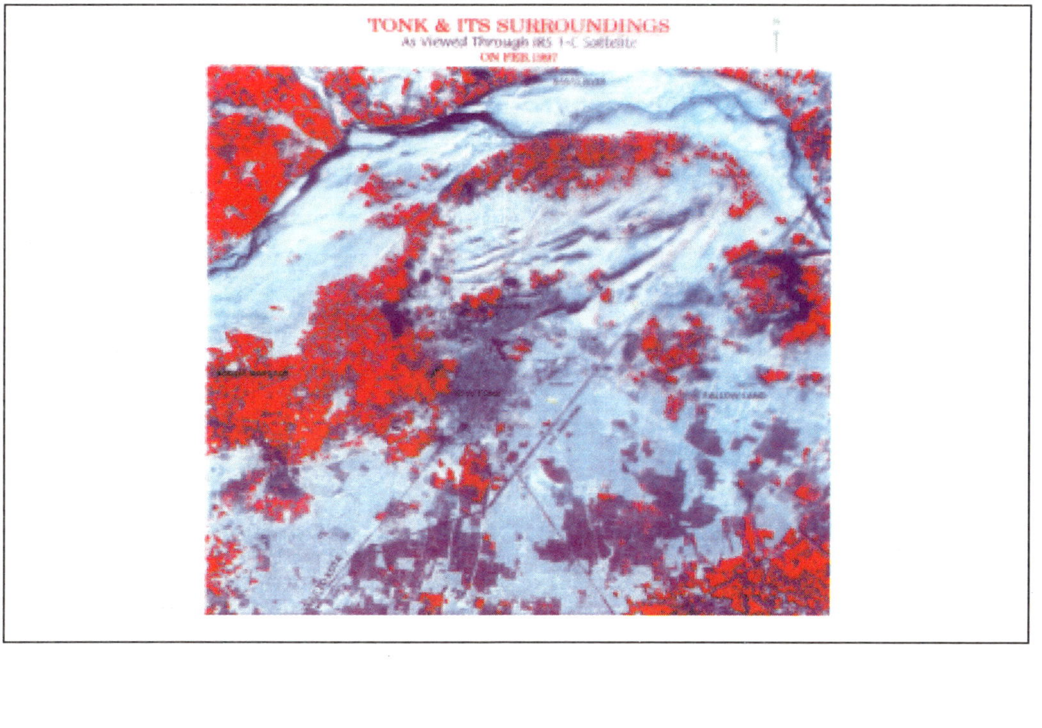
TONK & ITS SURROUNDINGS
As Viewed Through IRS 1-C Satellite
ON FEB.1997

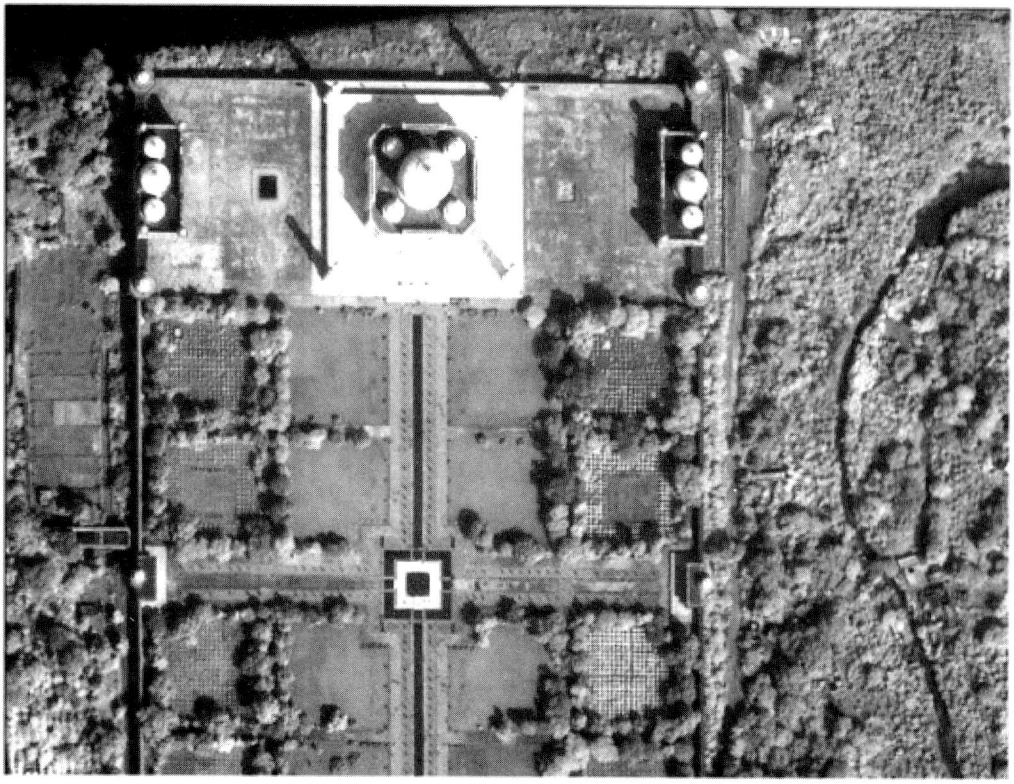

VILLAGE LEVEL PLANNING

Using Remote Sensing Inputs and GIS

Population pressure and haphazard growth of Human settlements (Urban/Rural) are causing deterioration in living conditions and worsening of environmental scenario in the Country. These can be improved if Land Development Planning is done properly and natural resources of the area are taken care in a scientific manner.

Development of resources requires accurate information on their present status, suitable strategy for their conservation and growth followed by regular monitoring.

Modern technologies like Remote Sensing (RS) and Geographical Information System (GIS) play a vital role in providing much needed information for sustainability of resources.

Remote Sensing by virtue of synoptic view of large area, repetitive coverage, data availability in multi-spectral range with variable resolution has been of immense use in inventory of resources, trends analysis, planning for development and monitoring the developmental efforts.

Remote sensing has immense utility in:

- Inventory for current status of natural resources.
- Planning for their development.

- Monitoring change as a result of developmental efforts.

Preliminary Interpretation

Visual interpretation techniques are employed to study digitally enhanced products on the basis of image characteristics:

- Size
- Shape
- Tone
- Texture
- Pattern
- Association
- Shadow
- Location

APPLICATION OF REMOTE SENSING

1. Land use/Land cover mapping
2. Geology and Geomorphology
3. Land degradation studies
   – Salt affected, Ravenous, Sand dune and Movement
4. Soil conservation planning
5. Water resources
   – Surface and Ground Water
6. Forest and Vegetation mapping
7. Agriculture
8. Human settlement
9. Irrigation
10. Monitoring/change detection

Objective of Study

- Human settlement planning (Rural)
- Village development
- Natural Resources Development

Village Development plan can be divided into two categories:

- Habitation expansion plan
- Overall development plan

Habitation expansion plan comprises of following stages:

- Inventory of current set-up
- Built up areas, their spatial distribution
- Residential
- Amenities – Hospital/Dispensary, School, Panchayat Ghar, Veterinary Hospital, Dairy Centre, Play Ground and others. Agriculture land – single/double crops
- Drainage/Water body
- Forest area
- Mining activity
- Wasteland

All these will be derived from Remote Sensing inputs.

Based on above information and after consulting secondary data on Ground Water, other Water Resources available (Surface/Sub-surface), Population, Population growth trends, Amenities requirements – interaction with local inhabitants and administrative machinery a plan can be designed for village development.

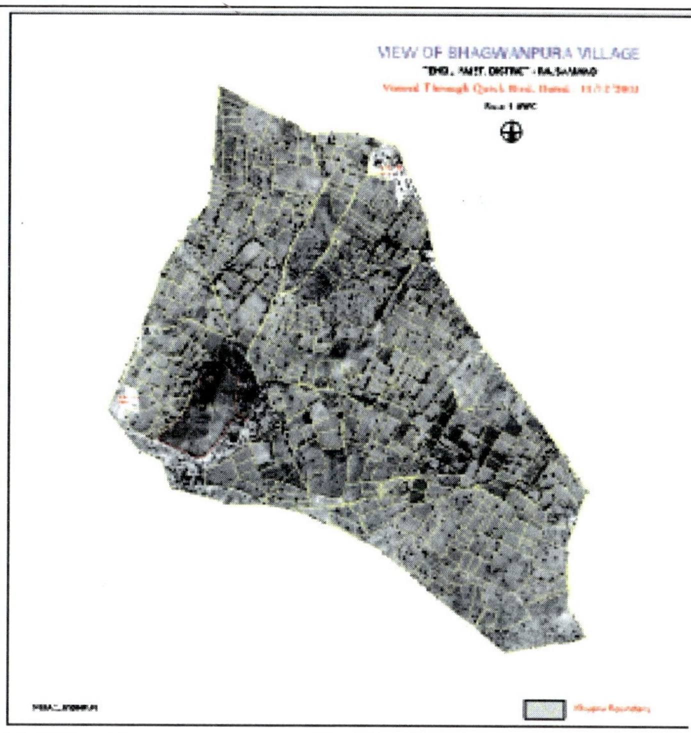
VIEW OF BHAGWANPURA VILLAGE

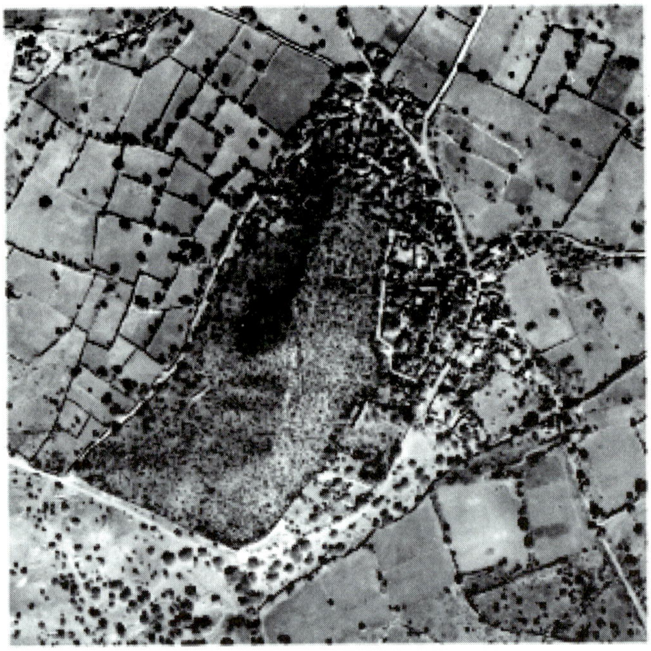

COMPONENTS OF A GIS

A GIS comprises three major components:

- The Hardware – used to store, process and display.
- The Software – used to control and perform operations.
- The Expertise – the human element required to drive the system to meet needs.

Components of a GIS

A working GIS integrates five key components: hardware, software, data, people, and methods.

Hardware

Hardware is the computer on which a GIS operates. Today, GIS software runs on a wide range of hardware types, from centralized computer servers to desktop computers used in stand-alone or networked configurations.

Software

GIS software provides the functions and tools needed to store, analyse, and display geographic information. Key software components are:

- Tools for the input and manipulation of geographic information.
- A database management system (DBMS).
- Tools that support geographic query, analysis, and visualization.
- A graphical user interface (GUI) for easy access to tools.

Data

Possibly the most important component of a GIS is the data. Geographic data and related tabular data can be collected in-house or purchased from a commercial data provider. A GIS will integrate spatial data with other data resources and can even use a DBMS, used by most organizations to organize and maintain their data, to manage spatial data.

People

GIS technology is of limited value without the people who manage the system and develop plans for applying it to real-world problems. GIS users range from technical specialists who design and maintain the system to those who use it to help them perform their everyday work.

Methods

A successful GIS operates according to a well-designed plan and business rules, which are the models and operating practices unique to each organization.

How GIS Works

GIS stores information about the world as a collection of thematic layers that can be linked together by geography. This simple but extremely powerful and versatile concept has proven invaluable for solving many real-world problems from tracking delivery vehicles, to recording details of planning applications, to modelling global atmospheric circulation.

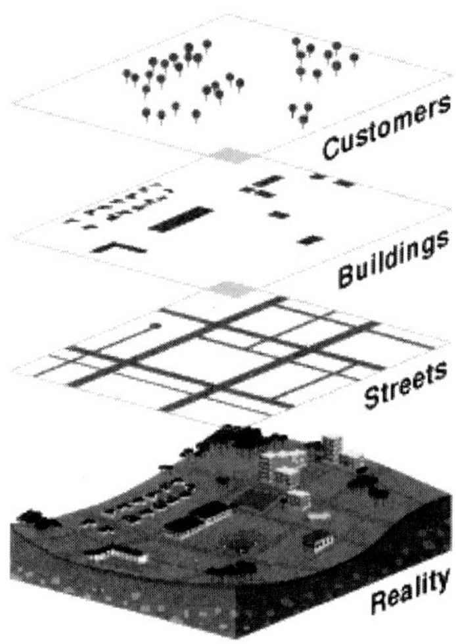

GEOGRAPHIC REFERENCES

Geographic information contains either an explicit geographic reference, such as a Latitude and Longitude or National Grid Coordinate, or an implicit reference such as an address, postal code, census tract (zone) name, forest stand identifier, or road name. An automated process called geocoding is used to create explicit geographic references (multiple locations) from implicit references (descriptions such as addresses). These geographic references allow you to locate features, such as a business or forest stand, and events, such as an earthquake, on the earth's surface for analysis.

Data Structure of the GIS Package

The data structure of the GIS has a major bearing on the accuracy of the representation of spatial data. Most GIS packages adopt either a raster or vector structure, or their variants, internally to organize spatial data and represent real world features.

The key element in both these models is that links are established between attribute information and spatial features. The precise techniques used to create these links vary from GIS to GIS. There are two types of GIS Packages available.

The Graphics Data Base

There are two fundamental data structures used within GIS Graphics databases. Each has inherent advantages and disadvantages to the user.

These two structures are:

- Cector encoding and
- Raster encoding

RASTER DATA

A Grid (or Raster) system stores data as a string of characters in which each character represents a location. It is a cell-based system in which a map is represented as an array of rectangular or square cells. In the Raster-based system, a geometric array of rectangular or square cells represents the data each with an assigned value. Polygons can be formed indicating areas of homogeneous characteristics.

The Raster system used to display satellite imagery resembles a television image composed of picture elements (pixels), which are essentially grid cells.

Advantages

- A grid pattern is particularly useful where spatial pattern (i.e. Land features) need to be recognized for comparison and analysis.
- A Grid based GIS can be easily used in a computer.

Disadvantages

Satellite imagery cannot be represented as vectors.

SAMPLE APPLICATIONS OF GIS

Digital Data Base of Colonies in GIS Environment

- Digitized map of colony.
- Linked with database in GIS environment.
- Query based record of all plots.
- Regular updating of all transformations.

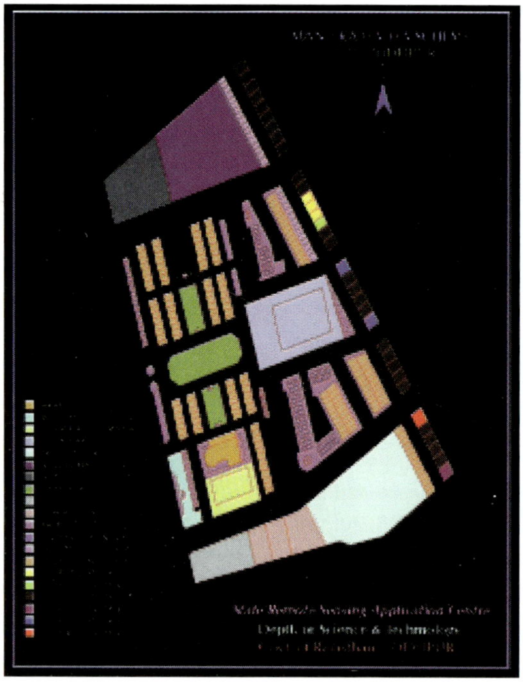

Manji Ka Hatha

Searching for Government Offices

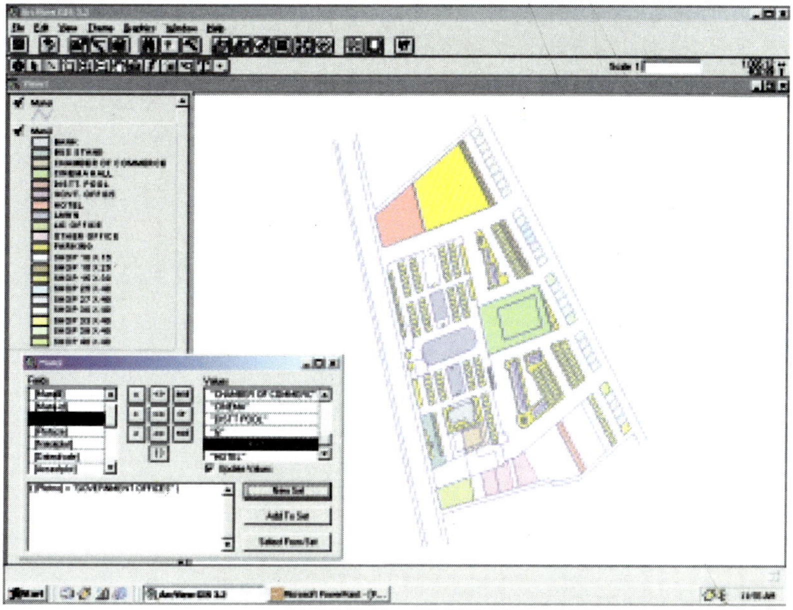

Searching for Specific Plot No

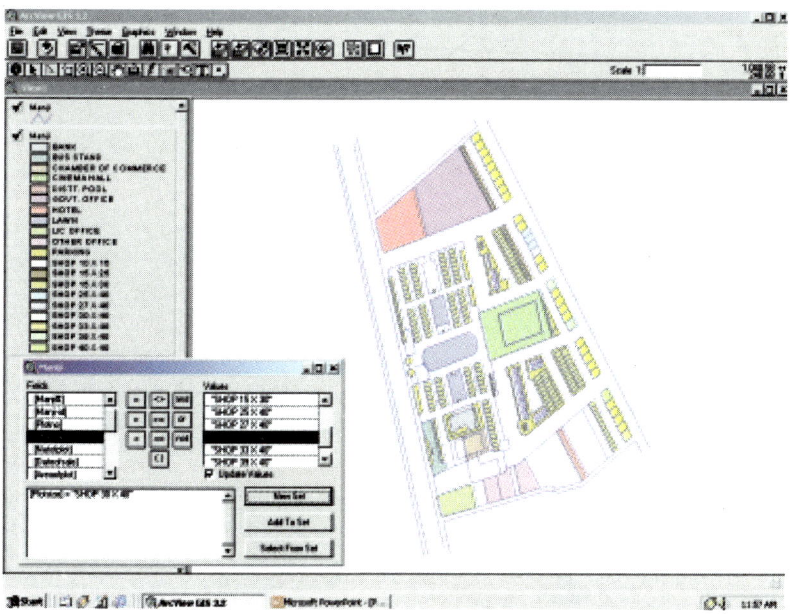

Searching for Shops of specific size

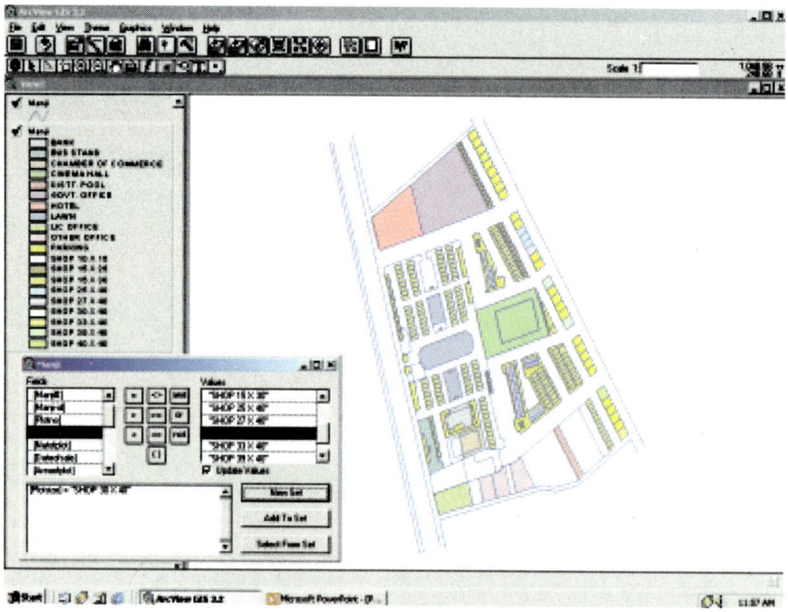

Searching for a plot owned by a Particular Person

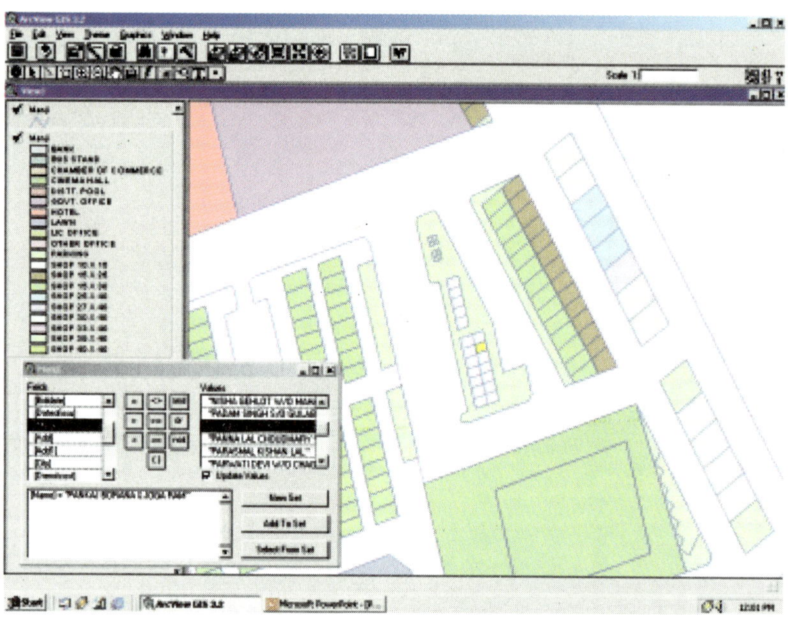

Searching for Cinema Hall

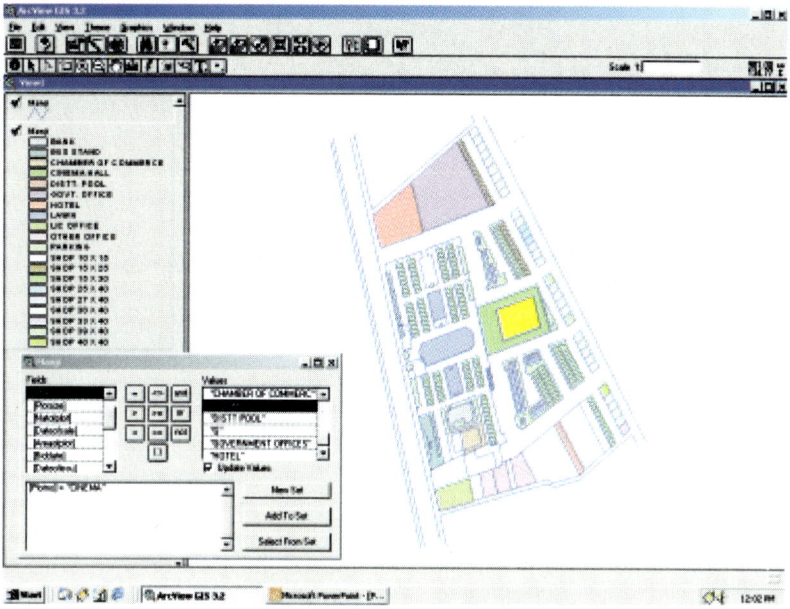

Identification of a Specific Plot

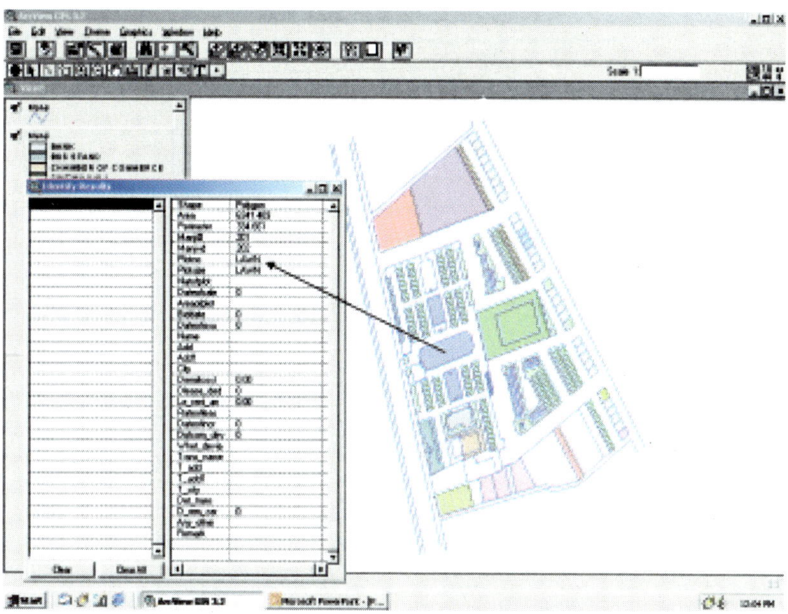

Bye Pass Study

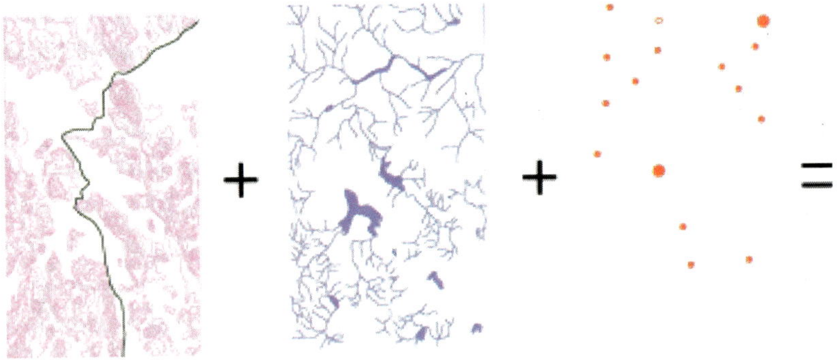

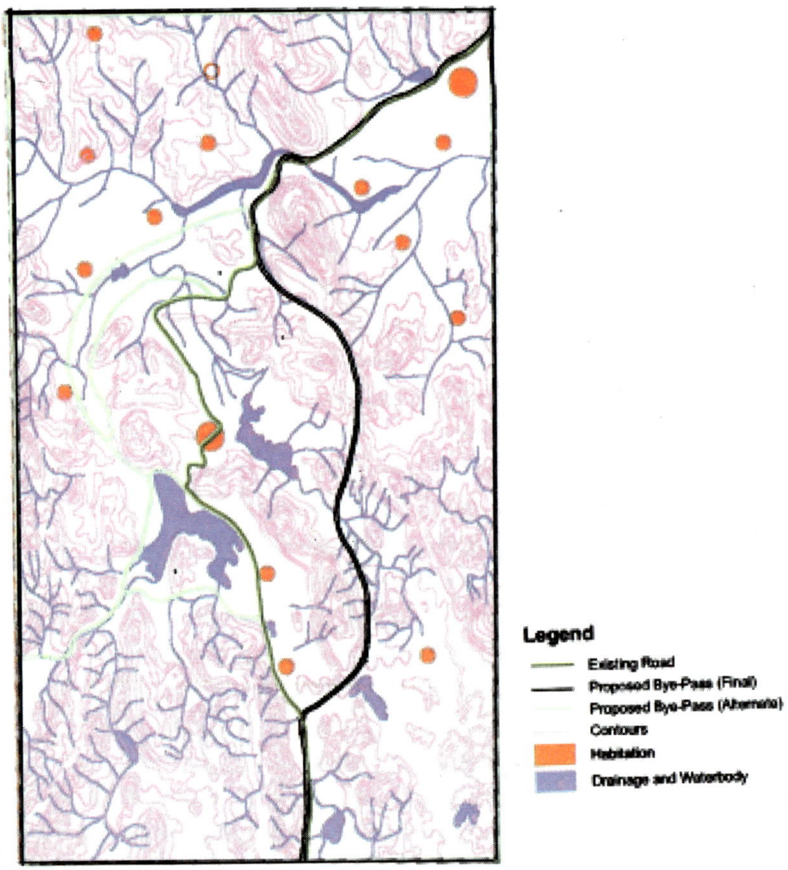

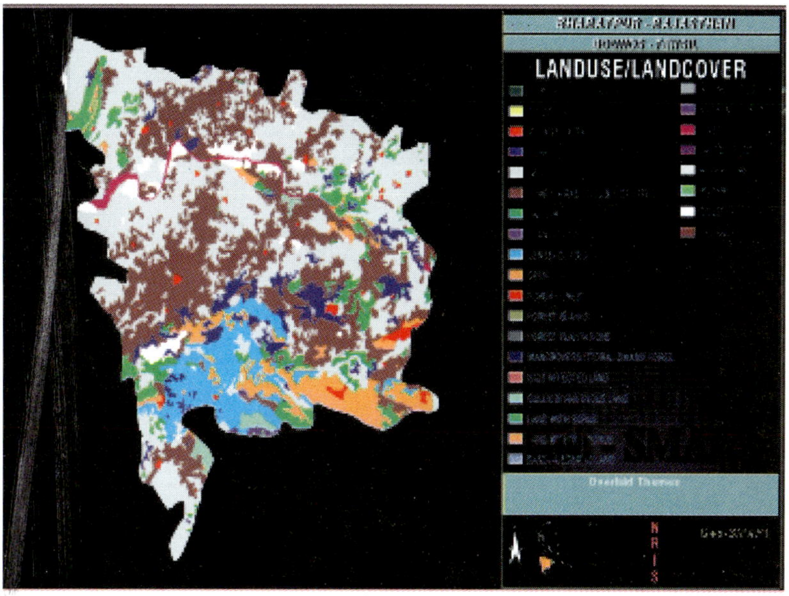
LANDUSE/LANDCOVER

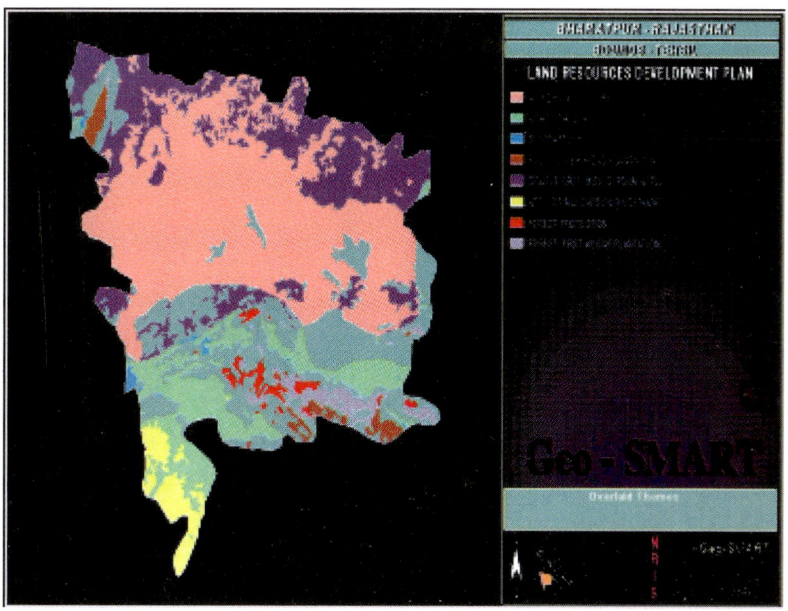
LAND RESOURCES DEVELOPMENT PLAN
Geo - SMART

Road Connectivity

Weir Tehsil, Bharatpur District

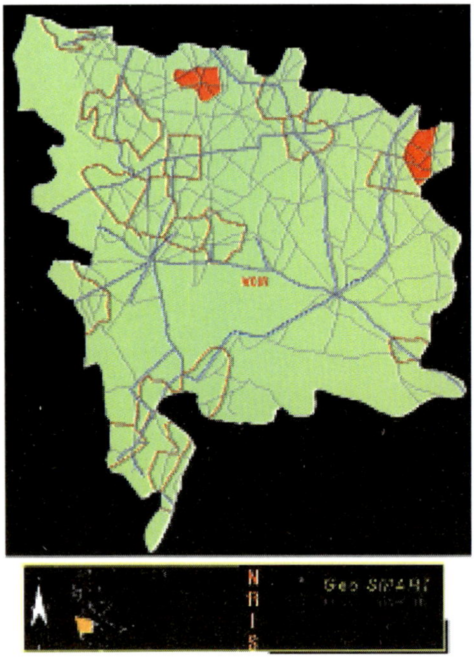

Priority villages for road connectivity	
Vill. Name	**Population**
Newara	2298
Nyampatpura	2053
Approximately 2km from the existing road.	

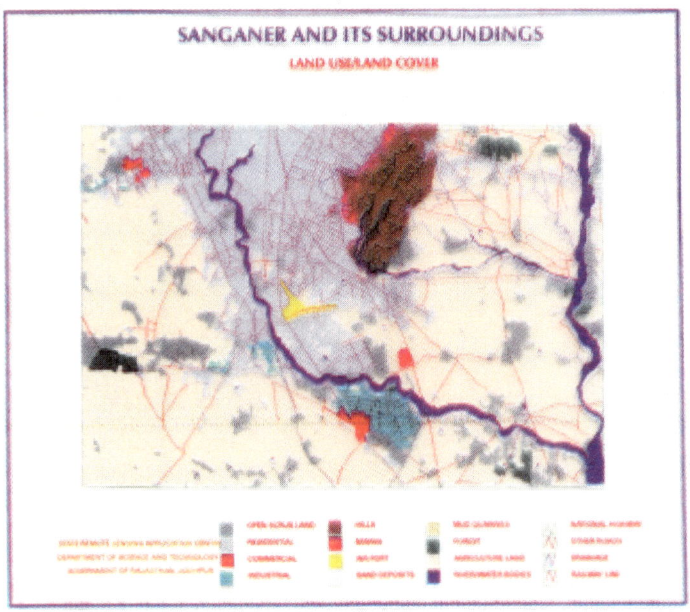

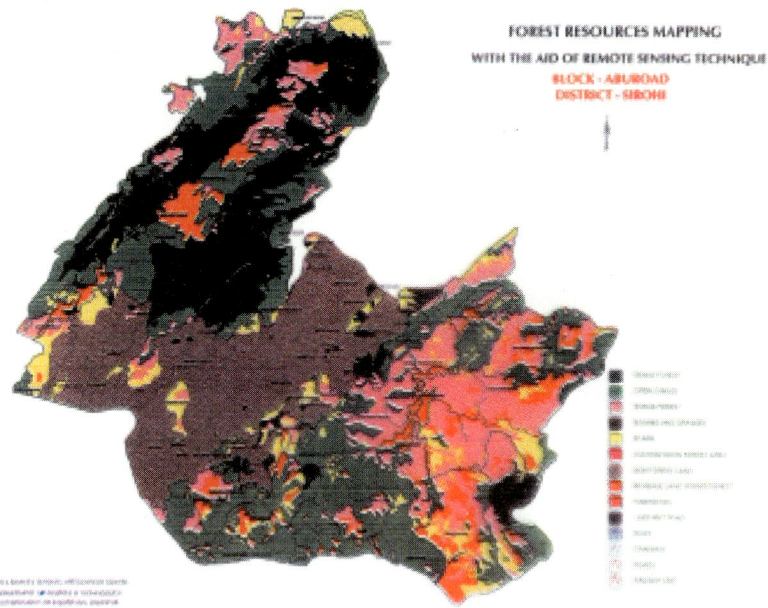
FOREST RESOURCES MAPPING
WITH THE AID OF REMOTE SENSING TECHNIQUE
BLOCK - ABUROAD
DISTRICT - SIROHI

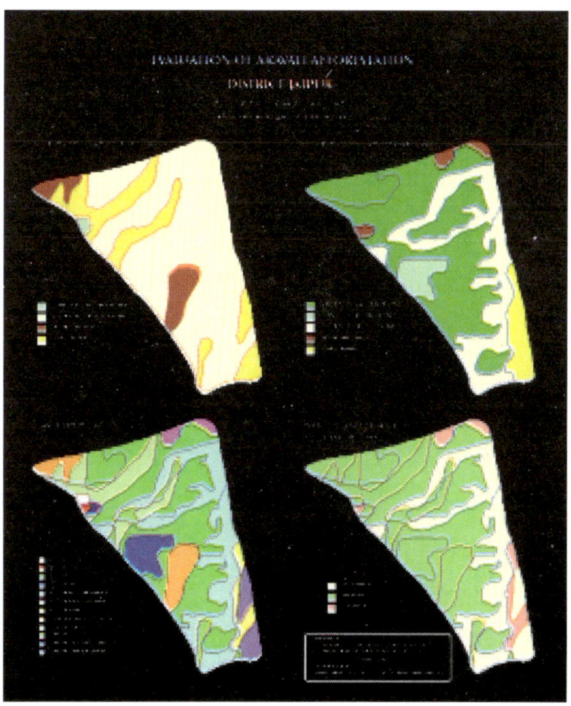

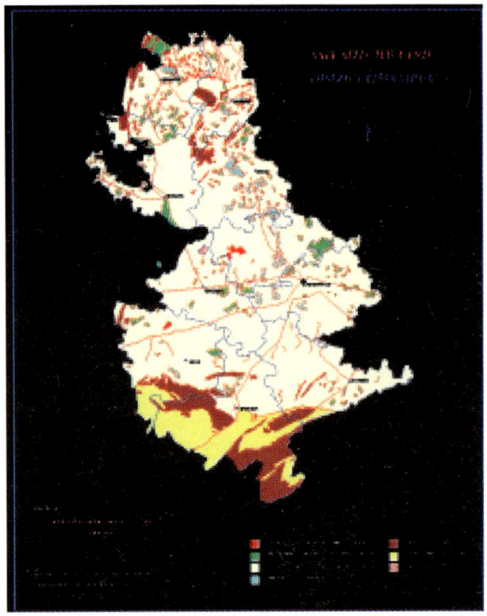

Geographic Data Concepts (Basic Map Concept)

Most spatial features can be displayed as:

- Points.
- Line
- Polygons
- A point is a small features (Like city, village) relative to the whole map.
- Lines are used to represent roads, pipelines, boundaries and other linear features.
- A polygon refers to any area or region.

Feature class	*Application and use*	*Example*
1. Arcs	Linear feature	Streets, contours, streams, sewer, power lines, gal lines.
2. Nodes	Points along linear feature	Valves on pipeline, interaction on streets, power poles, man hole covers.
3. Label points	Point location	Well sites, mountain peaks.
4. Polygon	Area Feature	Soil units, Land use parcels, buildings, footprints, forest land ownership.
5. Tics	Geographic registration and control	Registration for digitization/scanning.
6. Annotation	Feature labels	Streets names, place names on road maps.
7. Links	Rubber shipping and adjustments	Edge marking map, sheet marking, snapping data adjustment.

CAPTURING DATA

The geographic database is the most and long-lived component of the GIS, thus making data entry a very important consideration.

Because ARC/INFO integrates a variety of data types from a variety of sources, it provides multiple data entry options.

ARC/INFO offers efficient data entry methods for automating paper maps and other non-digital data sources. To take advantage of the vast collection of geographically referenced data that already exist in digital format, ARC/INFO provides the most comprehensive data conversion capability of any GIS on the market. ARC/INFO software's integrative capabilities also allow data sharing with other applications without the need for data conversion.

STORING DATA

Storing Geographic Data

A digital map database consists of two types of information: Spatial and Descriptive. These are stored as a series of files on the computer and contain either the spatial or descriptive information about the map features. These two types of data have to be linked and the spatial relationships between the map features have to be maintained.

This data integration opens the way for powerful and varied ways of looking at and analyzing your data. You can access information in the tabular database through the map, or you can create maps based on the information in the tabular database. To access and display this data, the computer must store both the graphic and tabular data in easily recognized and retrievable formats.

TOPOLOGY

The user interprets additional information from maps about the spatial relationships between features. For example, you can trace a route along a city street map to find your way from an airport to a hotel. You interpret these relationships by identifying connecting lines along a path, by defining the areas enclosed within these lines, and by identifying contiguous areas. In digital maps, such spatial relationships are depicted using topology.

Topology is a mathematical procedure for explicitly defining spatial relationships. For maps, topology defines connections between features, identifies adjacent polygons and can define one feature, such as an area, as a set of others (i.e. lines).

Creating and storing topological relationships has a number of advantages:

- Data is stored more efficiently.
- You can process data faster and process larger data sets.
- You can also perform analysis functions, such as modelling flow through the Connecting lines in a network, combining adjacent polygons with similar characteristics and overlaying geographic features.

The three major topological concepts are:

Arc-Node Topology—Arcs connect to each other at nodes (connectivity).

Polygon-Arc Topology—Arcs that connect to surround an area define a polygon (area definition).
Left-Right Topology—Arcs have direction left and right sides (contiguity).

DISPLAYING DATA

- Spatial data
- Non-spatial data

Spatial Data

Consisting of *Maps* and which have been prepared either by field surveys or by the interpretation of Remotely Sensed (RS) data. Some examples of maps are the soil survey map, geological map, land use map, village map etc.

Non-spatial Data

Attributes as complementary to the spatial data and describe what is at a point, along a line or in a polygon. The attributes of a soil category could be the depth of soil, texture, erosion, drainage etc. and for a geological category could be the rock type, its age, major composition etc.

Map features are logically organized into sets of layers or Themes of information. A base map can be organized into layers such as streams, soils, wells, administrative boundaries, and, so on.

Additionally, smaller areas, commonly corresponding to paper map sheets, can often be spatially combined into larger units called **Study Areas.**

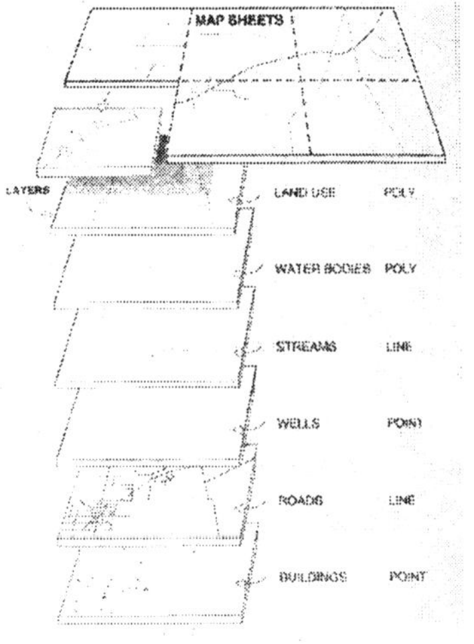

Arc/Info Coverage

Each layer in ARC/INFO is called a coverage. A coverage consists of topologically linked geographic and their associated descriptive data stored as an automated map; for example, Soils coverage.

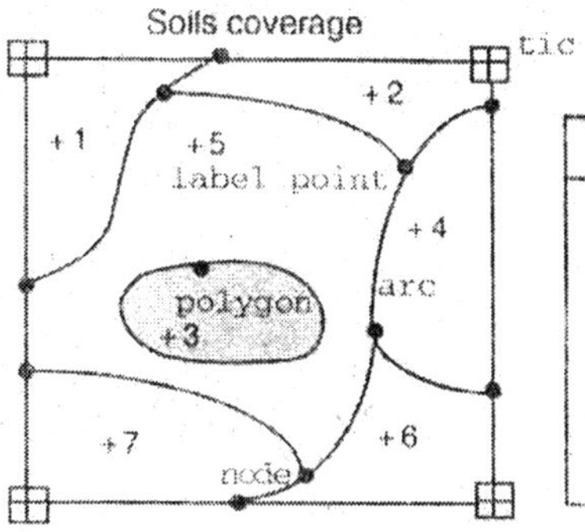

Soils attributes

ID	Soil	Class	Suitability
1	A3	113	HIGH
2	C6	95	LOW
3	B7	212	MODERATE
4	B13	201	MODERATE
5	Z22	86	LOW
6	A6	77	HIGH
7	A1	117	LOW

Point Features and Label Points

Points serve two functions in ARC/INFO. In a *point coverage*, they represent point features such as wells, students, or crime incidents. As *label points*, they serve to identify (label) each polygon in a polygon coverage. A label point is the link between a polygon and its associated attributes.

Tics

Tics are points that represent locations on the earth's surface for which real-world coordinates are known. Coverage use a common set of tics so that layers register to each other and to adjacent map sheets.

Summary Table of Coverate Features

Feature Class	Description	Attribute table	Example
Arc	A line defined by an ordered set of x,y coordinates. Represents linear feature, polygon boundaries.	AAT	
Node	The endpoints of an arc, or thepoint where two or more arcs connect.	NAT	
Lable point	A point defined by an x,y coordinate-a point features or the User-Ids of a polygon.	PAT	
Polygon	An area defined by the arcs that make up its boundary, including arcs defining islands inside. A lable point is used to assign a User-ID.	PAT	
Region	An area made up of polygon features.	PAT. subclass	
Tic	A geographic control point used to register and transform the coordinates in a coverage.	TIC	
Annotation	A text string describing a geographic feature. Shape points are used to position and draw annotation.	TAT .subclass	
Route	A linear feature composed of one or more arcs or parts of arcs.	RAT .subclass	
Section	An arc or portion of an arc used to define a route. The building blocks of routs.	SEC .subclass	

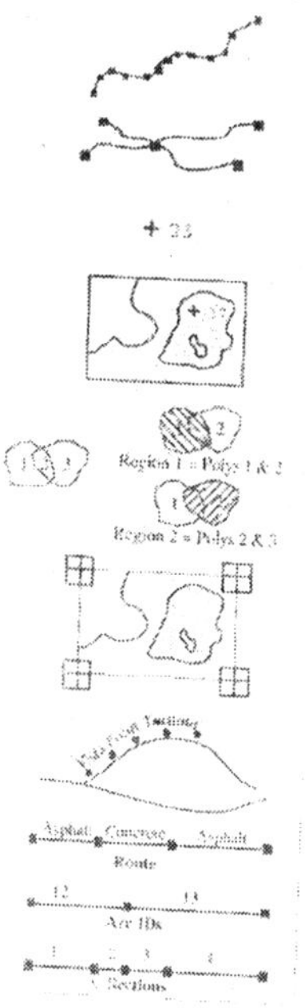

25

MATHEMATICAL MODEL FROM STATISTICAL PROCESSING OF GIS

RAJEEV JAIN

WHAT IS ARC/INFO?

The ARC/INFO software's geographic information system is a powerful computer technology that enables user to access, integrate, visualize, analyze, and output a world of information. ARC/INFO offers a complete solution for the automation, management, and display of geographic and associated data, which can be used by any organization that works with spatial data.

ARC/INFO software enables to integrate vector (x, y locations) map data, raster images such as photographs, scanned documents or satellite images, computer aided drafting (CAD) drawings, sound and video data, as well as wealth of tabular DBMS data into a single, integrated environment. ARC/INFO software is available for several UNIX workstations and the Intel NT and DEC Aloha platform.

What does ARC/INFO do?

- Topological map overlays
- Buffer generation
- Proximity analysis
- Spatial and logical query
- Hydrologic modelling
- Surface analysis
- Network modelling
- Raster modelling
- Sophisticated tabular analysis

WHAT IS ARC TOOLS?

Arc Tools is a general-purpose, menu-based interface to ARC/INFO designed to get new start quickly. Common operations such as mapping, query, data automation, and analysis are implemented through a series of AML-based tools and menus.

Arc Tools were developed using an object oriented programming methodology that allows Arc Tools to be used as a base to build custom applications. Individual tools can readily be plugged into other applications, and adding custom tools is simple and efficient. The object oriented methods employed to create Arc Tools represent a major break through in GIS application development.

WHAT IS ARC TOOLS?

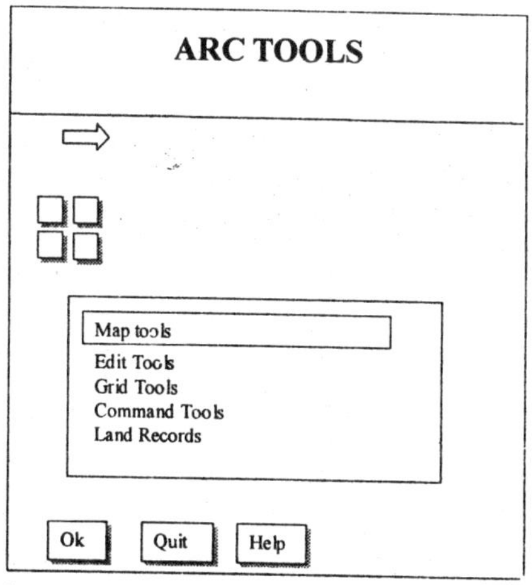

Arc Tools are functionally grouped into four tool sets: Map Tools, Edit Tools, Grid Tools, and Command Tools. This allows related tasks to be processed logically and quickly. Additionally there is a Land Records, an Arc Tools application that manages land records within the Arc Tools framework.

In each tool set a menu bar is displayed with pull down menus providing access to the component tools. The tools can be used together or one at a time. Many common operations have been included in more than one of the tool sets. For instance, the Arc Tools menu pull down is nearly identical in each of the tool sets. These menus provide general tools for data base management, modelling, changing workspaces, moving between tool sets, and accessing ARC/INFO programs, command lines and system files. The basic functionality of each tool set is based on the ARC/INFO program in which it operates.

Map Tools

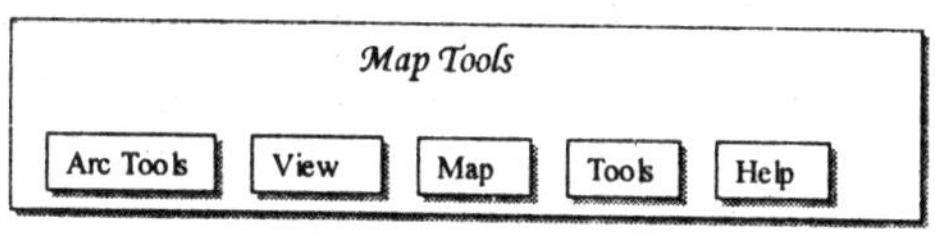

Map Tools operate from within the ARCPLOT program, providing environment for data display and query as well as map creation and output file generation. The Map pull down contains tools for managing the display and manipulation of map object properties associated with a particular map layout. Map layout can be created, edited and saved as a layout. The file can then be printed or plotted from within the Map Tools application.

Edit Tools

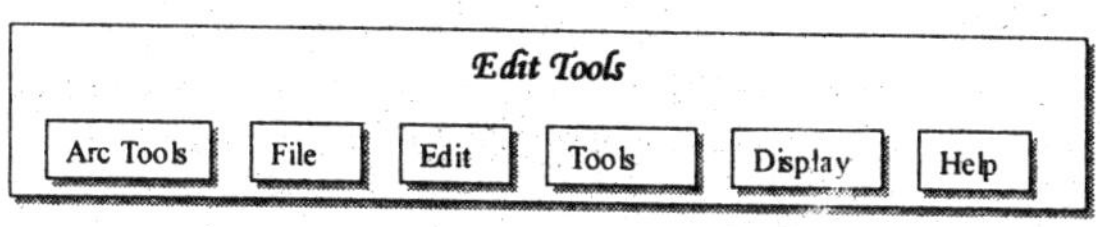

Edit Tools operate from within the ARCEDIT program, allowing the management of data automation and related processing tasks. The file pulldown cotains the tools to generated geographic fearture as well as tabular data. The edit pulldown provides access to a complete set of production tools to modify and manipulate coverage topology and associated feature attribute tables. The display pulldown tools manage the current graphic enviorment.

Grid Tools

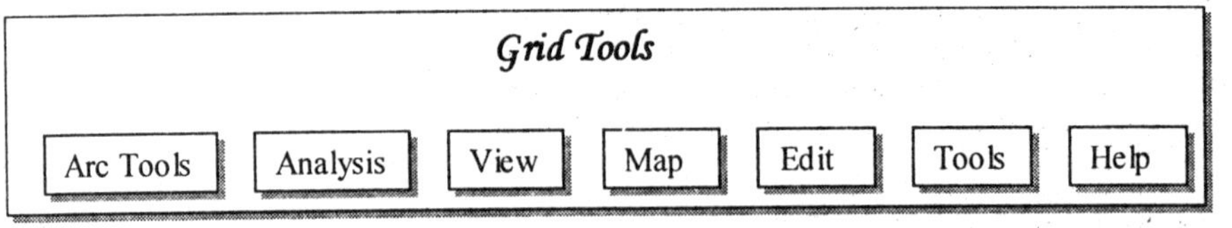

Grid Tools provide a sophisticated enviroment within the GRID program for performing spatial analysis and data display. A series of sequential steps with associated properties can be defined, edited and saved as a model.

Command Tools

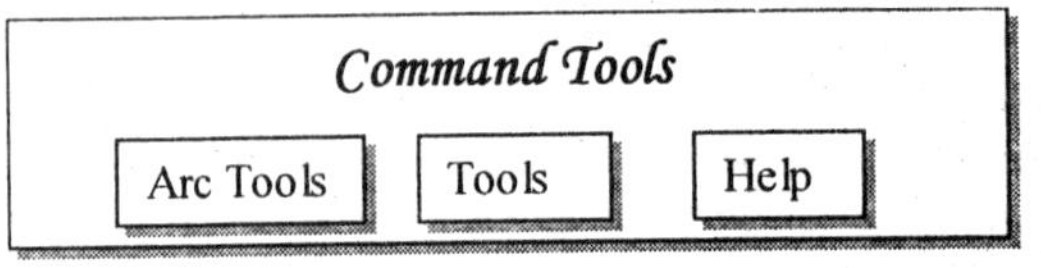

Command Tools operate within the ARC program. These Tools are also accessible from within the Map Tools, Edit Tools, and Grid Tools sessions: From the main menu bar each of these tool set, select Command tools..... on the Tools pulldown.

MAKING SPATIAL DATA USABLE

Key Terms – Some digitizing terms and the symbols used to represent them

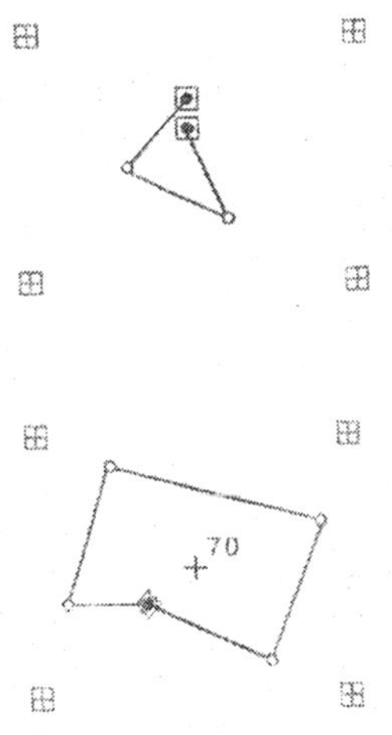

Terms	Symbol	Description
Arc	—	A line feature or the border of a polygon. An arc is defined by a from-node and a to-node and vertices in between.
Node	•	An arc endpoint, in many cases, the end of one arc marks the beginning of another, so it can be said that nodes also mark intersections.
Vertex	∘	A point within an arc that provides its shape.
Pseudo node	○	The point at which an arc connects to itself (a loop), or to only one other arc.
Dangling node	□	An arc endpoint not connected to another arc.
Label point	+	Used either to represents a point feature or to identify a polygon.
User-ID	70	A number assigned to each feature. These values should be unique. Once established, the user can after these values as needed.
Tic	⊞	A registration or geographic control point. These allow all coverage features to be registered to the same coordinate system.
RMS Error	0.004	The calculated difference between recorded and specified tic locations, expressed as a residual of the means squared, Values are usually just slightly greater than 0,000—the higher the value the greater the error.

Step 1: Steps for making spatial data usable

Making spatial data usable requires correcting the coordinate data to make it free of errors and topologically correct

Step 1: Construct a topology
Step 2: Identified digitizing errors
Step 3: Correct errors
Step 4: Reconstruct Topology

Topology makes explicit the relationship between the geographic features within a coverage. The process of constructing topology identifies errors that often exist in digitized data such as:

- Arcs that do not connect to other arcs
- Polygon that are not closed
- Polygons that have no label point or too many label points
- Incorrect User-ID

Constructing topology identifies these errors because it creates arc intersections, identifies arcs that make up each polygon and associates a label point with each polygon.

Until you construct topology, no polygon exist and arcs that cross each other might not connect with a node at their intersection.

ARC/INFO provides two commands that create topology for you automatically: BUILD and CLEAN.

Identify Digitizing Errors

Once you have constructed topology, you can identify most digitizing errors by making a plot of the digitized coverage to overlay and compare with the original source map.

ARC/INFO identifies and marks potential node errors with special symbols.

Here is a brief description of the two types of nodes and the potential errors they represent.

An island pseudo node Pseudo node marking a two-arc intersection

Pseudo nodes, drawn with a diamond symbol, occur where a single line connects with itself (an island) or where only two arcs intersect.

Pseudo nodes do not necessarily mean that there's an error or a problem. For example, pseudo nodes that represent island polygons (a spatial pseudo node) or the point where a road changes from pavement to gravel (an attribute pseudo node) are acceptable.

A dangling node, represented by a square symbol, connects to only one arc. It is the unconnected node of a dangling arc. Dangling arcs and nodes are created when digitized arcs stop short of, or extend past, an intended intersection point.

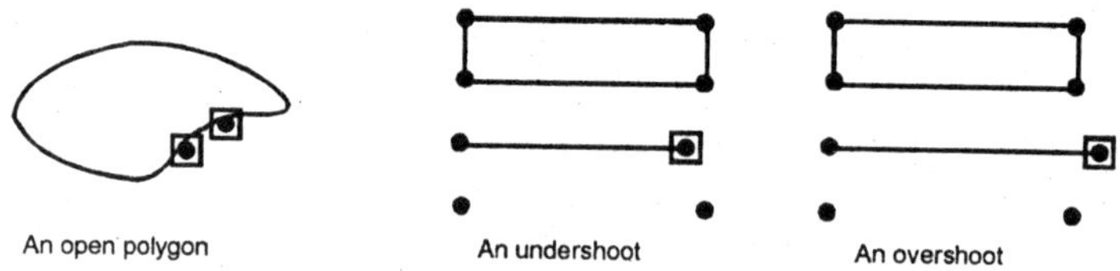

In some cases, dangling nodes are Acceptable.

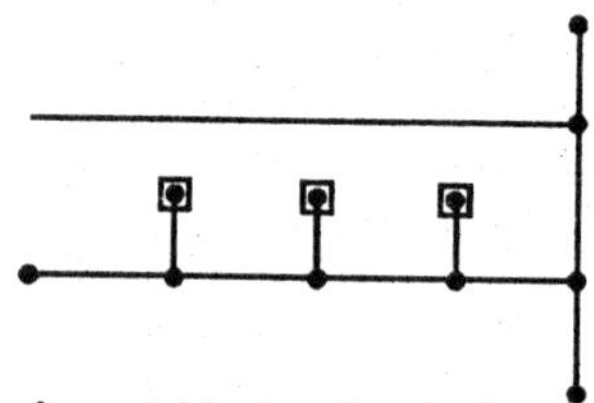

Acceptable dangling nodes

Common label errors in polygon coverage's include having a polygon that contains no label point and having more than one label point in single polygon. The later can be a case of too many label points or of not enough polygons. While it can happen that more than one label point is accidentally digitized for single polygon, it con also be that a polygon contains multiple label points because one or more arcs don't intersect with other arcs as they should (undershoots), resulting in one polygon where there should be more.

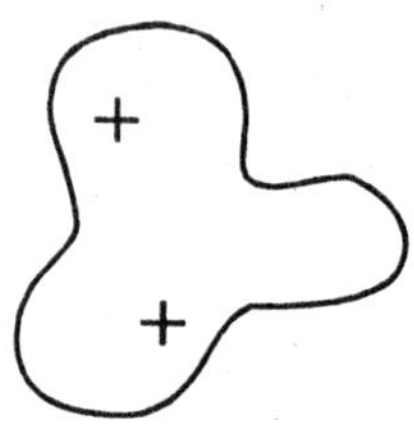

Polygon with two label points

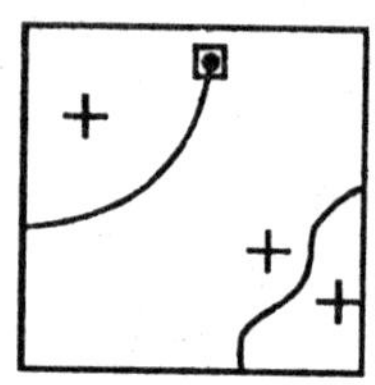

Polygon with two label points resulting from a dangling arc

Here are the node and label errors that are identified automatically.

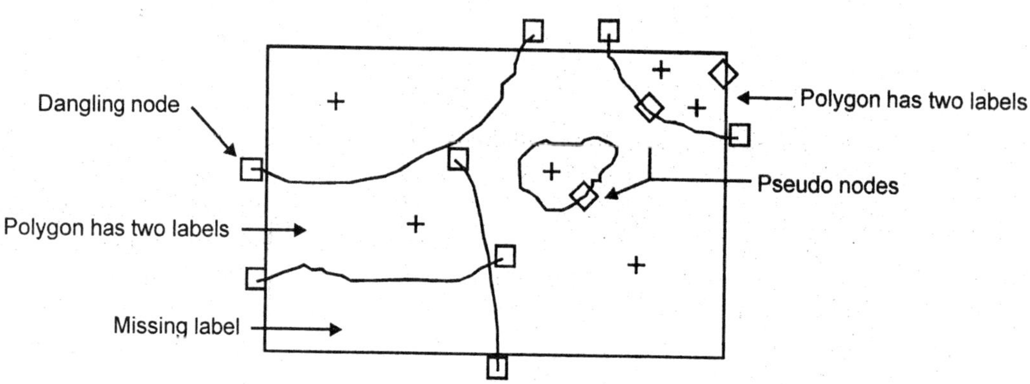

Here are the same errors, as they would be interpreted by you, after comparing them with the source map.

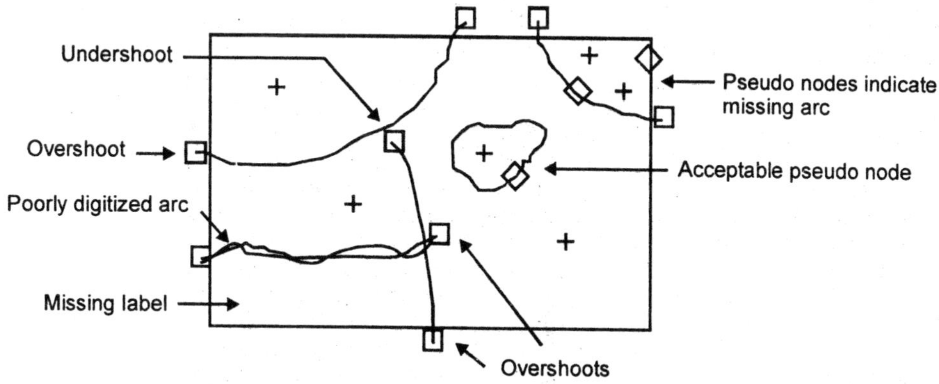

Correcting errors is one of the most important steps in constructing a database. Unless you correct the errors in our data. Subsequent calculations, analyses, and maps will not be valid.

For example, polygons without label points cannot have descriptive attributes attached to them, and area calculations cannot be accurately computed for unclosed pclygons.

Correcting errors simply means adding missing data, and removing and replacing inaccuracies with correct data.

Error	What should be marked
Missing arcs	Draw then in.
Missing label points	Mark position and correct User-ID value.
More than one label point in a polygon	Identify which ones to delete.
An overshoot	Indicate whether it should be deleted.
Incorrect User-ID values	Mark the correct value.

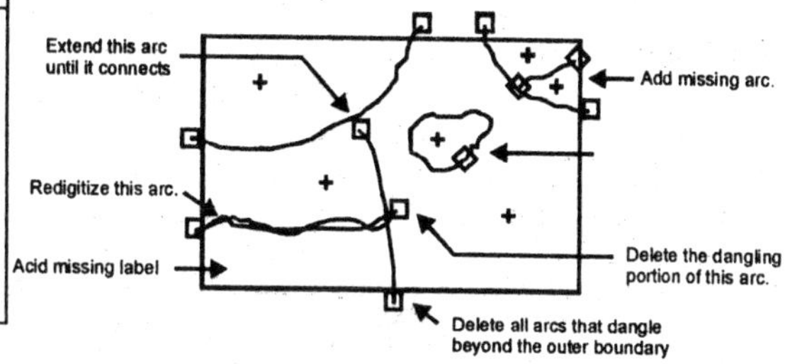

Here is the same set of data after correcting the spatial errors.

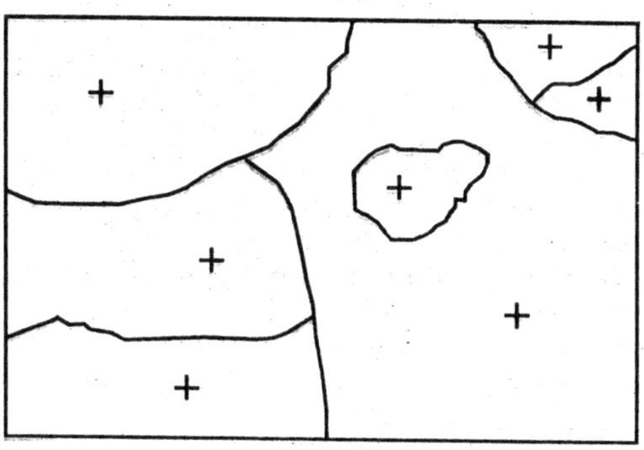

Reconstruct Topology

Editing the spatial characteristics of a coverage alters its topology. In such cases, you must always reconstruct topology to re-establish the spatial relationships. After you reconstructs the topology, repeat checking for any errors that remain. Then, if necessary, repeat Steps 3 and 4. Here is a chart showing the overall process:

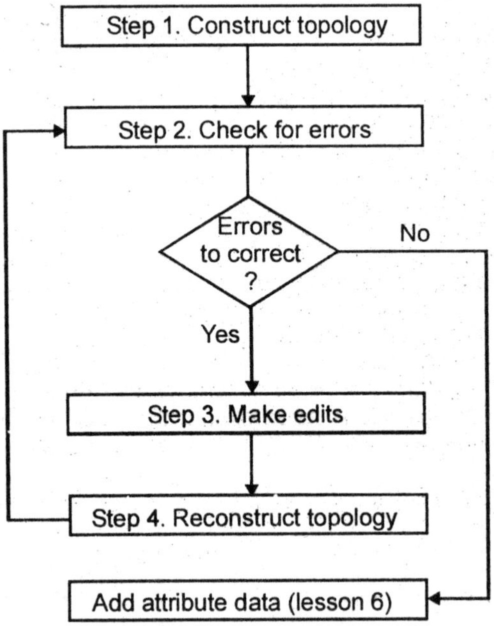

Potential Errors with Different Features

Feature	Potential errors	Terminology
Arc	Only one node at an endpoint	Dangling arc
	Node extends too far	Overshoot
	Node falls short	Undershoot
	No node where arcs intersect	Intersection
Nodes	Only one arc attached	Dangling node
	Only two arcs attached	Pseudo node
	One arc connected to itself	Pseudo node
Labels	None in a polygon	Missing lables
	More than one in a polygon	Too many labels

Potential Errors with Different Features

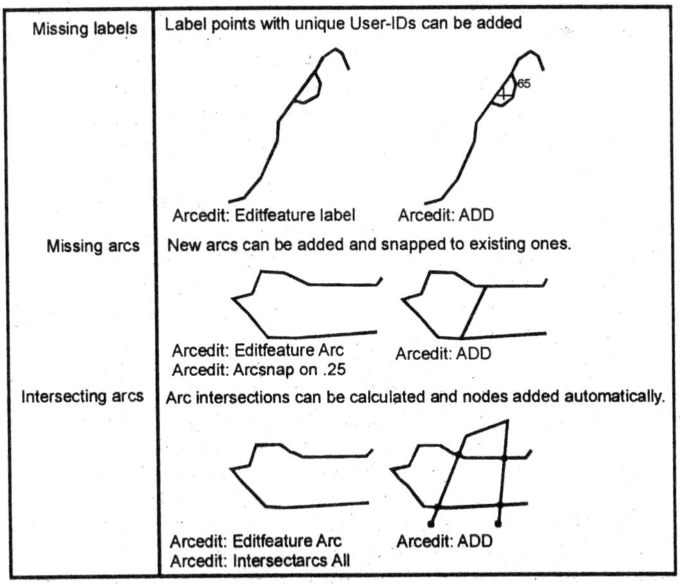

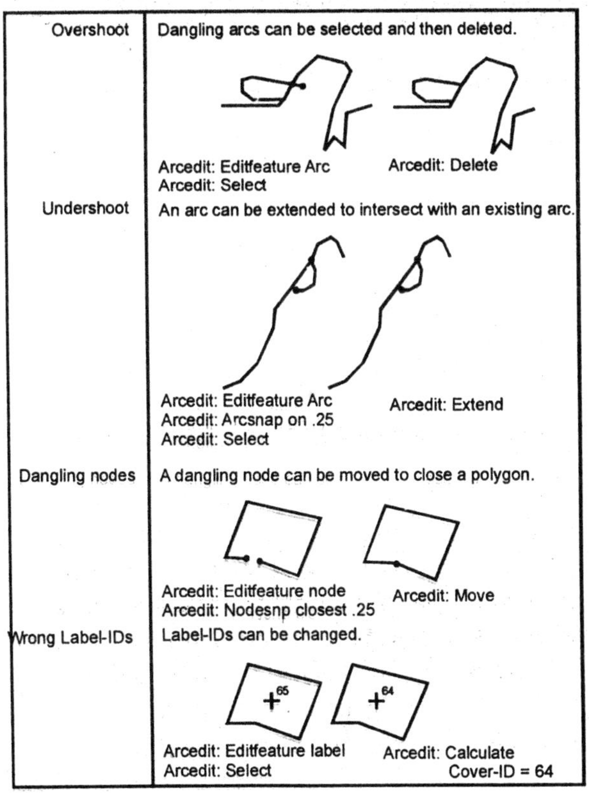

Construction Topology

To create spatial relationships between the features in a coverage, it is necessary to construct topology. ARC/INFO assigns an internal number to each feature. These numbers are then used to determine arc connectivity and polygon contiguity. Once calculated, these values are recorded and stored in a tabular format called a *feature attribute table.*

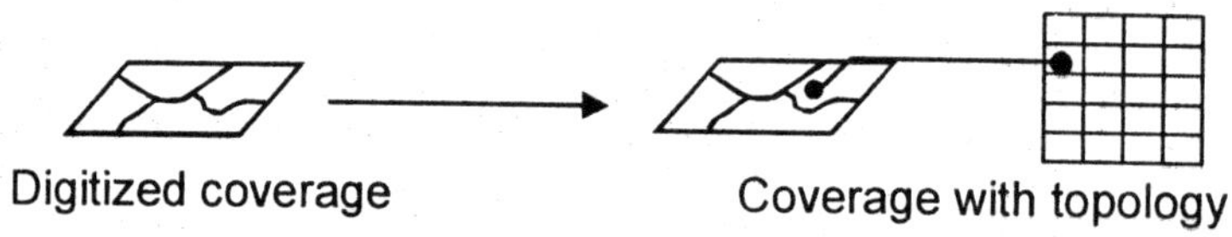

Feature Attribute Tables

Feature attribute tables are INFO files associated with each feature type. For example:

- constructing topology for a polygon coverage creates a polygon attribute table (PAT);
- for a line coverage, an arc attribute table (AAT); and
- for a point coverage a point attribute table (PAT).

Each table is composed of rows and columns. The columns are individual feature, such as polygon number 2.

Column of data

AREA	PERIMETER	Cover#	Cover-ID
-21.285	18.493	1	0
2.041	12.500	2	61
2.514	13.253	3	60
0.727	4.367	4	62
13.419	42.764	5	59
0.174	2.877	6	76
:	:	:	:

Row of data → Record for each feature

Item and attribute values

Standard items in an AAT

The AAT contains seven standard items created in the following order:

FNODE#	Internal node number for the beginning of an arc (from-node)
TNODE#	Internal node number for the end of an arc (to-node)
LPOLY#	Internal number for the left polygon (the Cover# in a corresponding PAT)
RPOLY#	Internal number for the right polygon (the Cover# in a corresponding PAT)
LENGTH	Length of each arc, measured in coverage units
Cover#	Internal arc number (values assigned by ARC/INFO)
Cover-ID	User-ID (Values assigned by the user)

Standard items in a PAT

AREA	Area of each polygon, measured in coverage units
Perimeter	Length of each polygon boundary, measured in coverage units
Cover#	Internal polygon number (assigned by ARC/INFO)
Cover-ID	User-ID (assigned by the user)

The PAT contains four standard items created in the following order:

The PAT for coverage of points always contains zero values for both AREA and PERIMETER

BUILD versus CLEAN

The commands BUILD and CLEAN construct topology in ARC/INFO. Although these commands perform similar functions—they both construct topology and create feature attributes tables—there are important differences. BUILD creates or updates a feature attribute table for point, line, polygon, node, or annotation coverage's. CLEAN, however, performs coordinate editing and created feature attribute tables for polygon and line coverage's only.

Whether you use BUILD or CLEAN for creating polygon or arc topology depends on how the data was originally digitized:

- BUILD recognizes only exiting intersections (discrete digitizing), whereas
- CLEAN creates intersections wherever lines cross one another Tolerances define the resolution of coordinates in a coverage. The TOLERANCE command can be used to set the different tolerances. For single precision coverages, these tolerance values are stored in the 'tol.adf' file, while for double precision coverages, these values are stored in the 'par.adf' file. Both files can be found within the coverage directory.

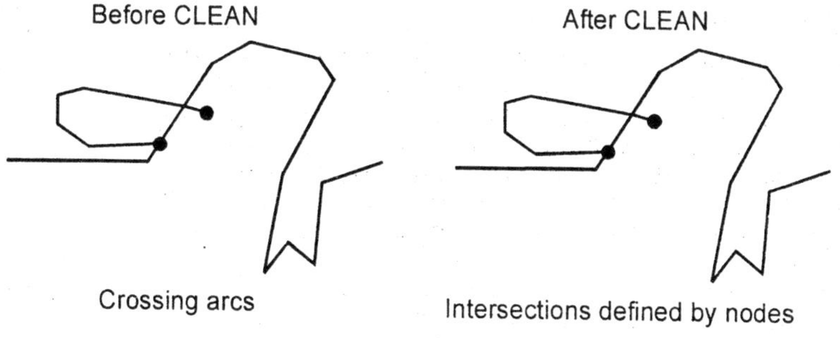

The fuzzy tolerance of a coverage is the minimum distance between arc coordinates in the coverage. The dangle length is the shortest acceptable length allowed for arcs which have the same polygon to their left and right sides. Both the fuzzy tolerance and dangle length are specified as inputs for ARC commands in which a geographic analysis is performed and topology is created (CLEAN, BUFFER and all overlay commands).

The tic match tolerance is the maximum allowable RMS error when a map of the coverage is registered on a digitizer using the COORDINATE command in ARCEDIT and ARCPLOT. The tic match tolerance is given in coverage units. Anytime the RMS error is higher than the specified tic match tolerance, you will be prompted to correct your map registration by the COORDINATE command.

Tolerance

EDIT, NODESNAP, WEED, GRAIN and SNAP tolerances are used in ARCEDIT. EDIT distance controls the search distance for selecting features; NODESNAP distance controls how close a node on an added arc has to be to an existing node to be snapped to that node; WEED tolerance controls the minimum allowable distance between vertices on an added arc; GRAIN tolerance controls the distance between vertices when you add an arc with a curve in it, add a circle or spline an arc during ADD; SNAP distance controls how close features must be to snap together.

Specifications for the Parameters for the Data File

Parameter		*Description*	*Examples*
Item name		Any name (up to 16 alphanumeric characters—names must begin with an alpha character)	LANDDG01-ID ROAD-CODE
Item width		Number of spaces (or bytes) used to store item values	3 bytes for codes ranging from 1 to 500
Item output width		Number of spaces used to display item values	3 spaces for codes ranging from 1 to 500
Item type		Data type of the item. The most common item types are:	
	C	Character—any combination of alphanumeric characters	Descriptive string: 'forest' Street address: '224 Elm Street'
	I	Integer—any characters that make up a valid integer	Class code: 200 Highway number: 91
	B	Binary integer—an integer number stored in binary format	Class code: 200 Unique ID: 11733
	N	Number—any characters that make up a valid decimal number	Percent: 10 Monetary value: 101.30
	F	Binary floating point—a decimal number stored in binary format	Area: 2344.45 Length: 46.4
	D	Date—stored as 8 bytes; displayed as 8 or 10 spaces	12/31/1989 12/31/1989
Number of decimals		For N or F, the number of digits to the right of the decimal place	Two places for monetary value: 4.22

Real-World Coordinate Systems

For any database to be useful for spatial analysis, all parts of the database must be registered to a common coordinate system. A coordinate system is composed of a spheroid (a mathematical description of the earth's shape) and a map projection (a mathematical conversion from spherical to planar coordinates).

Most maps display coordinate data by conforming to a recognized global coordinate system. The Universal Transverse Mercator (UTM), State Plane coordinate system, the Albers Conic Equal-Area, and Polar stereographic projections are examples of systems which use map projection to represent the earth's three-dimensional features on a flat surface.

Map projections ensure a known relationship between locations on a map and their true locations on the earth.

Latitude-Longitude

The most familiar locational reference system is the spherical coordinate system measured in Latitude and longitude. This system can be used to identify the locations on the earths surface.

Latitude-longitude is a geographic reference system—not a two-dimensional (planar) coordinate system.

This reference system measures angles from the centre of the earth, rather than distances on the earth's surface, it's not a planar coordinate system.

Planar Coordinate Systems

Planar coordinate systems (often called Cartesian coordinate systems) have several properties that make them useful for representing real-world coordinates on maps:

- There are two dimensions: X measures distance in a horizontal direction, and Y measures distance in a vertical direction.
- Measures of length, angle, and area are constant across the two dimensions.
- Various mathematical formulas exist to project the earth's spherical surface onto a flat, two-dimensional surface.

Map Projections

Because the earth is a spheroid, the spherical coordinate system has to be modified somewhat. A mathematical conversion must be used to create a flat map sheet from the three-dimensional surface. This transformation is commonly referred to as a *map projection.*

- Any representation of the earth's surface in two dimension always distorts shape, area, distance, or direction.
- Different projections produce different distortions.
- The characteristics of each projection make them useful for some applications and not useful for others.

Some commonly used map projections in the U.S. and North America

Projection	Characteristics	Maps
Mercator 	Shape: Preserved Area: Distortion increases toward poles Distance: Distortion increases from Equator or secant latitudes Direction: Straight lines represent constant compass bearing	■ Navigational maps ■ Time zone maps
Transverse Mercator 	Shape: Preserved Area: Distortion increases away from Central Meridian Distance: Distortion similar to area Direction: Some distortion	■ 7½' and 15' quadrangle maps for 22 states within the U.S. ■ Maps of North America
Albers Equal-Area Conic 	Shape: Some distortion Area: Preserved Distance: Some distortion Direction: Some distortion	■ U.S. maps ■ Some section maps of the U.S.
Lambert Conformal Conic 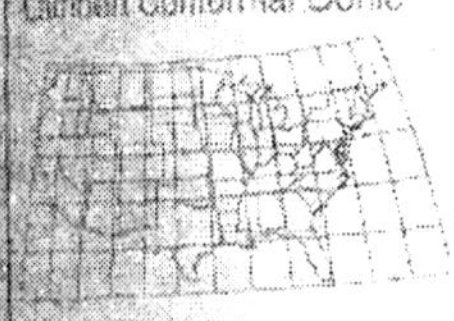	Shape: Preserved Area: Some distortion Distance: Some distortion Direction: Some distortion	■ 7½' and 15' quad maps for 32 states in the U.S. ■ State base maps
State Plane Coordinate System	Not a map projection—a defined set of coordinate systems which vary from state to state. East-West oriented State Plane Zones use the Lambert Conformal Conic projection while North-South oriented State Plane Zones use Transverse Mercator. Each zone has a set of parameters that define the projection for that zone. Most 7 ½ and 15' quadrangle maps for the U.S. use State Plane Zones.	
Universal Transverse Mercator *UTM is not used for very large areas.*	Not a map projection—a coordinate system defined by the transverse Mercator projection and a set of zones and offsets. Shape: Preserved Area: Distortion increases away from Central Meridian Distance: Distortion similar to area Direction: Some distortion	■ USGS 1:250,000 scale quadrangle map series (i.e., 1° latitude by 2° longitude)

Tic Features

Always establish tic points (representing known real-world locations) for a study area before the digitizing stage. Record these common tic locations on each map sheet to provide a common locational reference for each coverage.

Once a map is digitized, the tic points become registration or geographic control points for that coverage allowing all coverage features to be registered to a common coordinate system. Additionally, to ensure geographic control, adjacent coverage and other layers of the same area can be spatially registered using the same tics and the same map projection.

Transformation

When a map is digitized, the X and Y-coordinates are initially held in digitizer measurements (centimetres or inches). To make this information meaningful, and also to impose a scale factor, you must convert these measurements to the same real-world coordinate system and projection in which the original map was created. This process is known as *transformation*.

Points representing tics can be located on maps by identifying the X-and Y-coordinates from a known real-world coordinate system. These are usually given in both the map projection units, usually meters or feet, and degrees of latitude and longitude. Often real-world coordinates for the tics are taken directly from an original map. If these are not available, latitude and longitude values can be used to identify geographic registration points and later converted to real-world coordinates in the same projection used by the base map.

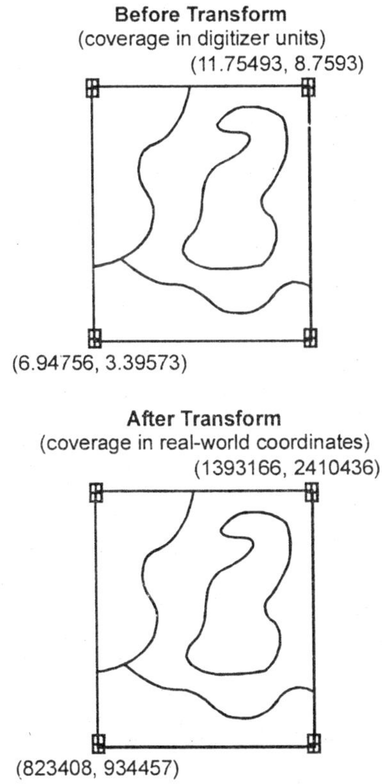

Using ARC/INFO to generate a coverage in real-world coordinates

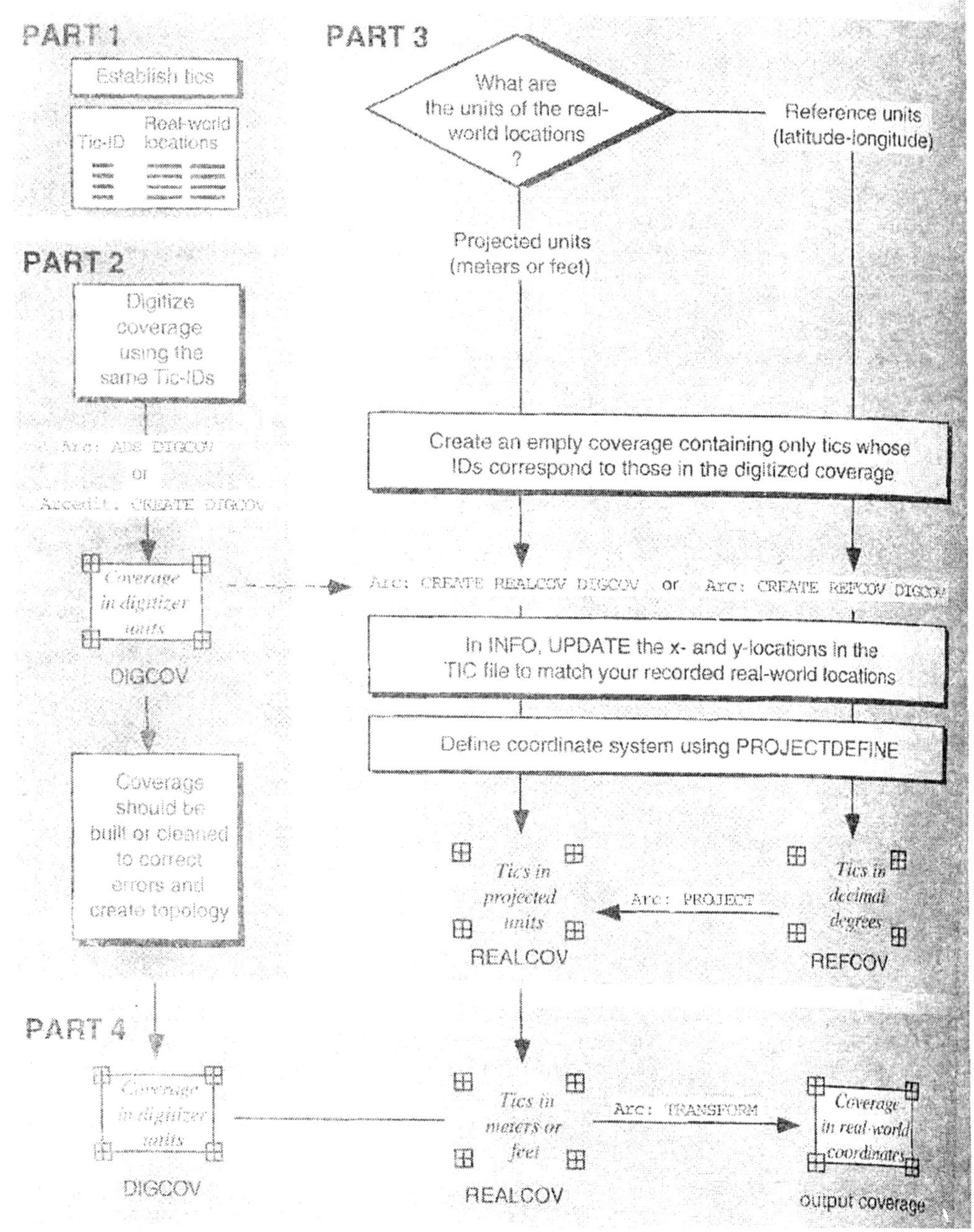

Flow Chart Showing Steps involved in Creating Map

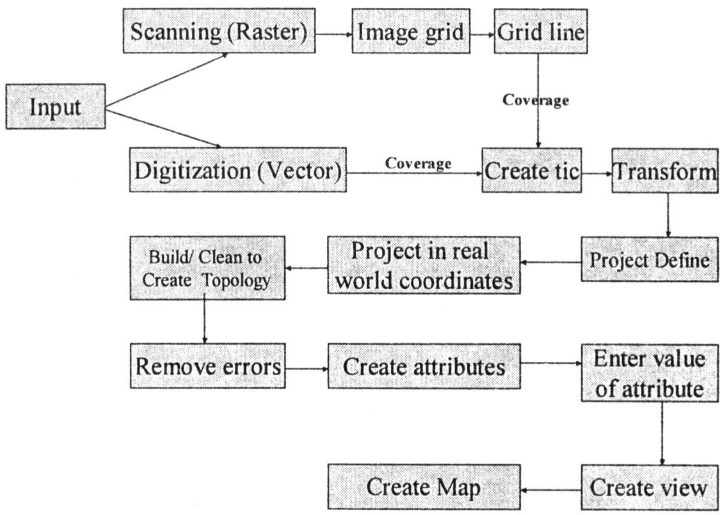

SAMPLE DATA

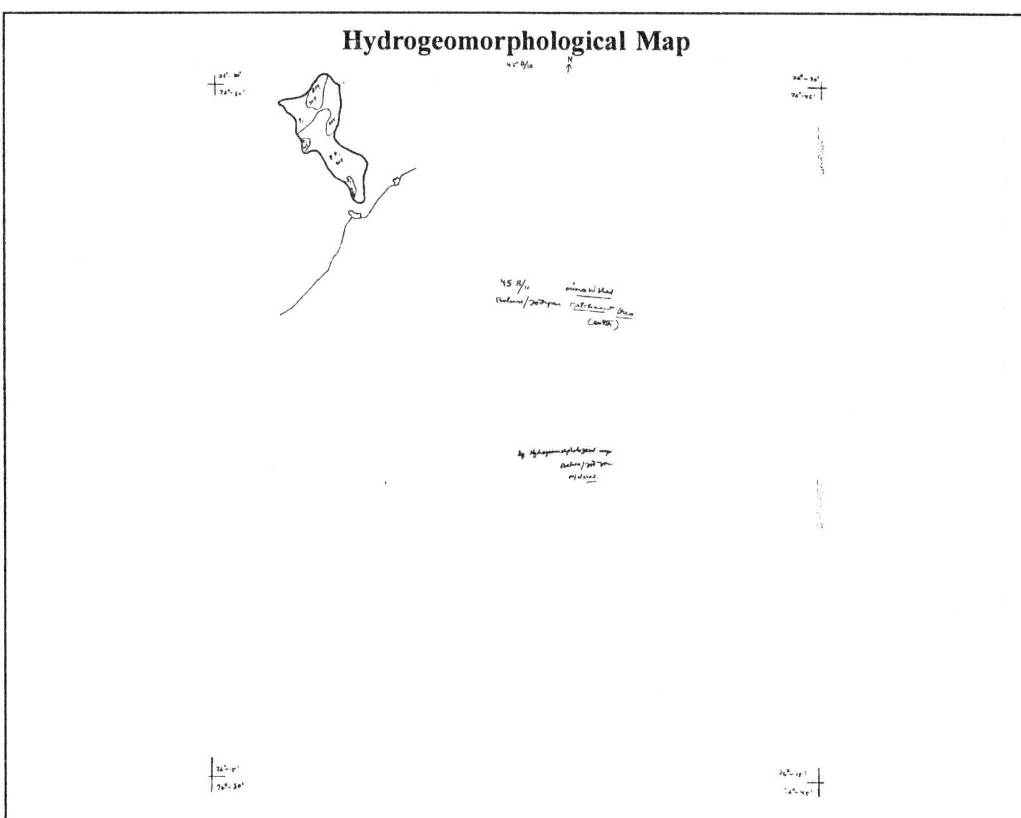

LULC Map

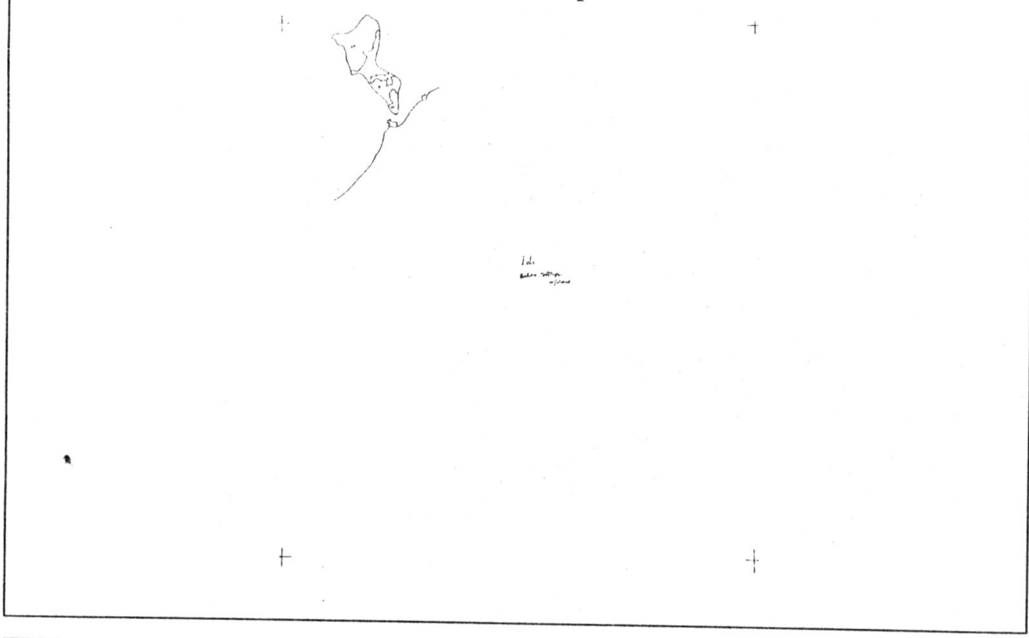

LRDP Map

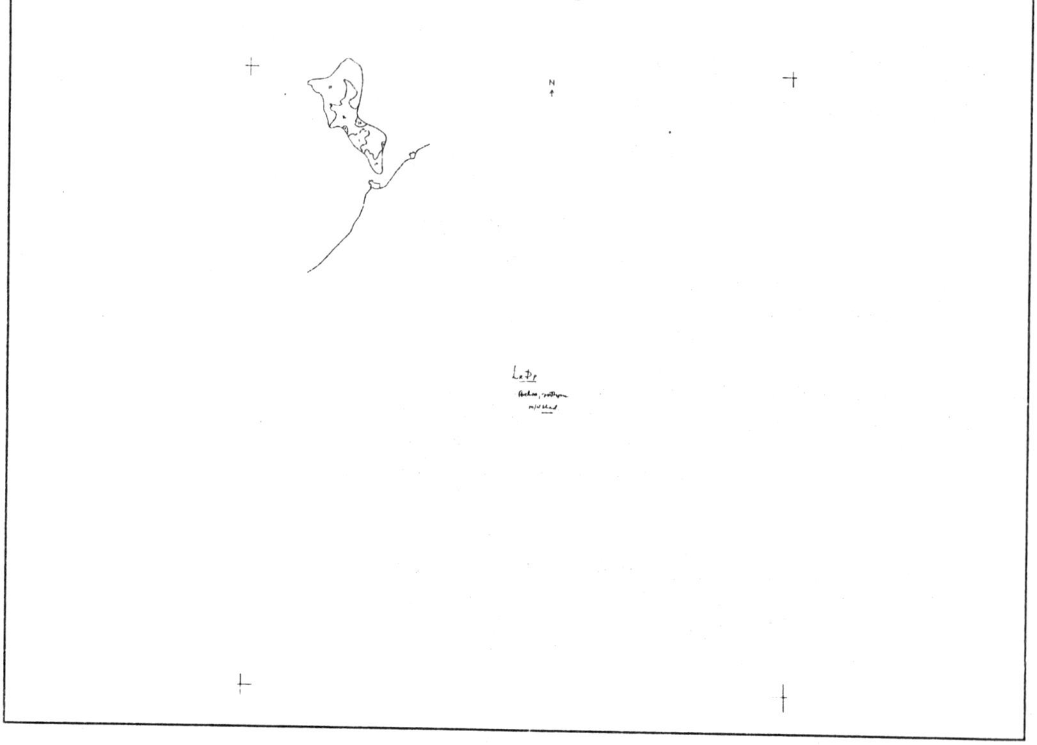

SOIL Map

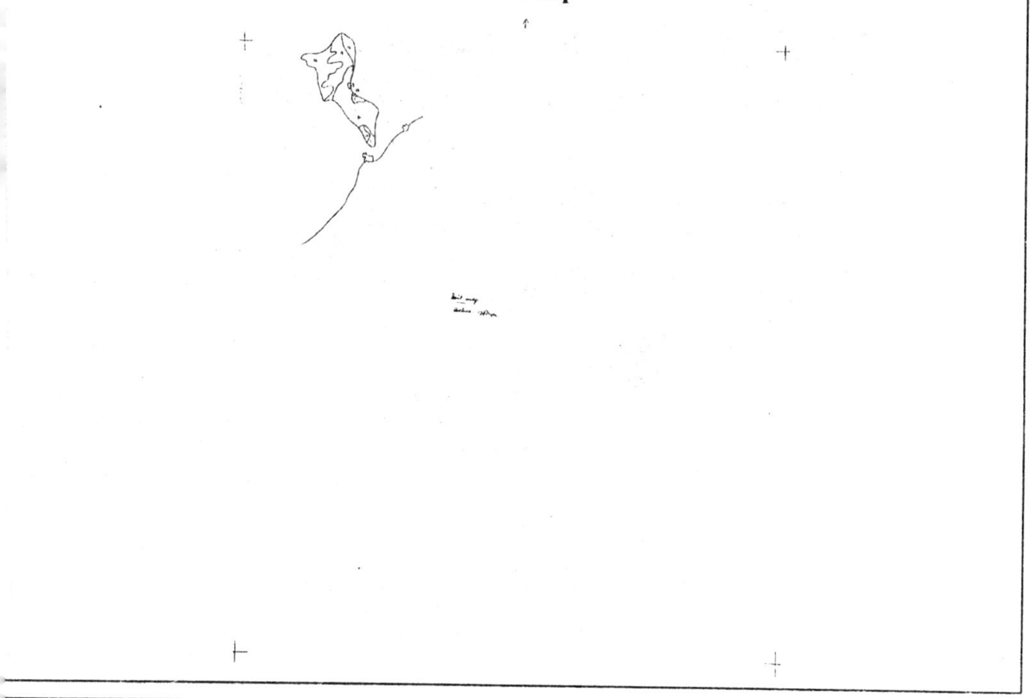

WATER QUALITY Map

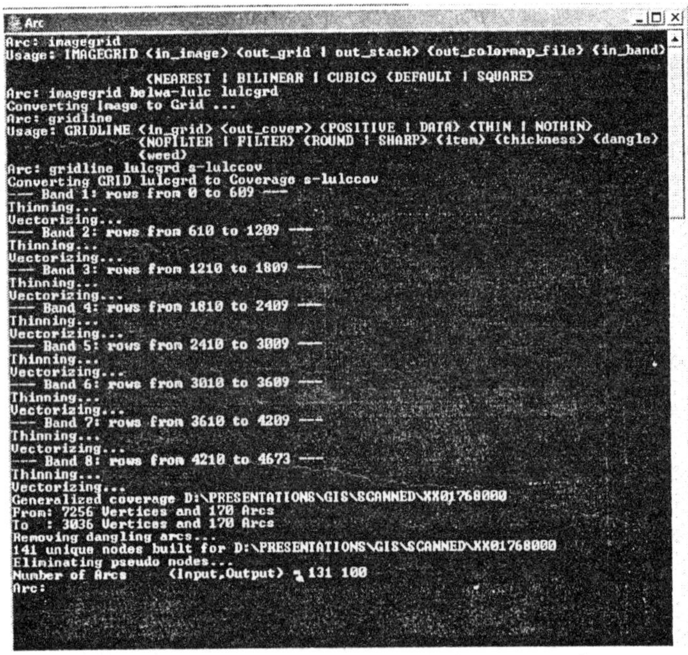

Coverage after conversion from Grid

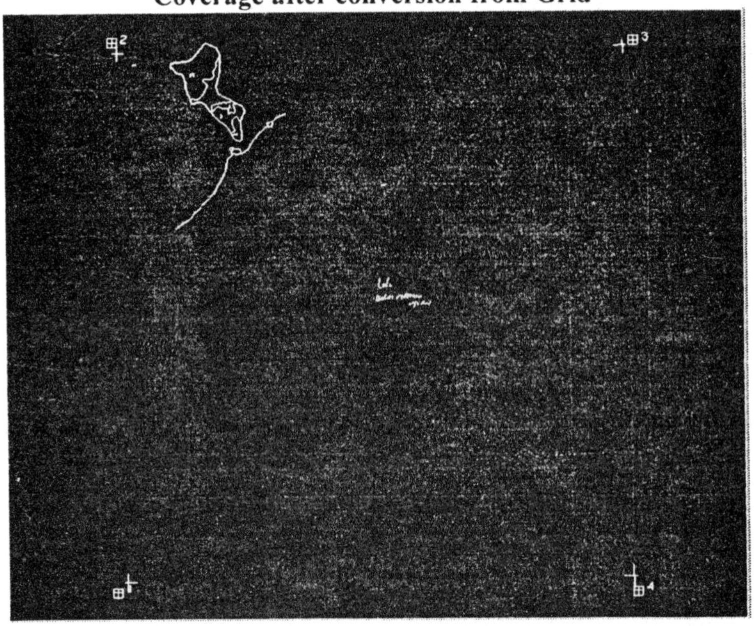

Tics are to be moved at known point Tics should be moved to Known Coordinate Points

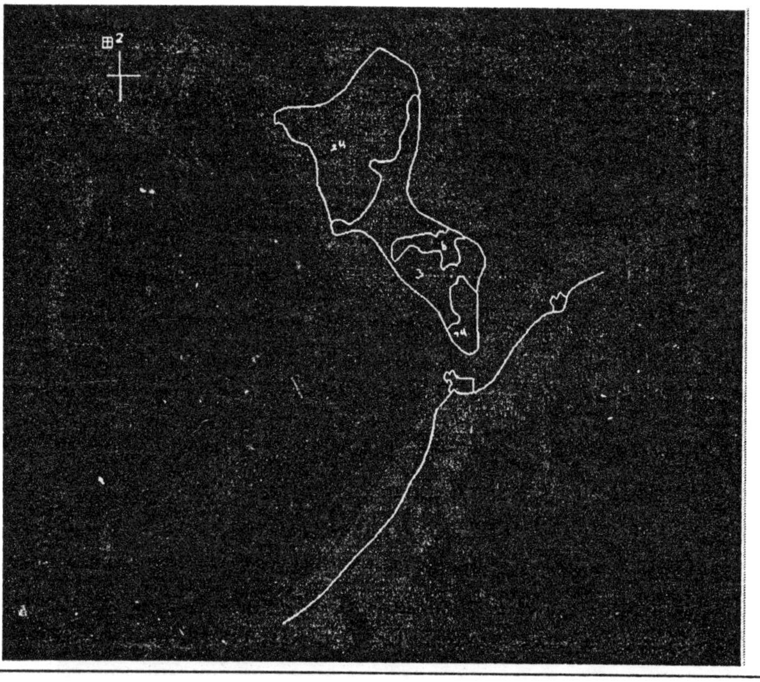

After Moving After Moving Tics

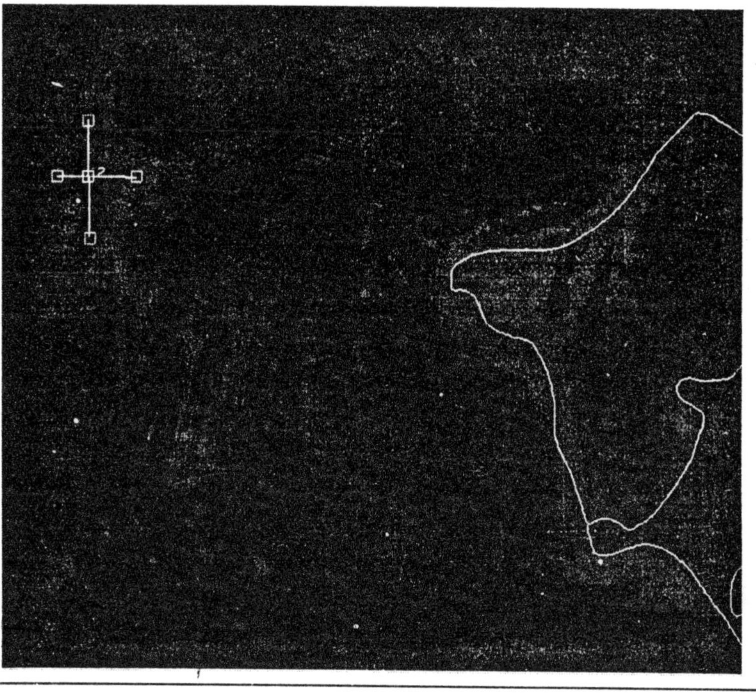

Errors are to be removed

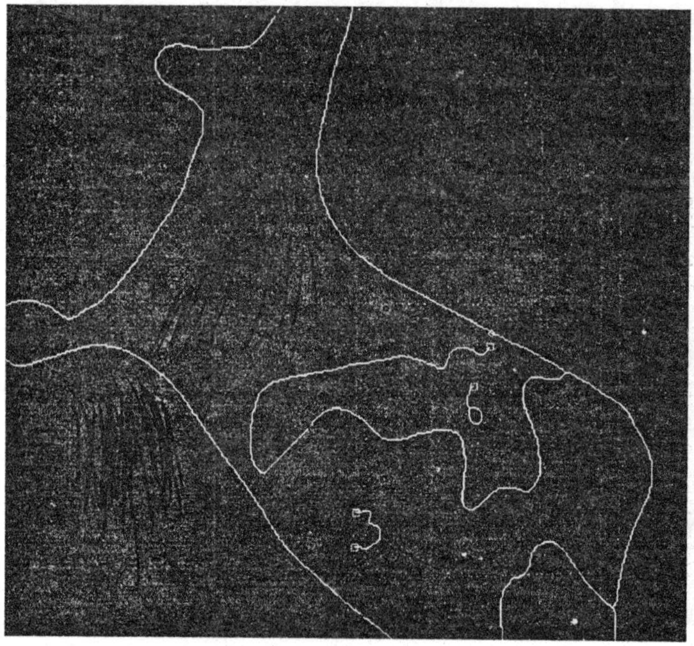

After Removing Errors

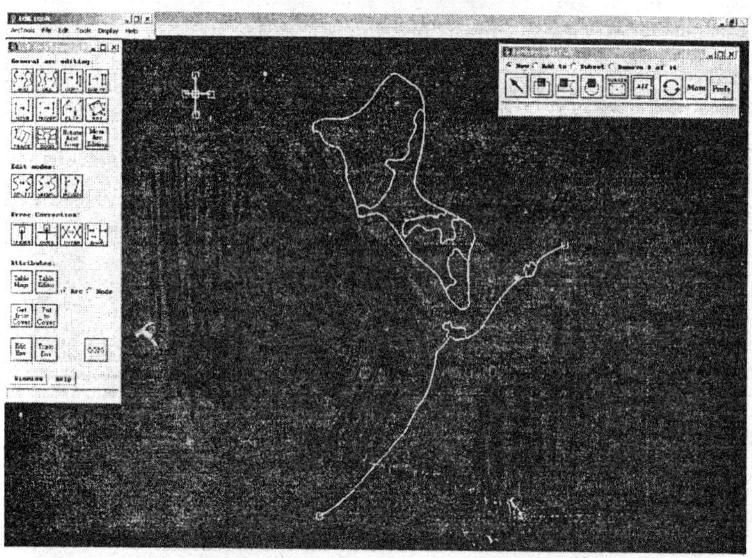

Clean or Build

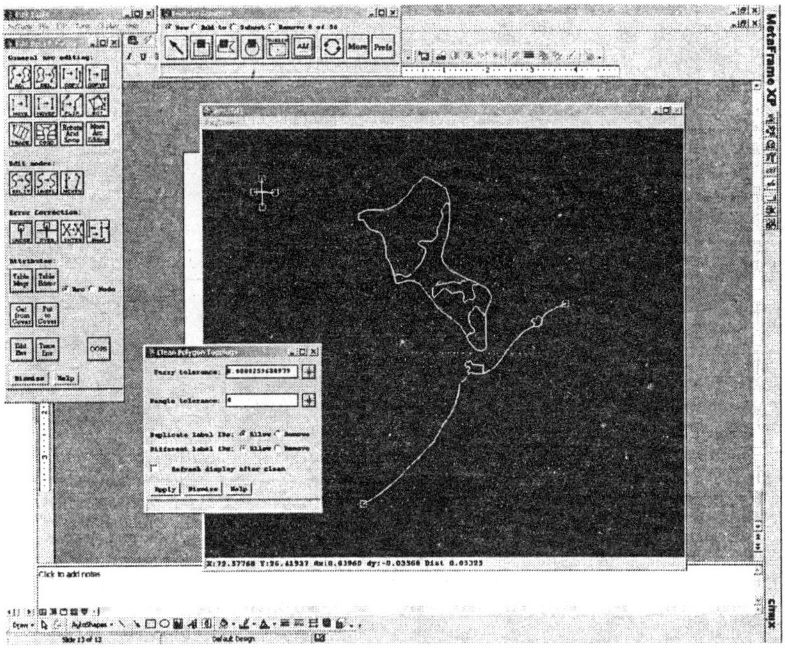

Create Fields

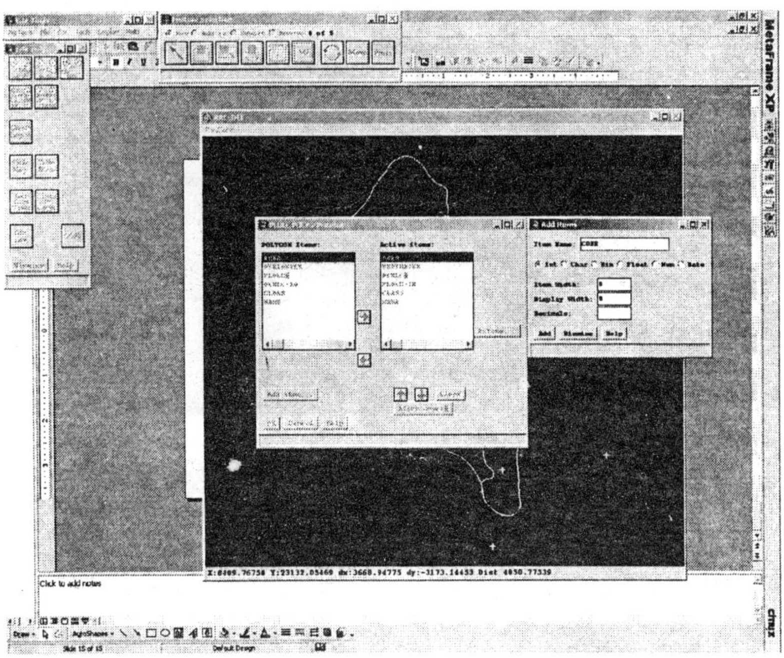

Create a tic coverage

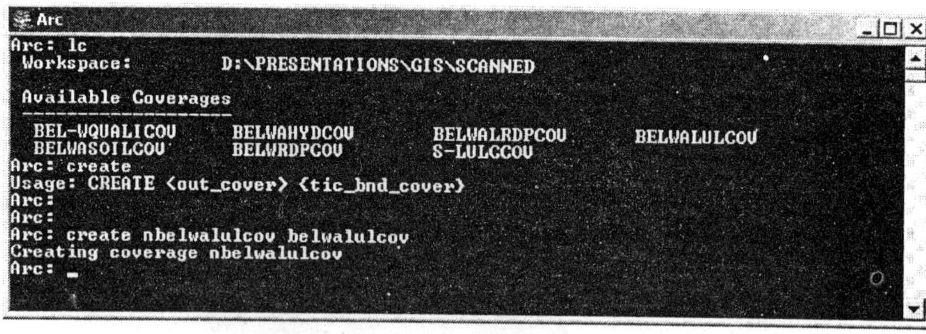

Enter Values for created tics

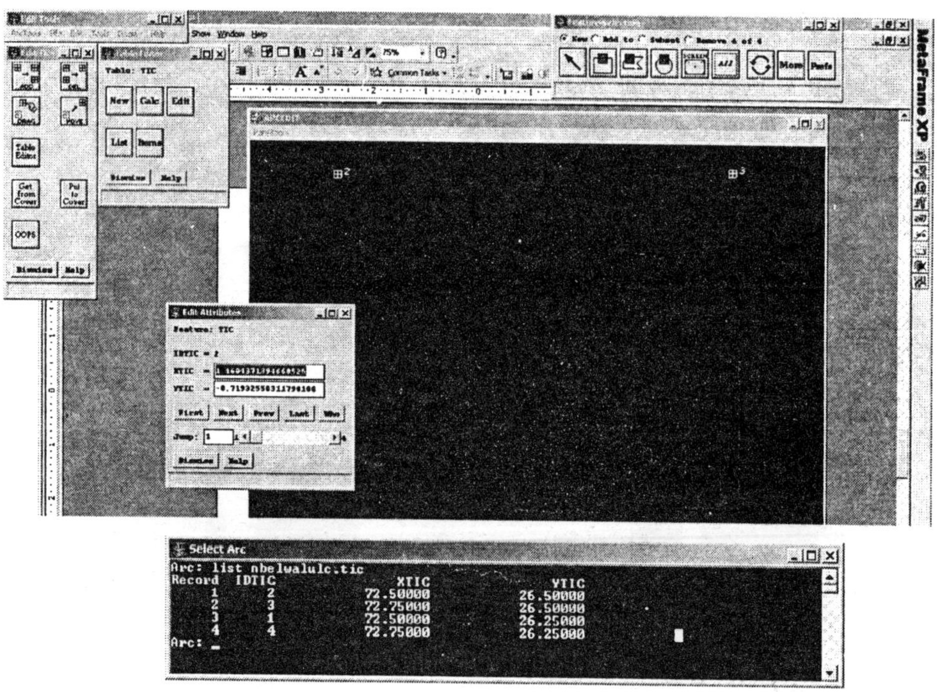

Transform the Coverage

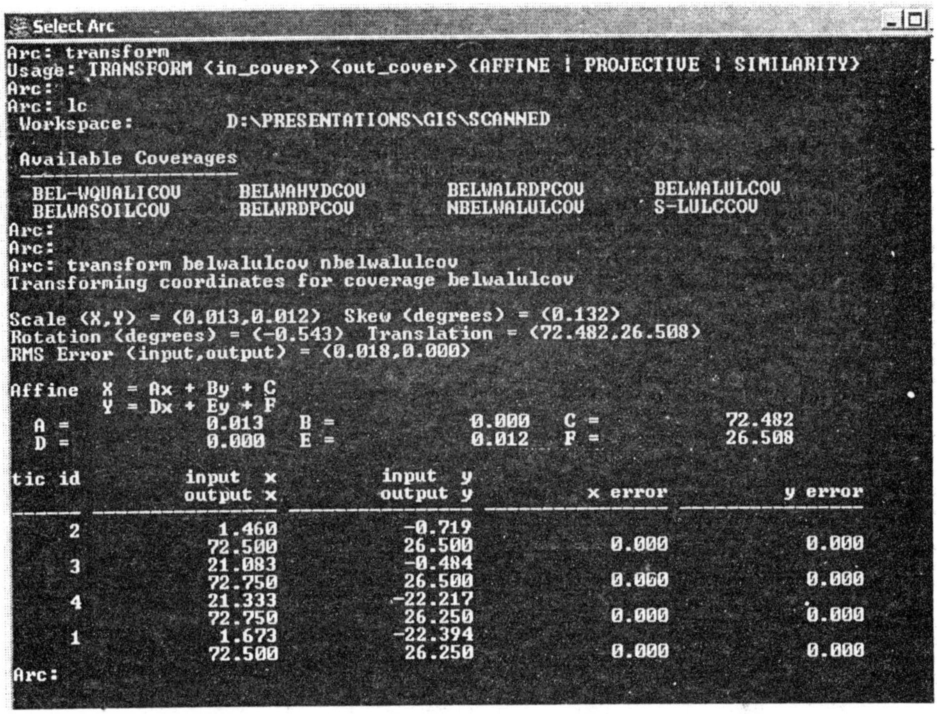

```
Arc: transform
Usage: TRANSFORM <in_cover> <out_cover> {AFFINE | PROJECTIVE | SIMILARITY}
Arc:
Arc: lc
 Workspace:          D:\PRESENTATIONS\GIS\SCANNED

 Available Coverages
 -------------------
  BEL-WQUALICOV      BELWAHYDCOV       BELWALRDPCOV      BELWALULCOV
  BELWASOILCOV       BELWRDPCOV        NBELWALULCOV      S-LULCCOV
Arc:
Arc:
Arc: transform belwalulcov nbelwalulcov
Transforming coordinates for coverage belwalulcov

Scale (X,Y) = (0.013,0.012)  Skew (degrees) = (0.132)
Rotation (degrees) = (-0.543)  Translation = (72.482,26.508)
RMS Error (input,output) = (0.018,0.000)

Affine  X = Ax + By + C
        Y = Dx + Ey + F
  A =          0.013    B =          0.000    C =          72.482
  D =          0.000    E =          0.012    F =          26.508

tic id         input  x        input  y
               output x        output y           x error          y error
------    ---------------  ---------------  ---------------  ---------------
     2          1.460         -0.719
               72.500         26.500             0.000            0.000
     3         21.083         -0.484
               72.750         26.500             0.000            0.000
     4         21.333        -22.217
               72.750         26.250             0.000            0.000
     1          1.673        -22.394
               72.500         26.250             0.000            0.000
Arc:
```

Define the Projection

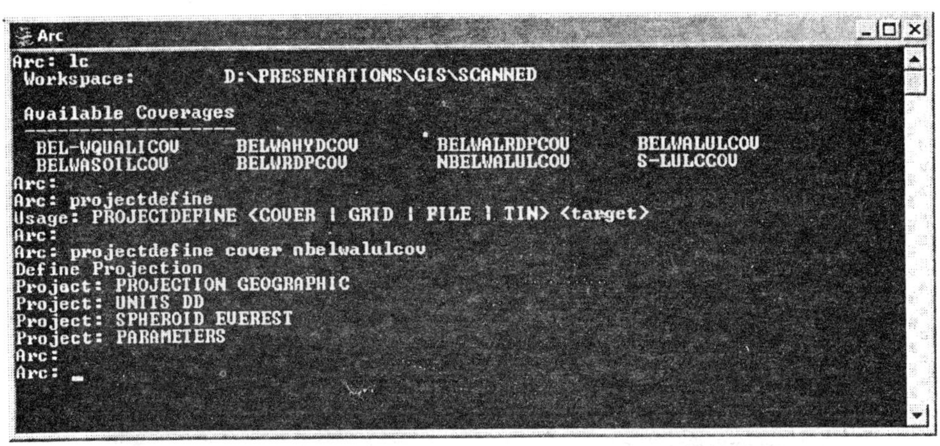

```
Arc: lc
 Workspace:          D:\PRESENTATIONS\GIS\SCANNED

 Available Coverages
 -------------------
  BEL-WQUALICOV      BELWAHYDCOV       BELWALRDPCOV      BELWALULCOV
  BELWASOILCOV       BELWRDPCOV        NBELWALULCOV      S-LULCCOV
Arc:
Arc: projectdefine
Usage: PROJECTDEFINE <COVER | GRID | FILE | TIN> <target>
Arc:
Arc: projectdefine cover nbelwalulcov
Define Projection
Project: PROJECTION GEOGRAPHIC
Project: UNITS DD
Project: SPHEROID EVEREST
Project: PARAMETERS
Arc:
Arc: _
```

Describe the Coverage

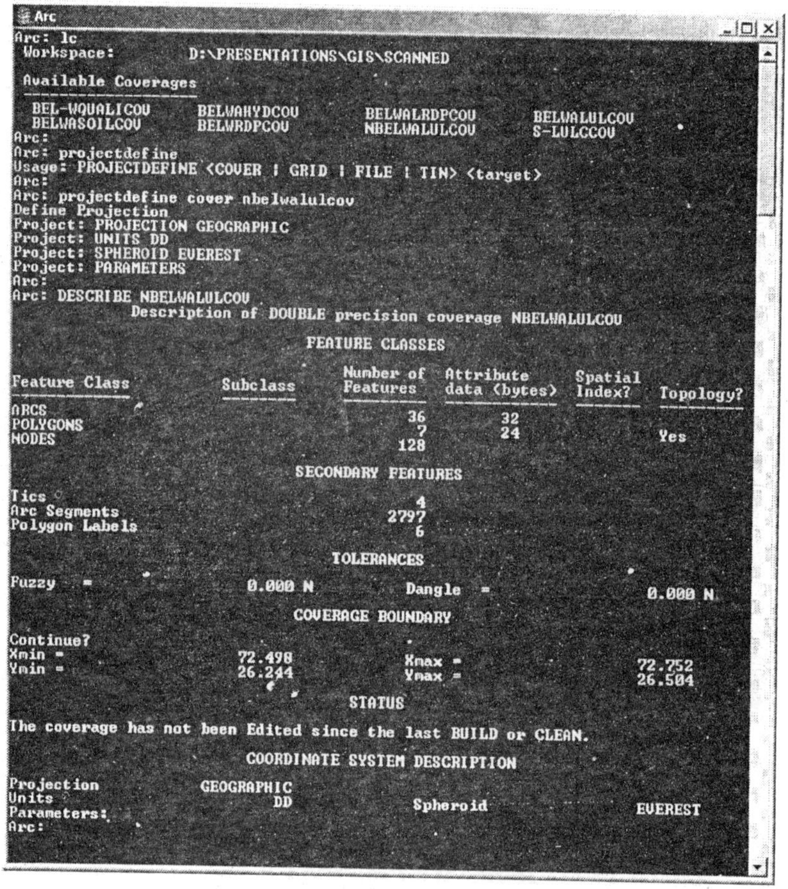

Project the coverage to real world Coordinates

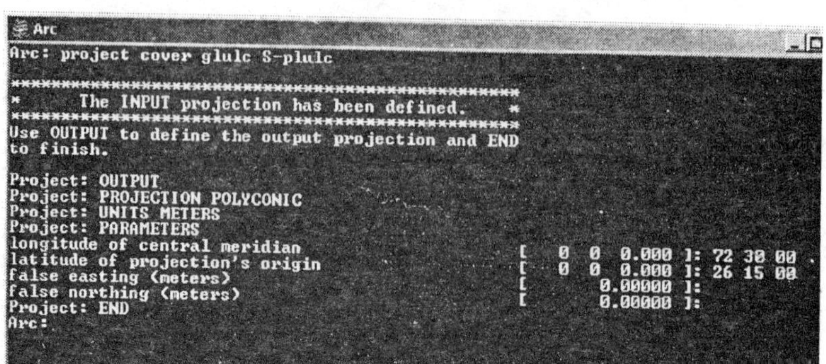

Display Errors

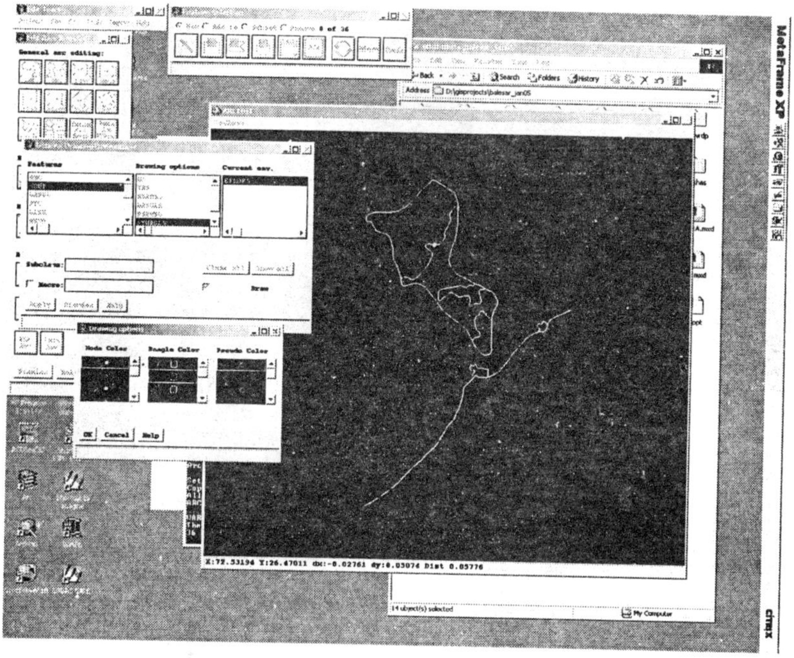

Add themes to a View

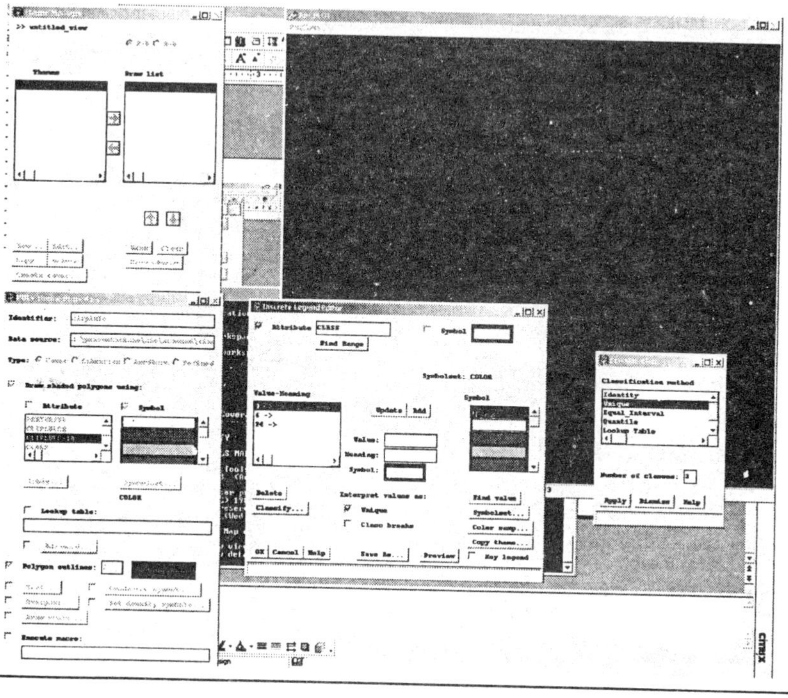

Create View

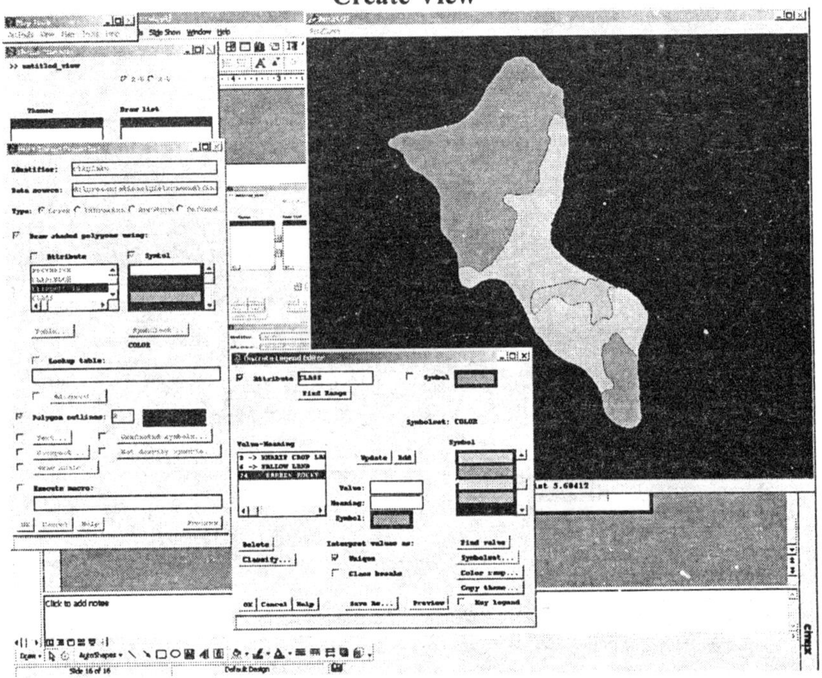

Calculate Statistics

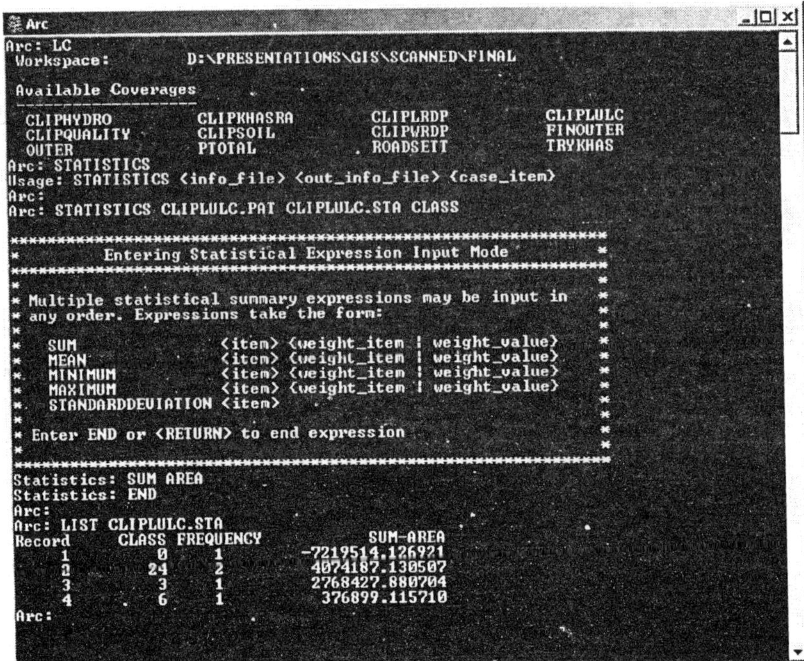

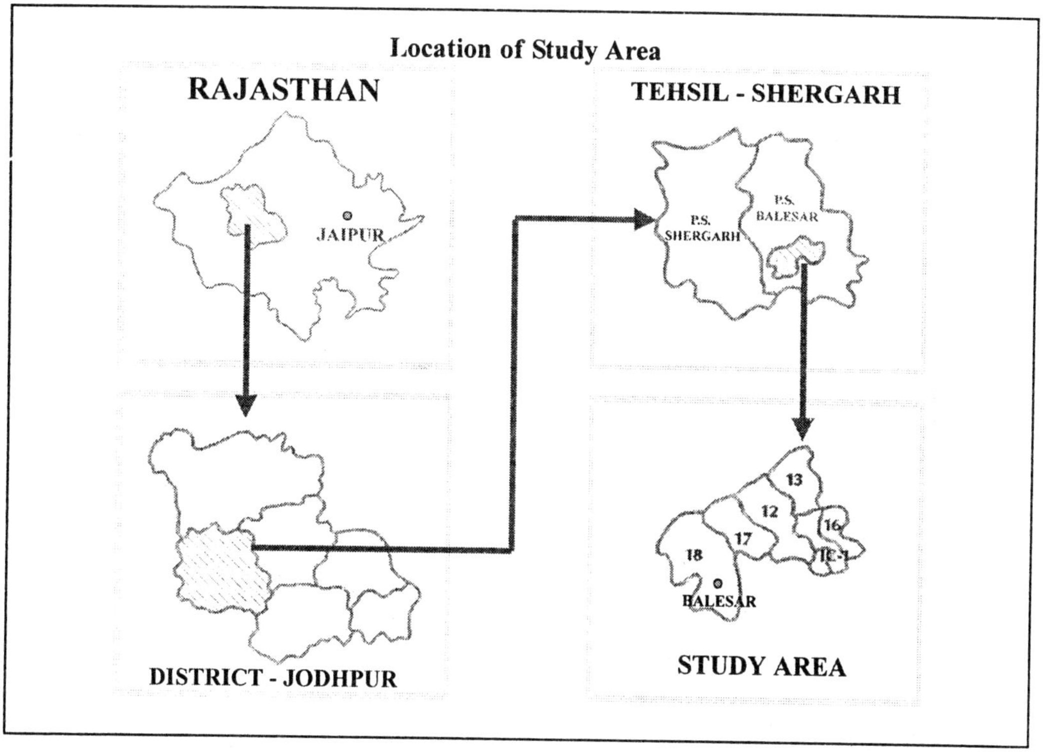
Location of Study Area
RAJASTHAN
JAIPUR
TEHSIL - SHERGARH
P.S. SHERGARH
P.S. BALESAR
DISTRICT - JODHPUR
13
12
16
17
18
IE-1
BALESAR
STUDY AREA

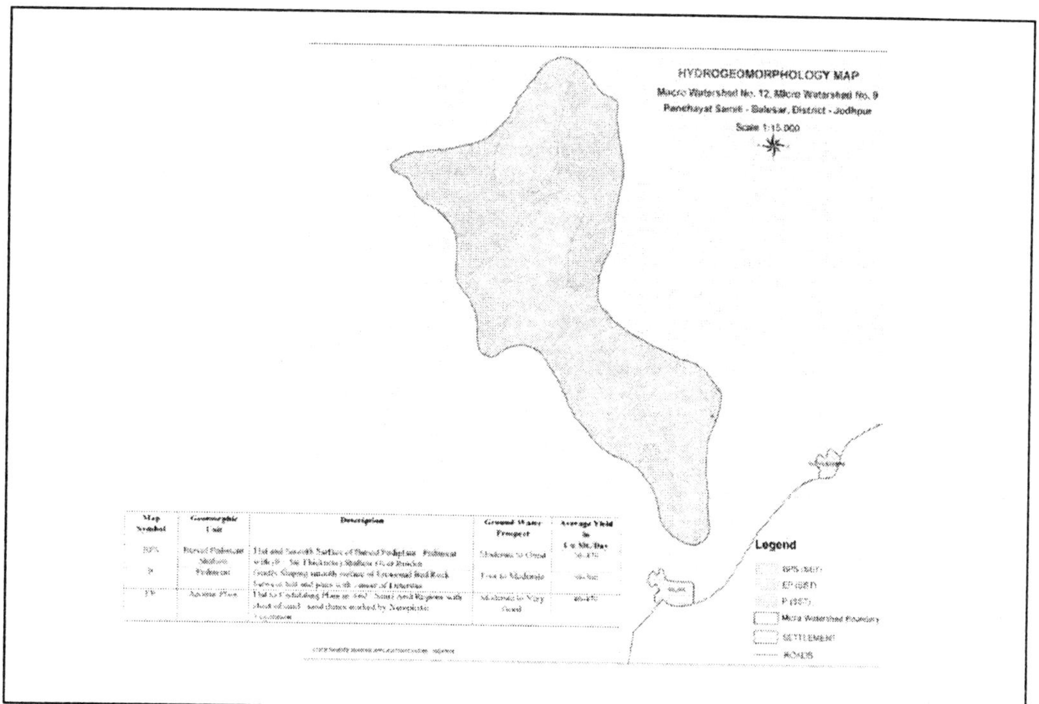
HYDROGEOMORPHOLOGY MAP
Macro Watershed No. 12, Micro Watershed No. 9
Panchayat Samiti - Balesar, District - Jodhpur
Scale 1:15,000
Legend
Micro Watershed Boundary
SETTLEMENT
ROADS

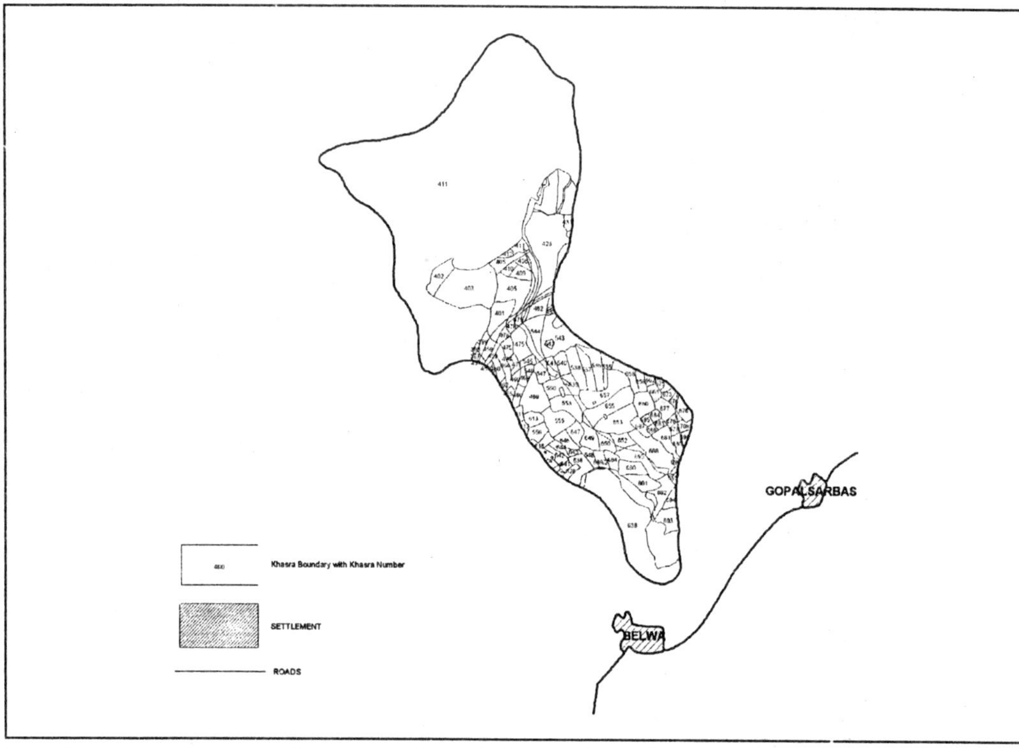
GOPALSARBAS
BELWA
Khasra Boundary with Khasra Number
SETTLEMENT
ROADS

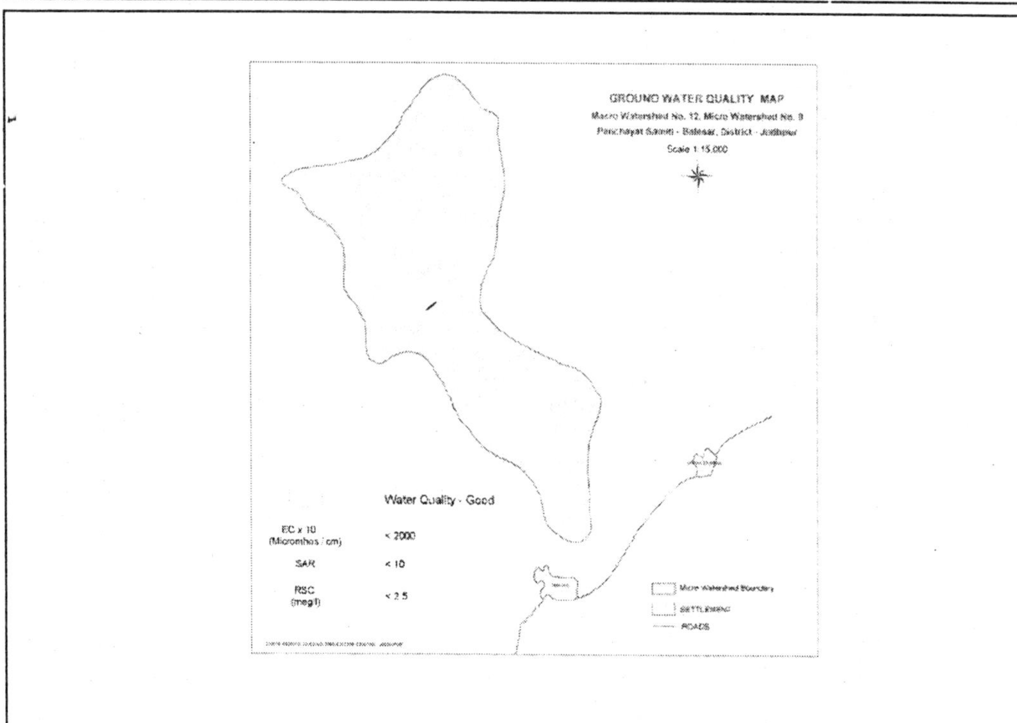
GROUND WATER QUALITY MAP
Macro Watershed No. 12, Micro Watershed No. 9
Panchayat Samiti - Balesar, District - Jodhpur
Scale 1:15,000
Water Quality - Good
EC x 10 (Micromhos / cm) < 2000
SAR < 10
RSC (meq/l) < 2.5
Micro Watershed Boundary
SETTLEMENT
ROADS

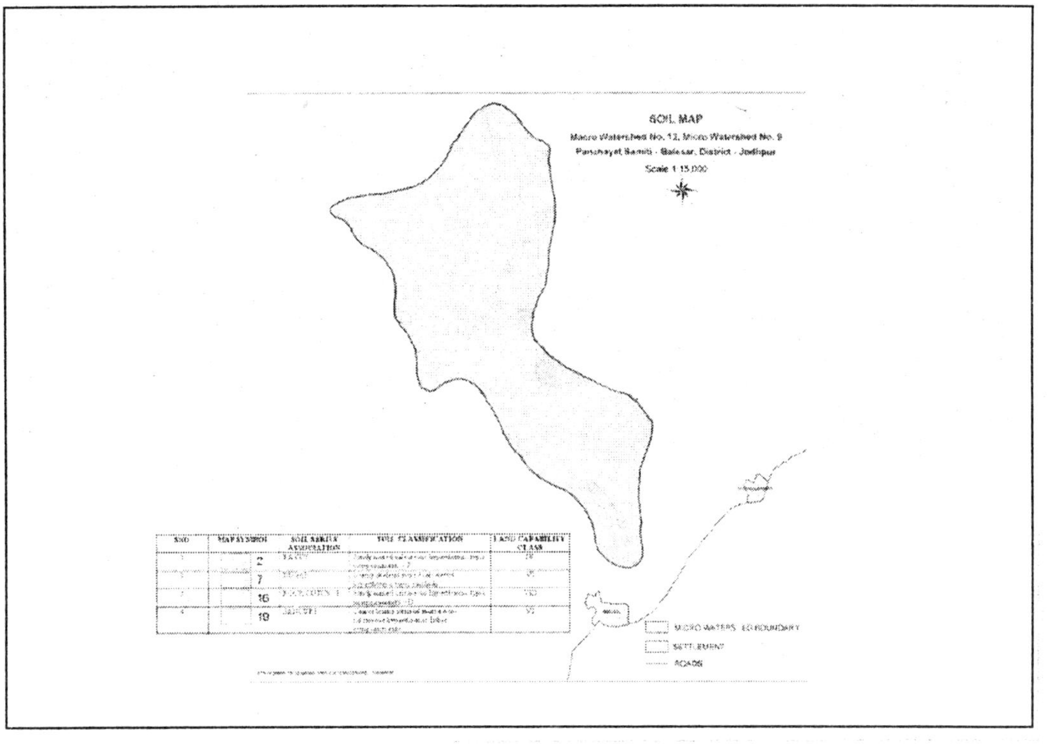
SOIL MAP
Macro Watershed No. 12, Micro Watershed No. 9
Panchayat Samiti - Balesar, District - Jodhpur
Scale 1:15,000
2
7
16
19

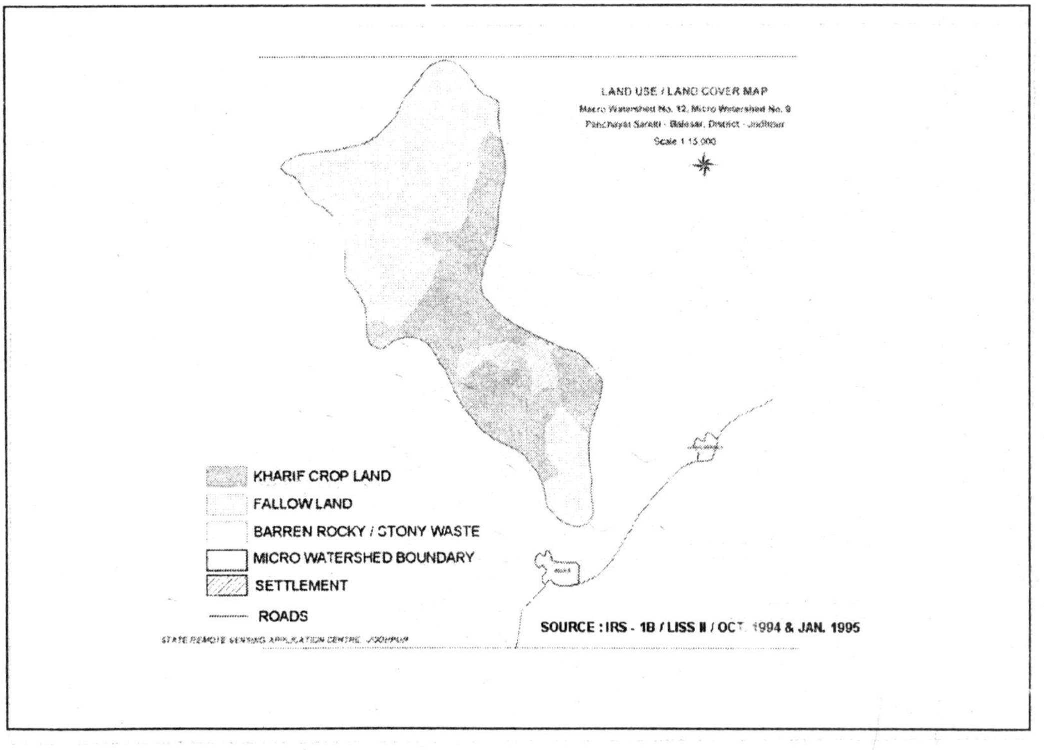
LAND USE / LAND COVER MAP
Macro Watershed No. 12, Micro Watershed No. 9
Panchayat Samiti - Balesar, District - Jodhpur
Scale 1:15,000
KHARIF CROP LAND
FALLOW LAND
BARREN ROCKY / STONY WASTE
MICRO WATERSHED BOUNDARY
SETTLEMENT
ROADS
SOURCE : IRS - 1B / LISS II / OCT. 1994 & JAN. 1995
STATE REMOTE SENSING APPLICATION CENTRE, JODHPUR

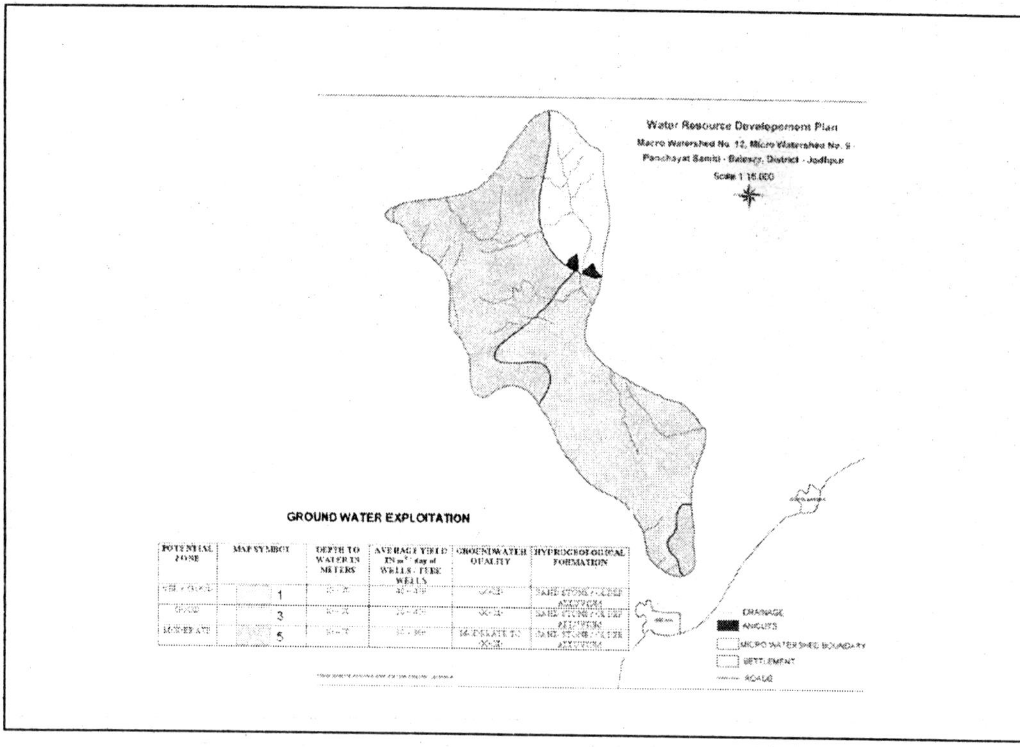
Water Resource Developement Plan
Scale 1 15,000
GROUND WATER EXPLOITATION
POTENTIAL ZONE
MAP SYMBOL
DEPTH TO WATER IN METERS
AVERAGE YIELD IN m² / day of WELLS - TUBE WELLS
GROUNDWATER QUALITY
HYDROGEOLOGICAL FORMATION
1
3
5
DRAINAGE
ANICUTS
MICRO WATERSHED BOUNDARY
SETTLEMENT
ROADS

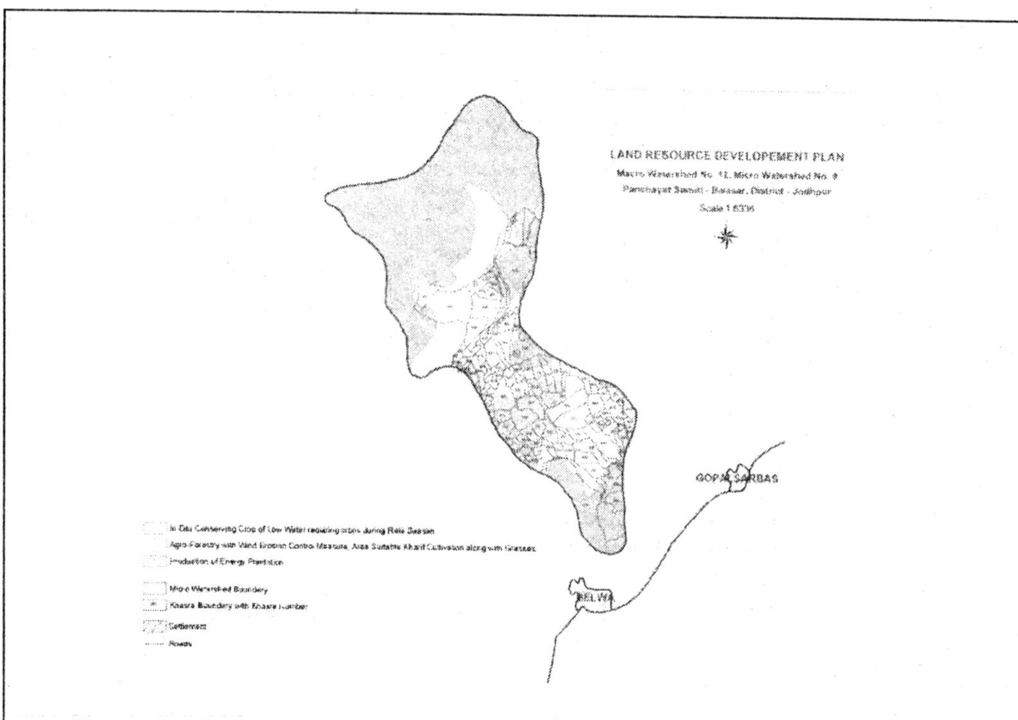
LAND RESOURCE DEVELOPEMENT PLAN
Macro Watershed No. 12, Micro Watershed No. 9
Panchayat Samiti - Balesar, District - Jodhpur
Scale 1:6336
GOPAL SARBAS
BELWA
Production of Energy Plantation
Micro Watershed Boundary
Khasra Boundary with Khasra Number
Settlement
Roads

26

VISUAL INTERPRETATION OF REMOTE SENSING DATA

MILAP PUNIA

REMOTE SENSING DATA PRODUCTS AND ANCILLARY DATA

Data Types

Satellite data products can be classified on the basis of:

(i) level of processing, and
(ii) product media

Level of Processing

There are six types under this category. These are:

Raw Data

Browse Data: used only for quick appraisal of quality of data on a desired date or selecting a cloud free data.
Partially Corrected Data: radio metrically corrected and partially corrected for geometric corrections.
Standard Data: corrected for geometric and radiometric distortions. Standard products can also be obtained as quadrant, full scene, or stereo pair in case of IRS PAN data.
Geocoded Data: standard product made to match with SOI topographic sheet in terms of area scale and alignment.
Special Data: generated after further processing on standard product by extracting enhancing, mosaicing and merging.

Product Media

There are two types under this category.

Digital: Satellite data in digital form can be supplied in Cartridge, DAT, CD-ROM, Floppies.

Photographic: Satellite data products in the form of hard copy products can be supplied as Prints and films in B/W and colour.

Referencing Schemes

Referencing scheme which is unique for each satellite mission is a means of conveniently identifying the geographic location of points on the earth. This scheme is designated by path and rows. The path row concept is based on nominal orbital characteristics.

Path: An orbit is the course of motion taken by satellite in space and the ground trace of the orbit is called as Path. Though the number of orbits and paths are same the designated path number in the referencing scheme and the orbit number are not the same. For example, in case of IRSIC satellite completes 341 orbits in a 24 day cycle with an orbital period of 101.35 minutes. This way satellite completes 14 orbits per day. On day one (D1) the satellite covers orbit numbers1, 318, 294, 270, 246, 222, 198 etc. assuming that the cycle starts with path1.

Row: Along a path the continuous stream of data is segmented into a number of scenes of convenient size. While framing the scenes equator is taken as the reference line for segmentation. The scenes are framed in such a manner that centre of any one of the scenes lies on the equator. Centre of the next scene lies northward or southward depending upon the number of scan lines defined under the scene. The lines joining the centre of the corresponding scene centers of different paths parallel to equator are called as rows the uniformly separated scene centers are such that same rows of different paths fall at the same latitude.

Interpretation Aids

On the basis of tasks to be carried out, Visual interpretation aids can be classified into three categories, viz. Aids for viewing including enhancement transfer of detail and measurement. Interpretation aids can also be classified as equipment for monoscopic viewing and stereo viewing. It should, however, be pointed out that the categorisation is not rigid as certain equipment can be classified into more than one category particularly some of the projection devices which can be used for both viewing and transfer of details. Some of the equipments have become obsolete due to availability of high quality and cost effective FCC data products. In context of land use/cover the aids used only for transferring of details have been discussed here. Basically the equipment in this category are meant for delineating the details from image to a map on the same scale or different scale. These devices employ the means for simultaneous viewing of both the image and the map. There are following three types of equipments which are commonly used for mapping, i.e. HME, LFOE, PROCOM and light tables. First two types come under the optical projection types instruments.

High magnification enlarger (HME): HME is developed at SAC is a versatile aid for visual interpretation of remotely sensed data in the form of transparencies, i.e. B/W or FCC dia-positives. Enlargements upto 20 times are possible in this instrument thus enabling a

1:1 million dia-positives such as LANDSAT TM transparency to be enlarged up to the scale of 1:50,000 corresponding to SOI topographic sheets. Even high magnifications have also been attempted with this equipment comparison. Projection of the image is carried out on a table attached to the equipment on which a map can be fixed for registration with image and also a tracing film for the preparation of a thematic map.

Large format optical enlarger (LFOE): Large format optical enlarger has also been developed at SAC. It is used for two and four times enlargement of 240 mm diapstives.1:1M scale images such as LANDSAT TM and MSS data can be enlarged upto the scale of 1:250,000 scale corresponding to SOI topographical scale at this scale. IRS LISS II images at 1:500,000 scale can be enlarged to a scale of 1:250,000 using a LFOE having capability of two times enlargements. It can also project multi-spectral images in 70 mm format for easy comparison.

Light tables: In addition to the instruments used for optical projection an interpretation aid used for delineation of features from hardcopy paper prints are light tables. Interpretation corresponding to the scale of paper print can be carried out. Handling of images for interpretation is relatively easier using light tables.

Exercise: Path/Row Identification

Path/row identification is must before procuring any satellite data products. This is done as following:

Step – 1: Procure the reference map of the satellite of which data is to be used.
Step – 2: Identify your area based on latitude and longitude using a topographical map.
Step – 3: Find out the location of the same area on the reference map on which lat-long coordinates are also given.
Step – 4: Note down the path and row number covering the area of interest.

Interpretation of Land Use/Land Cover Map

(For detail discussion on classification system, interpretation key, ground truth data collection etc. please refer lecture no. 2.)

Exercise: Delineation of Different Land Features

Interpretation of Geomorphological Map

The map represents various geomorphological units and lineaments mapped using satellite data.

(For detail discussion on classification system, interpretation key, ground truth data collection etc. please refer lecture no. 3.)

Exercise: Geomorphology Mapping

Step 1: base map preparation (source: survey of india topographical map)

- delineation of streams
- delineation of cultural features, e.g. roads (both metalled and unmetalled)
- delineation of railway line
- delineation of habitations (major villages, towns etc.)
- delineation of water bodies (tank, lake, pond etc.)

Step 2: delineation of geomorphological features

(source: satellite image of the study area)

Step 3: cross-checking with SOI topographical map

- strike of the cultural lineaments representing carttracks, roads, rails etc.

Step 4: Final geomorphological map

Demonstration of Software Packages (ERDAS)

Introduction to ERDAS

Exercise – 4: Image extraction and enhancement
Exercise – 5: Digital classification
Exercise – 6: Map projection and Geo-referencing

ARC/INFO GIS Demonstration

ARC/INFO is one of the first GIS packages that was available commercially and is a package being used all over the world. ARC/INFO has been developed by Environmental Systems Research Institute (ESRI), Redlands, USA.

ARC/INFO brings together a strong geographic analysis and modelling capability with a complete interactive system for entry, management, and computer display of spatial data. ARC/INFO facilitates an easy to learn command language for operation of each programme. This command language, together with extensive macros building capability, allows the user to build, adopt and expand the basic tools for customizing applications. Figure 1 shows ARC/INFO software's architecture.

ARC/INFO Data Structure: ARC/INFO is a vector based GIS package consists of spatial data base build around relational DBMS. It organizes geographical data using relational and topological model. Spatial data bases are simply structured with co-ordinates and topology (vector form). The coordinates (x, y) data are used to identify geographic features locations and topological data are used to identify ARC, Node, and Polygon relationships. The cartographic data are then linked by way of common feature numbers to the attribute data managed by relational DBMS.

Concept of the Package: The fundamental element in ARC/INFO is a coverage which is the database of the thematic data. A coverage could consist of map features-point data, lines or polygon data; user-defined annotation-text associated to features and attributes organised as tables. Each coverage has a definite extent and a set of tic-points associated which help in registration of the coverage and the local transformation for display etc. Spatial data is organised as a COVERAGE-separately for points, lines and polygon maps. Attribute data is organised as tables as part of the INFO module and is linked to the coverage by internal and system generated link/key fields.

ARC/INFO Functionalities: ARC/INFO has a wide range of functionality which has been developed based on a tool-box concept—where each function can be visualised as a tool and having a specific utility. Thus, based on the user requirement a specific tool or function could be utilised. The major modules are discussed below:

ARCEDIT: Database creation in ARC/INFO is possible through the process of digitisation using the ARCEDIT module. ARCEDIT is a powerful editing utility having capabilities for feature based editing. These modules include the functions for co-ordinate entry using different devices-digitisers, screen cursors etc.

INFO: INFO is a complete relational database manager for the tabular data associated with geographic feature in map coverages. ARC/INFO keeps track of, and updates map feature attribute table which are stored as INFO data files. ARC/INFO maintains the relationships between each map and corresponding tabular record. INFO can be used to manipulate and update each feature's attribute by performing logical and arithmetic operations on the rows and columns of the table.

ARCPLOT: This module has got capabilities for generating cartographic quality outputs from the database. This includes utilities for interactive map composition, editing map compositions, plotting and printing etc. The map composition functionalities include the incorporation of coverage features as per required scale, generalization, symbolization, transformation, etc. Placements of non-coverage features-like legends, free text, logos, graphic shapes and so on can also be done.

TIN: The TIN module of ARC/INFO can be used to create, store, manage and perform analysis pertaining to the third dimensional data. The TIN module is a part of the overall ARC/INFO package and has utilities for data conversion, modelling display and querying capabilities for the third dimensional data. The modelling capabilities include calculation of slope, aspect, volumes, etc.. Additional functions like VIEWSHED for determining spatial visibility zones, Line of sight are also provided. Display capabilities mainly provide a visualisation of the three dimensional surface either on terminal or can be plotted as a hard copy at a specified angle, orientation, vertical exaggeration etc.

NETWORK: NETWORK module of ARC/INFO performs two general categories of functions—address geocoding and geographic network analysis. Geographc network analysis can be done using the ROUTE module for optimal path determination and the ALLOCATE module for resource allocation analysis. Based on impedances and resource demand, the ROUTE and ALLOCATE model the line network to determine the path and resource allocation. Network functions analyse and model real world networks such as city streets, water ways, telephone lines, providing a tool for vechicle routing, optimum facilities siting, zoning and redistricting, optimum route selection, and time/distance flow analysis.

GRID, is a grid based module of ARC/INFO and is available on mainframe platforms. GRID has an interface to ARC/INFO and thus coverages can be converted to GRID and also from GRID to ARC/INFO. GRID supports powerful modelling tools.

Apart from the above, ARC/INFO also provides gateways to and from other systems-ERDAS system, grid data, AUTOCAD-DXF format, IGES format and a flat file format.

ARC/INFO Platforms: ARC/INFO is available on a wide range of platforms—starting from a PC and on to workstations. ARC/INFO is also available on a variety of operating systems—DOS, Windows 95, Windows NT, and UNIX.

REFERENCES

1. ARC/INFO manuals.
2. ERDAS manuals.

27

CARTOGRAPHIC VISUALIZATION

MILAP PUNIA

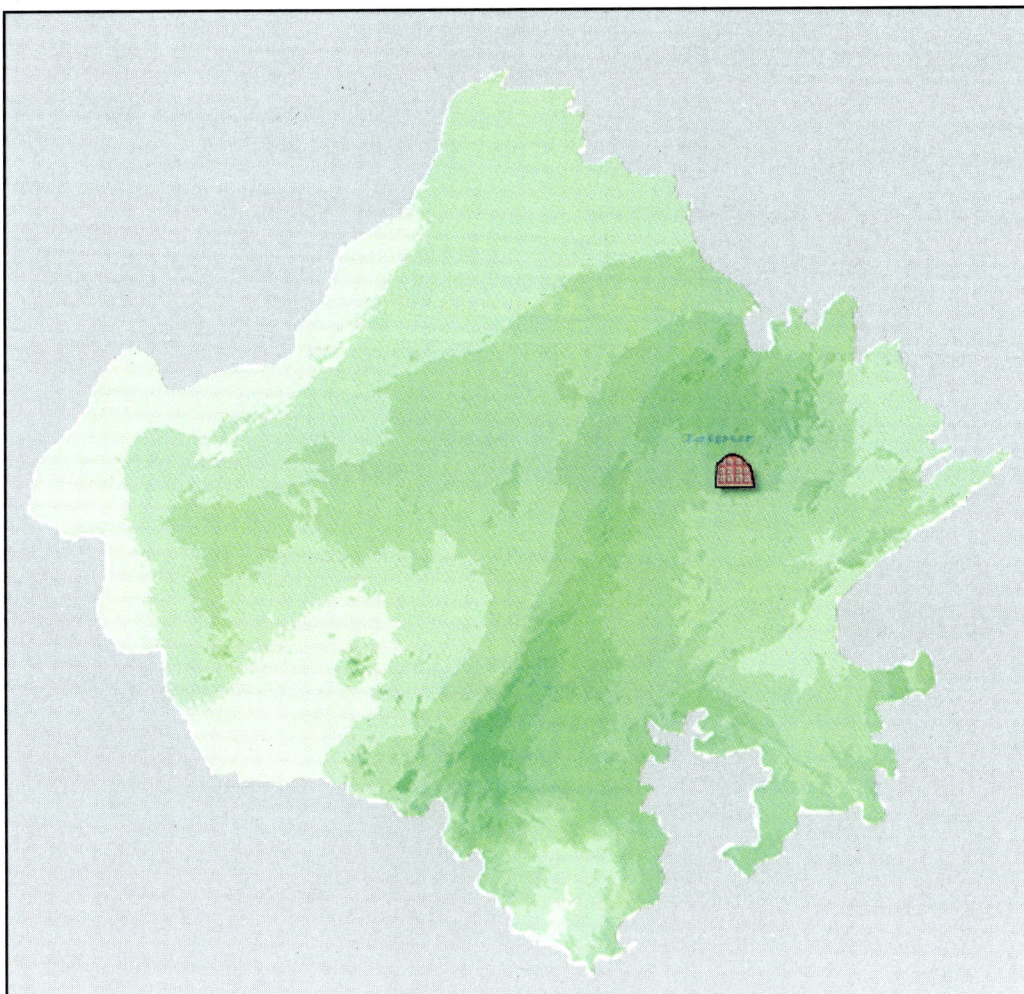

- The nature of maps and of their use in science and society is in the midst of remarkable change-change that is stimulated by a combination of new scientific and societal needs for georeferenced information and rapidly evolving technologies that can provide that information in innovative ways.
- A key issue at the heart of this change is the concept of visualization.
- Cartography has always been about visualization—in the sense of making aspects of our world visible.

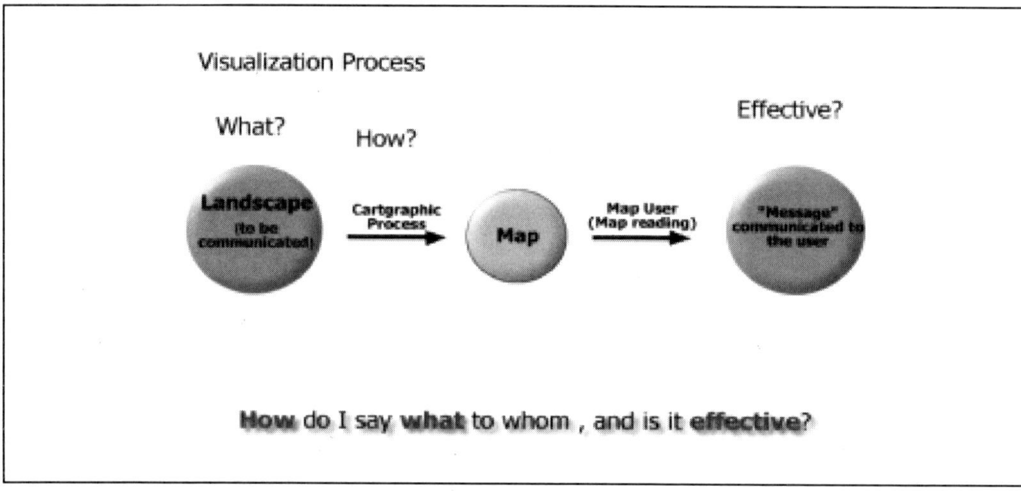

WHAT IS GEOVISUALIZATION?

(Geo) Visualization

Use of computer-supported, interactive, visual representations of (geospatial) data to strengthen understanding

the purpose of visualization is *insight*, not pictures
(or maps)
insight = discovery, and explanation, leads to decision making

Geovisualization

The use of visual geospatial displays to explore data and through that exploration to generate hypotheses, develop problem solutions, and construct knowledge.

Geovisualization Integrates...

...... approaches from
geographic information systems
image analysis
exploratory data analysis
scientific visualization
information visualization

Cartography

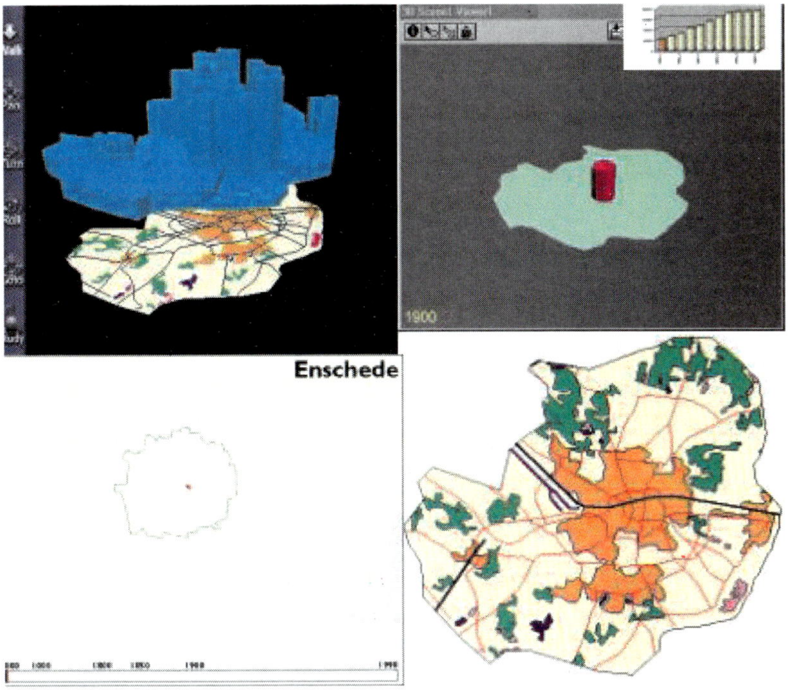

Visualization and Maps

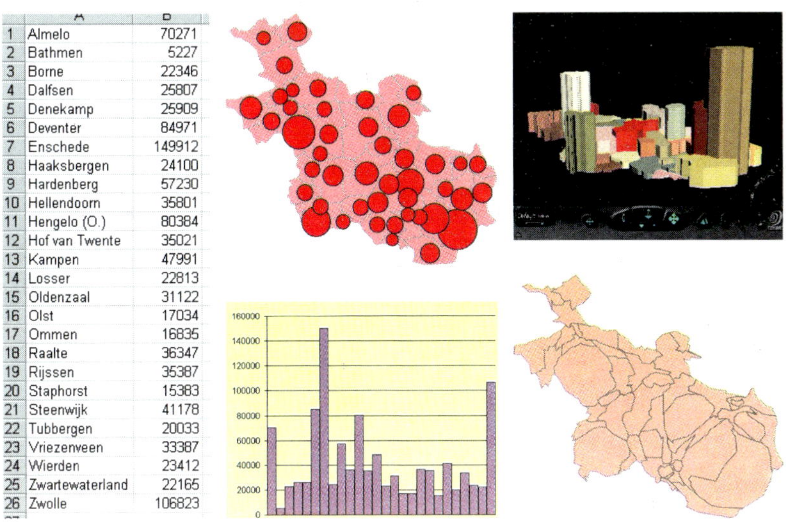

	A	B
1	Almelo	70271
2	Bathmen	5227
3	Borne	22346
4	Dalfsen	25807
5	Denekamp	25909
6	Deventer	84971
7	Enschede	149912
8	Haaksbergen	24100
9	Hardenberg	57230
10	Hellendoorn	35801
11	Hengelo (O.)	80384
12	Hof van Twente	35021
13	Kampen	47991
14	Losser	22813
15	Oldenzaal	31122
16	Olst	17034
17	Ommen	16835
18	Raalte	36347
19	Rijssen	35387
20	Staphorst	15383
21	Steenwijk	41178
22	Tubbergen	20033
23	Vriezenveen	33387
24	Wierden	23412
25	Zwartewaterland	22165
26	Zwolle	106823

Multiple Linked Interactive Views

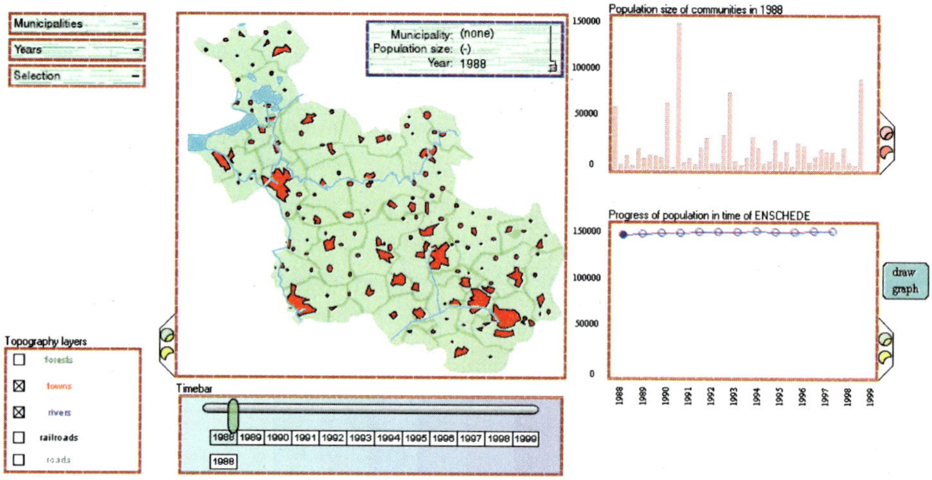

Image Analysis

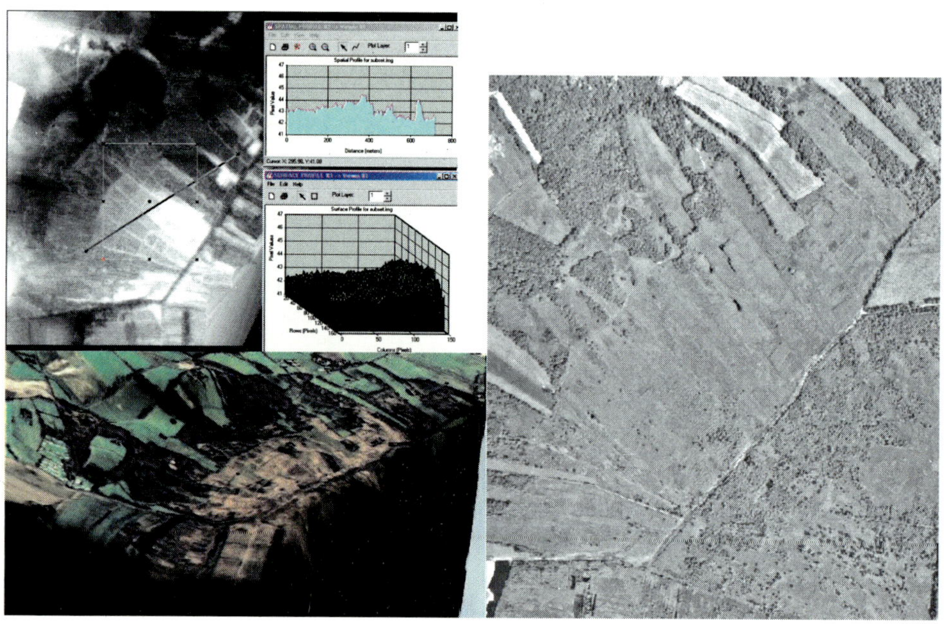

GEOVISUALIZATION AND MAPS

Maps are used to stimulate (visual) thinking about geospatial patterns, relationships and trends:

- view geospatial data sets in a number of *alternative* ways.
- multiple representations without constraints (traditions).

28

USING GIS TO UNDERSTAND CANCER DISTRIBUTION

R.K. GUPTA

TITLE

- Analyzing factors associated with cancer occurrence: A geographical systems approach"
- Mutlu Hayran
- Turkish Journal of Cancer, volume 34, issue 2, April 2004.

OBJECTIVES

- Use GIS to analyze cancer incidence data
- Look at prostate cancer distribution in Pennsylvania
- Link the geography of risk factors with the results of the disease

DATA

- Sources
- Prostate cancer diagnosis from 1993 to 1997
- Diagnosis date
- Stage at diagnosis
- Vital status
- Age
- Race
- Location
- Distance to nearest hospital
- Distance to toxic release sites

METHODS

- Gather data
- Calculate frequencies
- GIS mapping
- Look for patterns
- Smoothed maps

- Statistics
  - Analyze clusters
  - Autocorrelation

RESULTS

- Range 55.4 to 193.93 per 100,000 people
- Average of 134 per 100,000 people
- Fewer incidences in the northeast
- More incidences in the south east
- Important variables
  - Less than two bedrooms
  - Closeness to toxic site

RESULTS

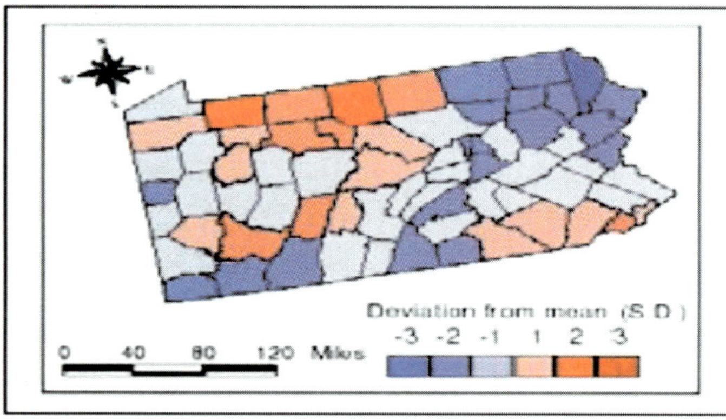

Smoothed map for age-adjusted prostate cancer incidences

RESULTS

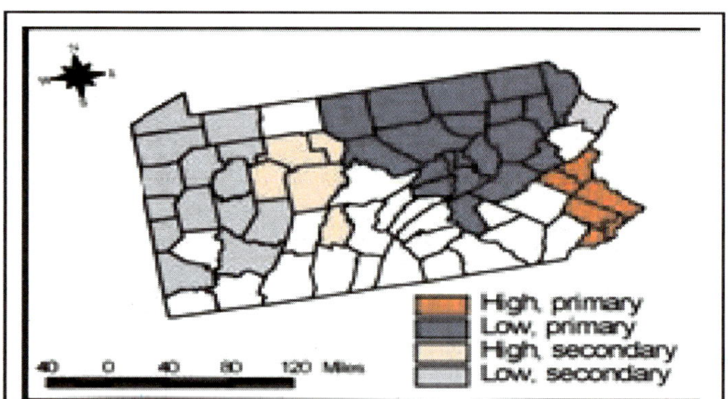

High and low clusters detected using spatial scan statistics

DISCUSSION

- Less cancer in less than two bedroom homes
  - Poorer?
  - Less likely to go to hospitals?
- More cancer near toxic sites
  - More urban
  - More hospitals
- Supported by other studies

IMPLICATIONS

- GIS useful to study of geographic distribution of diseases
- Helpful in finding patterns
- Good for generating hypothesis
- Locate areas that need help

29

TOPOLOGY BUILDING OF THE CARTOGRAPHIC INFORMATION

MILAP PUNIA

...going for field work

DTM
Accuracy Assessment

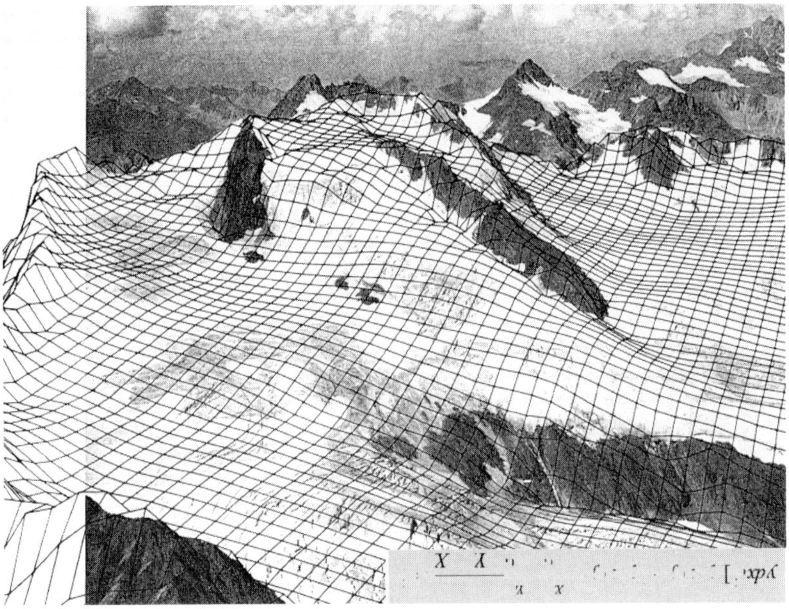

Accuracy - why to bother?

(a) evaluating an existing DTM
(b) designing a new DTM

Available Global DTM Data Sets

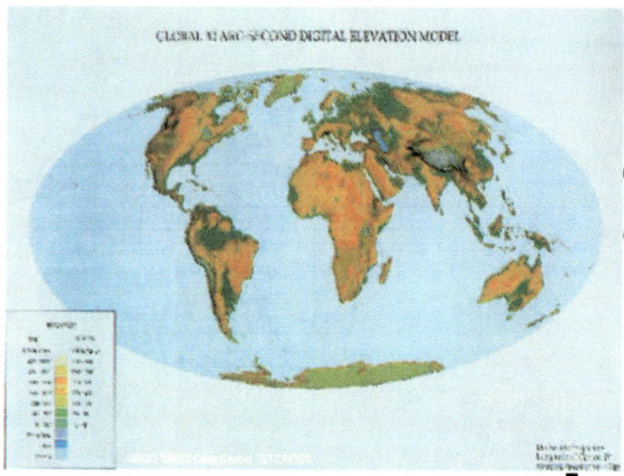

spacing: 30 arcsec (1km)

DLR Shuttle Radar Topographic Mission

SRTM Data Products

DTED (Digital Terrain Elevation Data)

tile size: 1 deg. × 1 deg.

posting: 3 arcsec (30 m)

datum: mean sea level, WGS 84

SRTM

Global coverage
60° North to 58° South

American C-band:
6 cm wavelength,
complete coverage;
not available since 11/9

German X-band:
3 cm wavelength,
gaps between swaths;

accuracy ~3m
in open terrain

San Francisco

DTM Data Acquisition Techniques

- **Ground Survey**
  - – field surveying
  - – airborne laser scanning
  - – interferometric SAR
- **Photogrammetry**
  - – (semi-automatic) digitizing
  - – image matching
- **Digitizing Maps**
  - – (semi-automatic) line following
  - – scanning

Data Processing Methods

Different sampling patterns

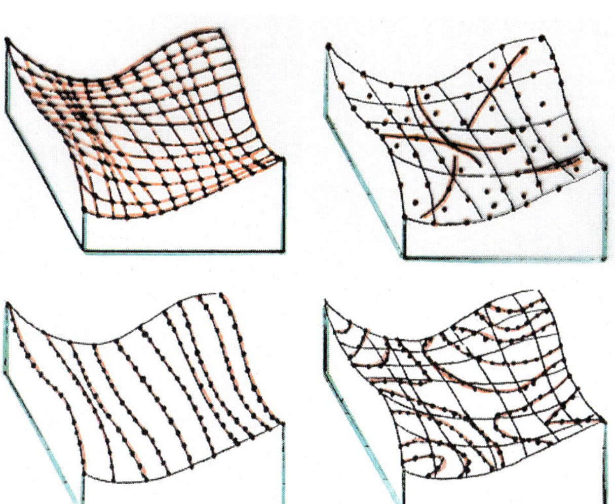

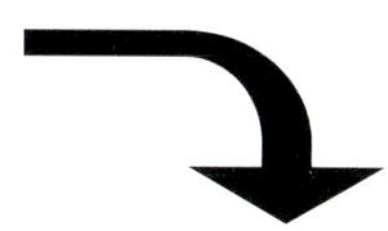

Grid DTM by Interpolation

Linear interpolation
Cubic interpolation
Moving average
Finite elements
Linear prediction
Kriging
……

- (irregular) grid
- profiles
- contours
- morphologic

Accessing Accuracy of existing DTM

(a) Trust the meta data
(b) Empirical Approach

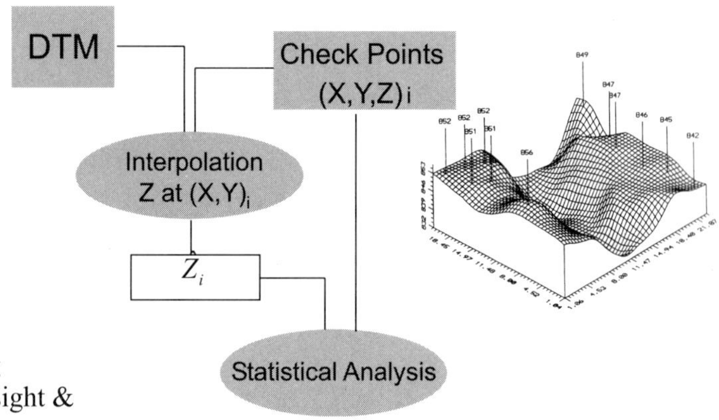

simple & convincing
but no theoretical insight &
not for designing

$$\hat{\sigma} = \sqrt{\frac{\sum \left(Z_i - \hat{Z}_i\right)^2}{n}}$$

Predicting Accuracy of new DTM

Which choices to meet specifications?

- grid or TIN?
- point spacing? breaklines?
- precision of elevation values?
- which interpolation method?
- how to validate end products?

Proper design:

Meet specifications

at minimum costs

From Application to DTM Spec's

Example 1: contour map to be produced from DTM

Contour lines 1:25,000
contour interval 10m
90% of the checked elevations in error less than CI/2

From experience: σ_{DTM} < **CI/5**

Example 2: Predicting the water level from the discharge of a river

Using Manning's formula for the discharge at the cross section of a river

$$Q = R^{2/3} \cdot S^{1/2} \cdot A \cdot n^{-1}$$

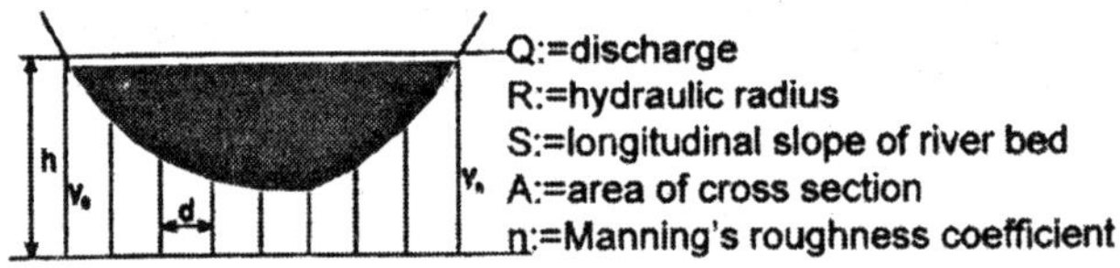

we can calculate A.
From A and the DTM we can calculate h, the water level

$$Q = R^{2/3} \cdot S^{1/2} \cdot A \cdot n^{-1}$$

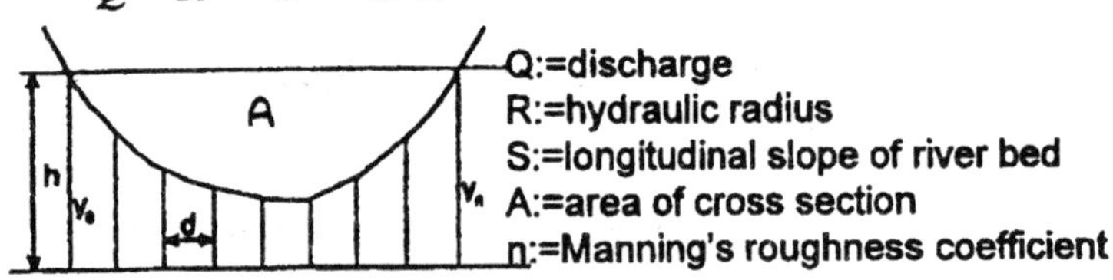

h by numeric integration from the DTM:

$$(x_m - x_0)\,\mathbf{h} = A + (y_1 + y_3 \ldots\ldots + y_{2m-1})(x_m - x_0)/2$$

$$\rightarrow \mathbf{h} = A/(x_m - x_0) + 1/m\Sigma y$$

Tolerable error of h $\Rightarrow$ tolerable error of DTM

Error propagation (assuming $\sigma_y = \sigma_{DTM}$):

$$\sigma_h^2 = \sigma_{DTM}^2/m$$

90% confidence interval: [h 1.6 σ_h]

e.g., $1.6\sigma_h < 30$ cm $\rightarrow \sigma_{DTM}$ **< 20 SQR(m)**

Example 3: orthophoto

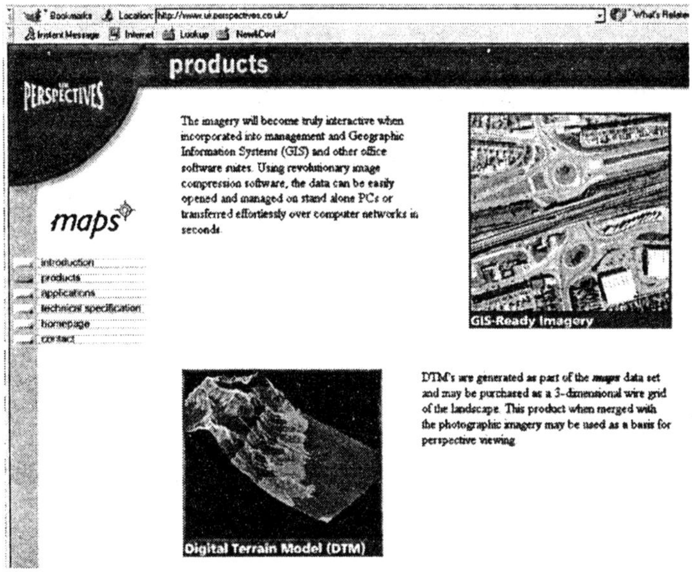

Orthophoto
25cm resolution
better than 2m

DTM accuracy?

E.g., wide angle photograph:

$$\tan\alpha_{max} = \frac{230}{2}\sqrt{2} \;/\; 150 = 1$$

⬇

$$\sigma_{DTM} < \sigma_{ortho}$$

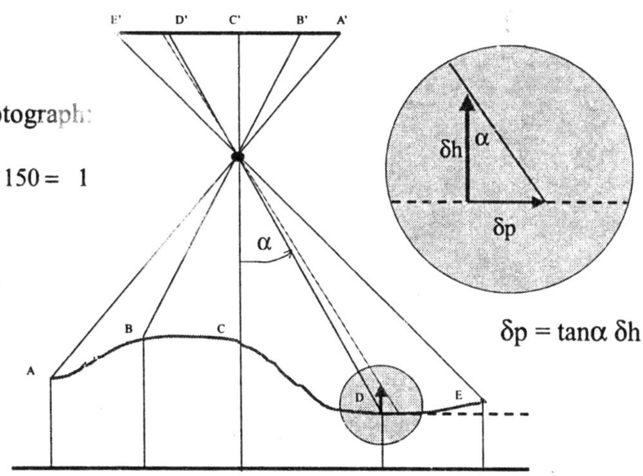

$$\delta p = \tan\alpha \; \delta h$$

From DTM Spec's to Spec's of DA

σ_{DTM} **< Value** *or* e_{DTM} **< Value**

(a) ground survey or Data Acquisition (DA) from images

- **point spacing?**
- **precision of elevation values?**

(b) digitizing contours

- **spacing of grid DTM?**
- **method of interpolation?**

To predict the accuracy of a DTM

from data acquisition and processing parameters

We must mathematically model
terrain and the processes leading to the DTM

Digital Terrain (Relief) Model

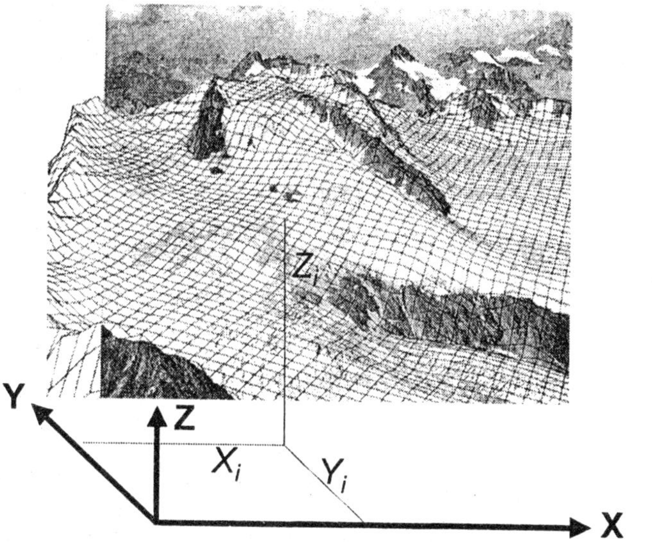

$$Z_i = f(X_i, Y_i)$$

elevation as single-valued function of horizontal position

Accuracy of DTM

Accuracy of the data $Z_i = f(X_i, Y_i)$

- point definition
- measuring precision

Accuracy of interpolation

- spacing between points
- interpolation method

Mathematical Model

Accuracy Measure

$$Z = f(x, y)$$

$$e(x, y) = f(x, y) - \hat{f}(x, y)$$

$$\boxed{\sigma^2 = \frac{1}{LX \cdot LY} \int_0^{LY} \int_0^{LX} e(x, y)^2 \, dxd}$$

$$\hat{\sigma} = \sqrt{\frac{\sum \left(Z_i - \hat{Z}_i\right)^2}{n}}$$

terrain relief

DTM error

error variance

vertical RMSE

Empirical Approach

$$\hat{\sigma} = \sqrt{\frac{\Sigma(Z_i - \hat{Z}_i)^2}{n}}$$

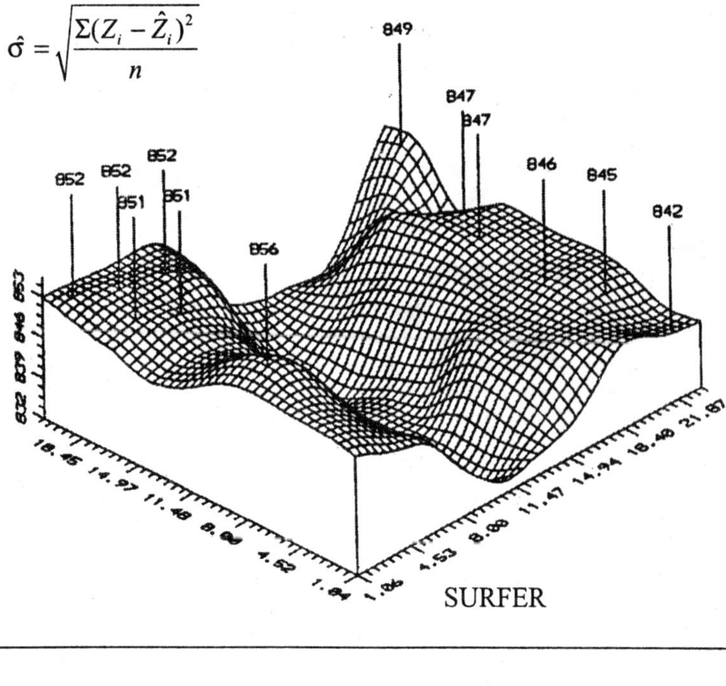

RMSE from checkpoints

- area of 8 by 12 km^2
- purely random checkpoint distribution
- percentage error to be smaller than 10%

☛ up to 1000 checkpoints !

Analytical Approach
(A)

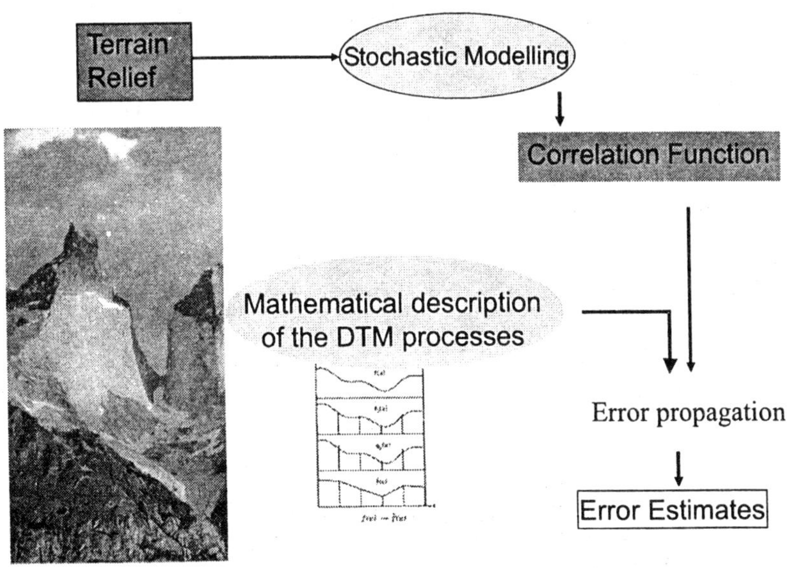

$$H = Z + \delta$$

Correlation Function

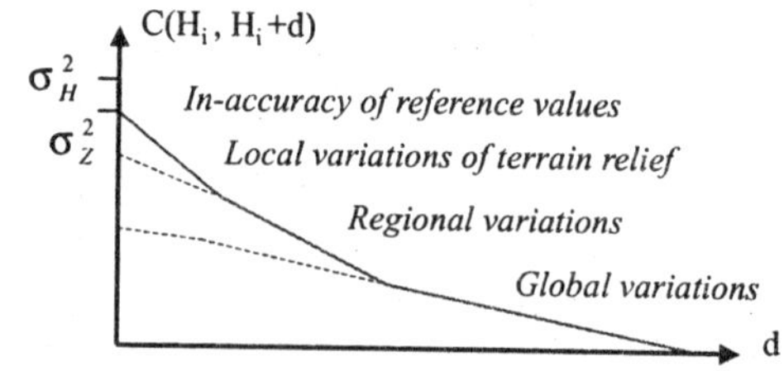

Error estimation

$$e_p = Z_p - \hat{Z}_p = Z_p - A^T F \qquad F = [\,H_1\ H_2\ \ldots\ldots\ H_n\,]$$

$$\sigma_e^2 = A^T C_{HH} A + \sigma_Z^2 - 2A^T C_{ZH}$$

$$C_{HH} = \begin{bmatrix} C(H_1,H_1) & C(H_1,H_2) & \ldots\ldots & C(H_1,H_n) \\ & C(H_2,H_2) & & \\ & & & C(H_n,H_n) \end{bmatrix}$$

$$C_{ZH} = [\,C(Z_p,H_1)\ \ C(Z_p,H_2)\ \ldots\ldots\ C(Z_p,H_n)\,]$$

Analytical Approach
(B)

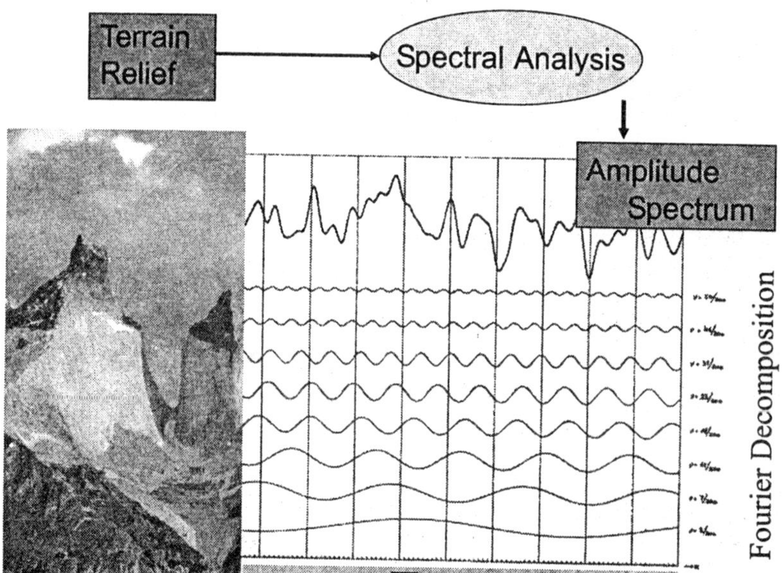

$$f(x) = c_0 + \sum c_i \sin(i2\Pi v_0 x + \varphi_i)$$

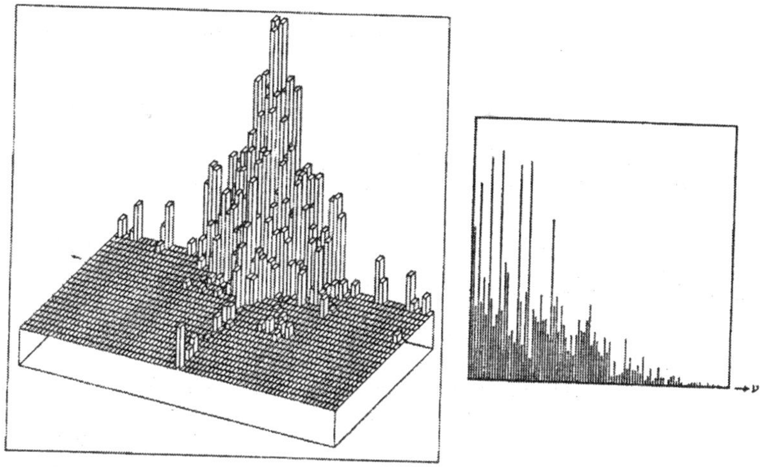

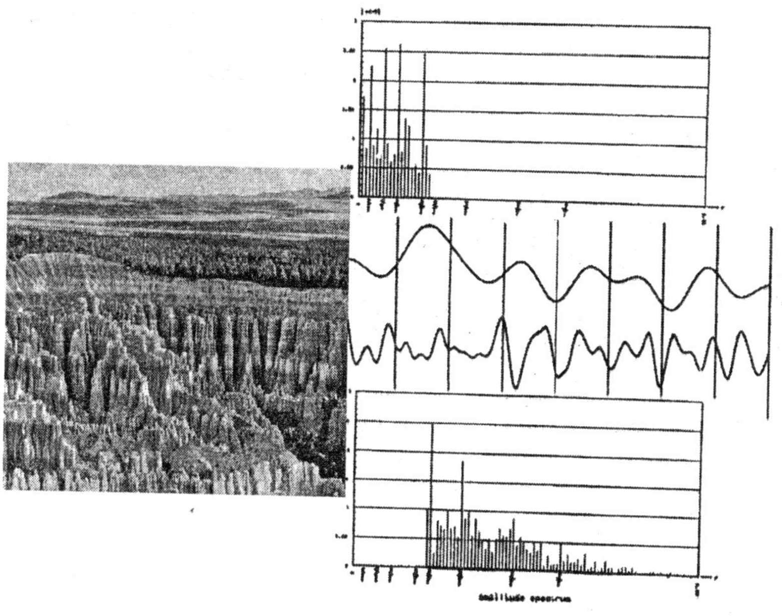

DTM Processes

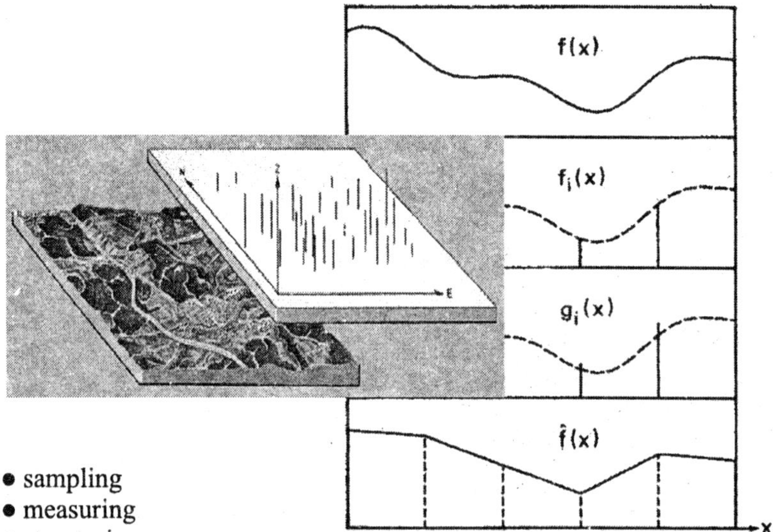

- sampling
- measuring
- structuring
- interpolation

$f(x) \rightarrow \hat{f}(x)$

Linear System

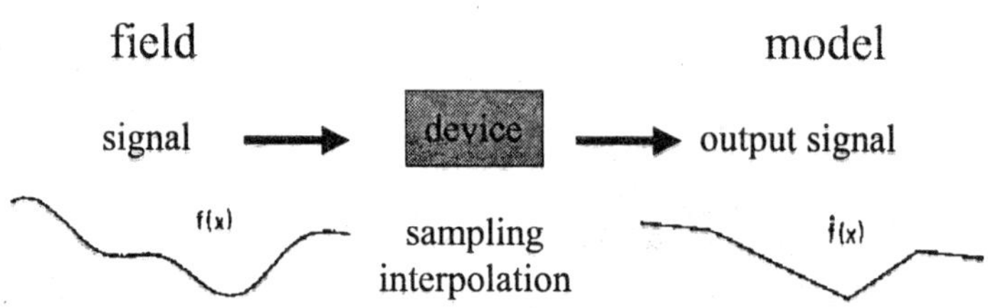

Linear: if $f(x)=c_1 f_1(x) + c_2 f_2(x)$

then $\hat{f}(x)=c_1 \hat{f}_1(x) + c_2 \hat{f}_2(x)$

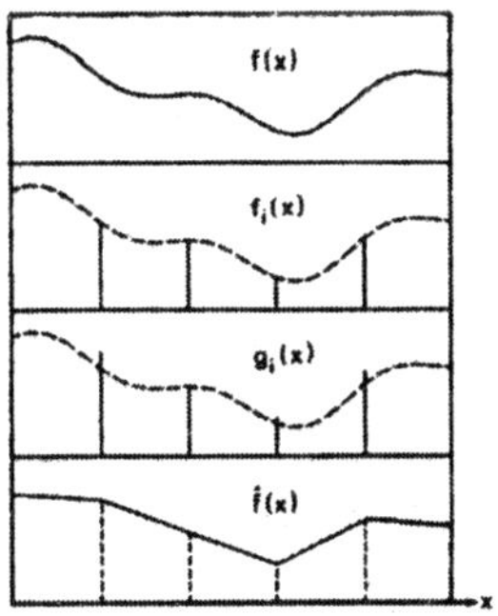

$$f_i(x) = \sum_{l=-\infty}^{\infty} f(l\Delta x)\,\delta(x - l\Delta x)$$ sampling theorem

$$g_i(x) = f_i(x) + m_i(x)$$

$$\hat f(x) = \int_{-\infty}^{\infty} a(t)\, g_i(x-t)\, dt = A^T F$$

$$e(x) = f(x) - \hat f(x)$$

$$\sigma^2 = \frac{1}{L}\int_0^L e^2(x)\, dx$$

$$\hat f(x) = a(x) * g_i(x)$$

$$\hat F(\nu) = A(\nu) \bullet G_i(\nu)$$

transfer function

illustration

Sampling

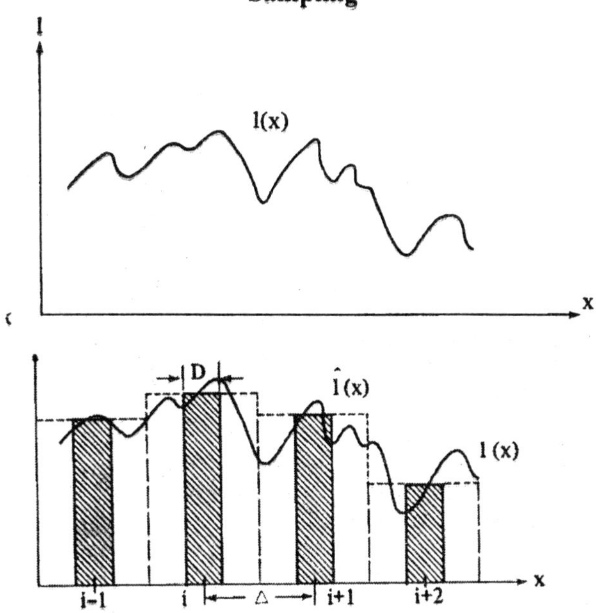

sampling $f_i(x) = f(x)\ i(x)$

interpolation $f(x) = a(x)\ f_i(x)$

$i(x)$ impulse function

$a(x)$ weighting function

convolution Fourier Transformationmultiplication

$$F(\nu) = A(\nu)\ F_i(\nu)$$

$A(\nu)$ transfer function of interpolation

Sampling Theorem

signal processing

$$\lambda = 1/\nu$$

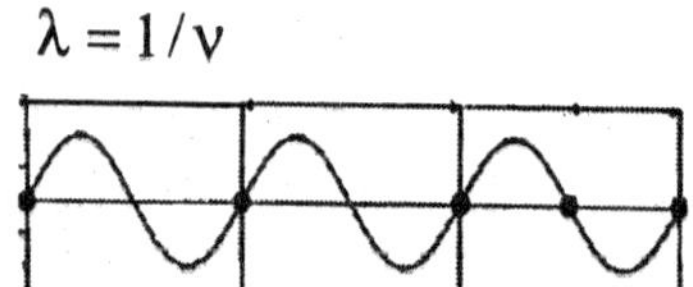

$$\Delta x \prec \frac{1}{2\nu_N}$$

band limited signal

Nyquist frequency ν_N

Grid
Complete Reconstruction: Whittakers interpolation
(sinc function as weighting function,
infinitely many sampling intervals)

Raster
Kell factor: 0.35 * Nyquist wavelength

Example: Contour Lines as Data Source

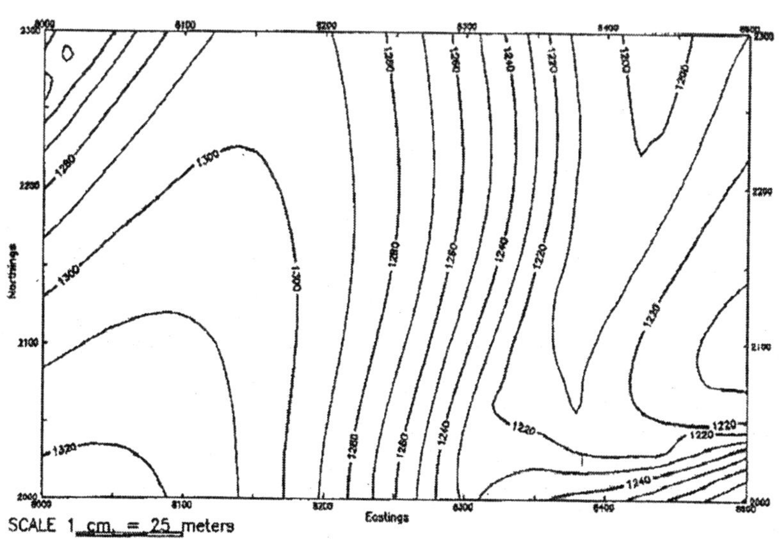

EXAMPLE: USGS 10m or 30m DEM?

Offered grid DTM is derived from contour lines 1: 24,000

Will 10m or 30m grid have better accuracy?
What is the rmse you can expect from 1:24,000 contours?
Does the choice 10m or 30m depend on the contour interval?
If not, what should it depend on?

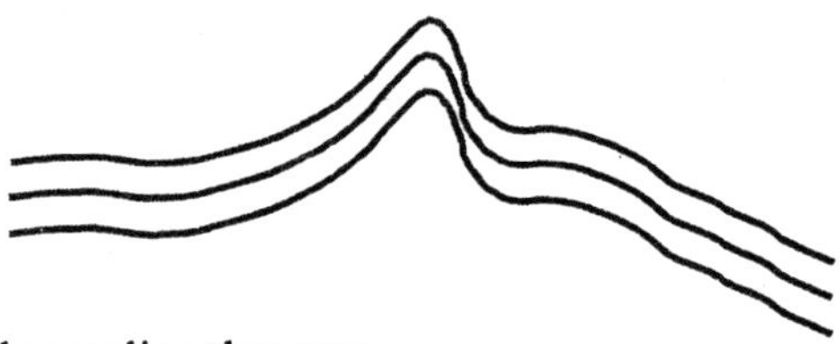

refined sampling theorem

EXAMPLE: USGS 10m or 30m DEM?

sampling spacing is relevant

Sampling by digitizing contour lines

- RMSE : 1/3 - 1/2 contour interval
- resolution of contour lines:

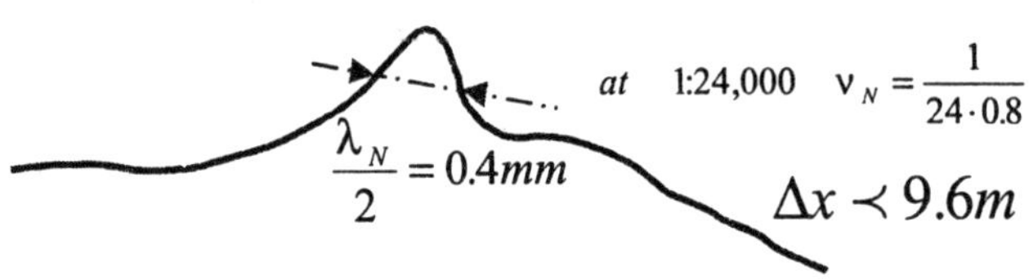

refined sampling theorem

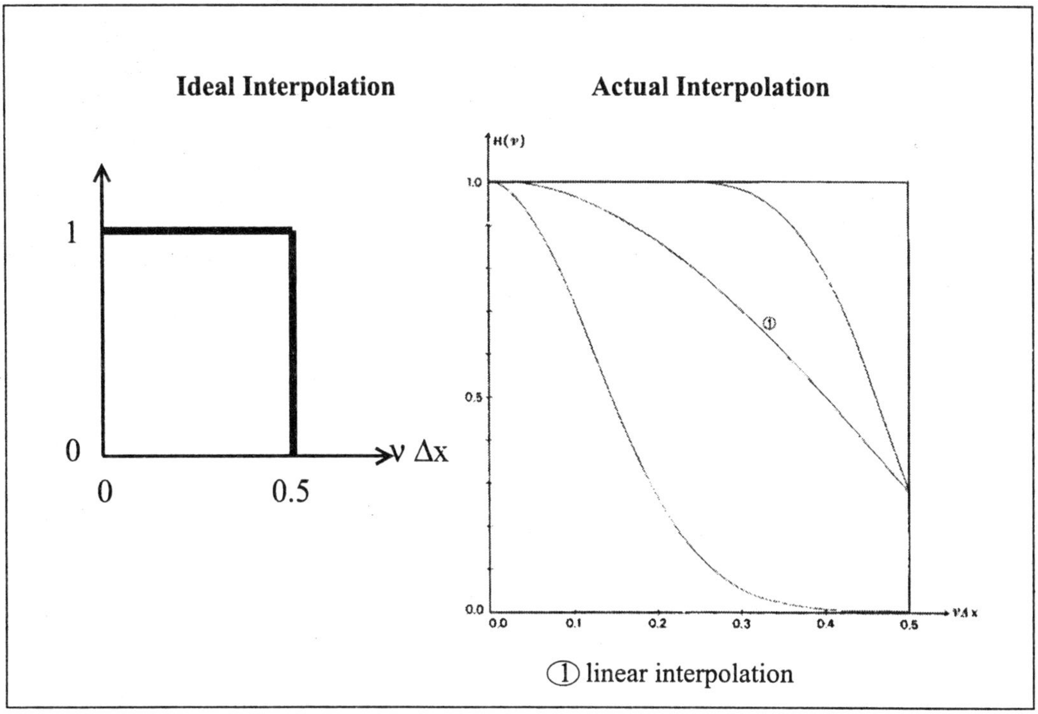
Ideal Interpolation
Actual Interpolation
1
0
0
0.5
ν Δx
H(ν)
1.0
0.5
0.0
0.0
0.1
0.2
0.3
0.4
0.5
① linear interpolation

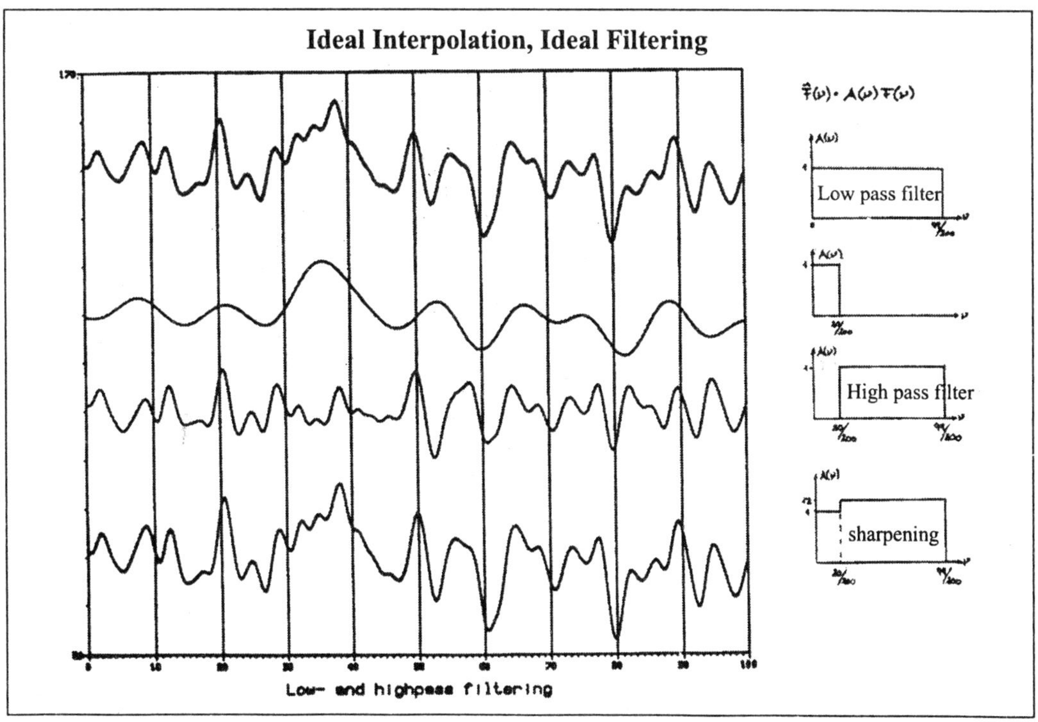
Ideal Interpolation, Ideal Filtering
Low pass filter
High pass filter
sharpening
Low- and highpass filtering

Actual Interpolation/Filtering

② Moving average

③ Linear Prediction

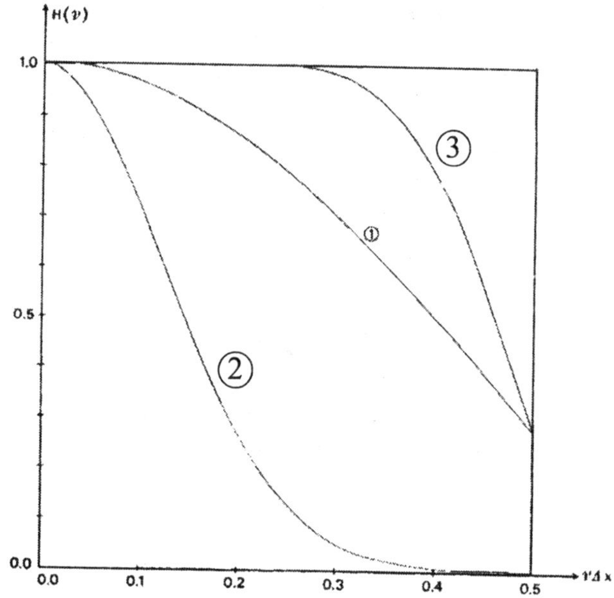

Actual Interpolation/Filtering

Moving average

$$w(t) = \frac{\sin \omega.t}{\omega.t} e^{-at^2}$$

$$\omega = \pi \; ; \; a = \frac{\pi}{100}$$

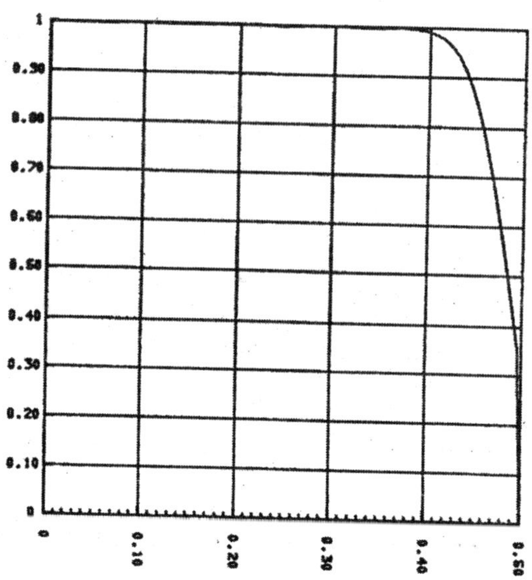

Bilinear Interpolation

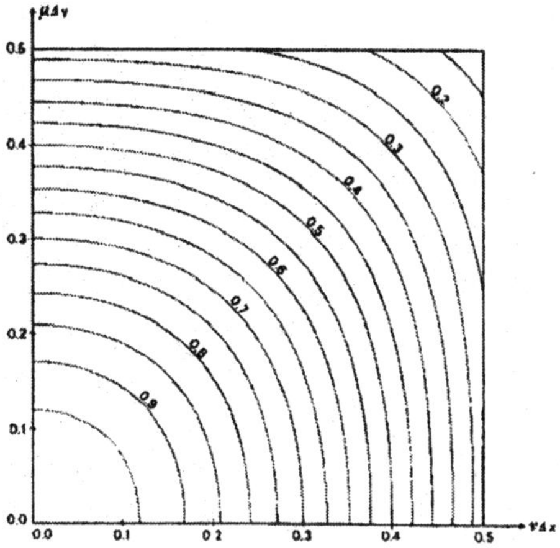

Bicubic Interpolation

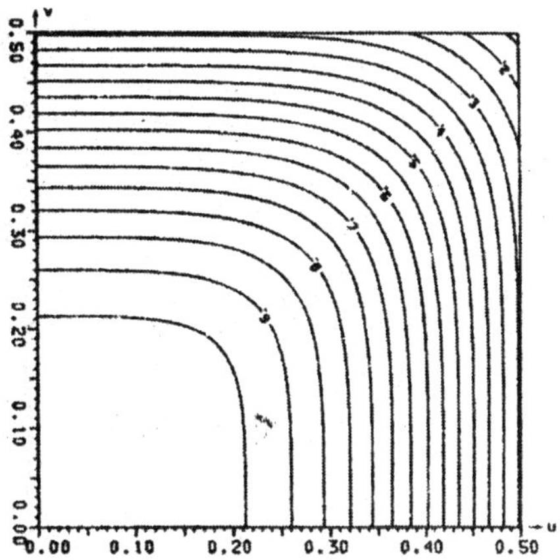

$$\hat{F}(\nu) = A(\nu) \bullet G_i(\nu)$$

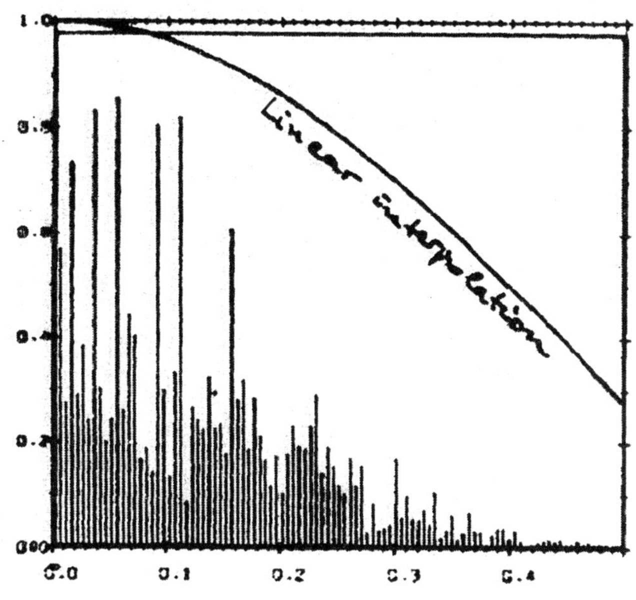

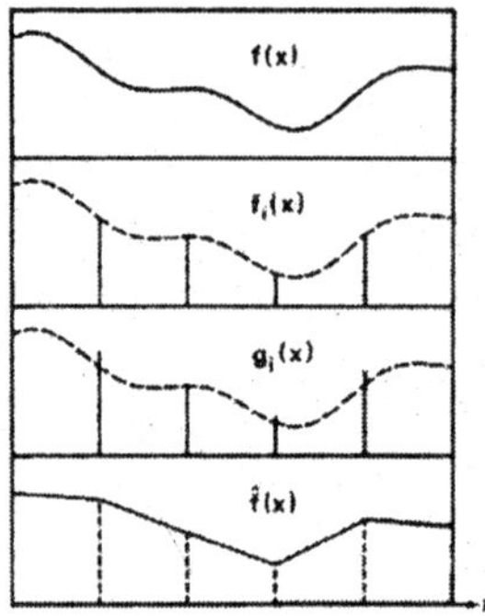

$$f_i(x) = \sum_{l=-\infty}^{\infty} f(l\Delta x)\,\delta(x - l\Delta x)$$

$$g_i(x) = f_i(x) + m_i(x)$$

$$\hat{f}(x) = \int_{-\infty}^{\infty} a(t)\, g_i(x-t)\, dt = A^T F$$

$$e(x) = f(x) - \hat{f}(x) \qquad f(x) - \hat{f}(x)$$

$$\sigma^2 = \frac{1}{L}\int_0^L e^2(x)\,dx \qquad \frac{1}{L}\int_0^L e^2(x)\,dx$$

Variance estimation in the spectral domain

$$\hat{f}(x) = a(x) * g_i(x)$$

$$\hat{F}(\nu) = A(\nu) \bullet G_i(\nu)$$

$$\hat{\sigma}^2 = 2\sum_{k=1}^{N/2} \{1 - H(\nu_k \Delta x)\}^2 |F(\nu_k)|^2 + 2\sigma_M^2 \int_0^{1/2} H(u)\,du \;; \quad u = \nu\Delta x$$

$$\hat{\sigma}^2 = \hat{\sigma}_S^2 + \hat{\sigma}_R^2$$

$$\{1 - H(\nu\Delta x)\}^2 = \left\{1 - \frac{1}{\Delta x} A(\nu)\right\}^2 + 2\sum_{k=1}^{\infty} \frac{1}{\Delta x^2} A^2\left(\nu + \frac{k}{\Delta x}\right)$$

Analytical Approach

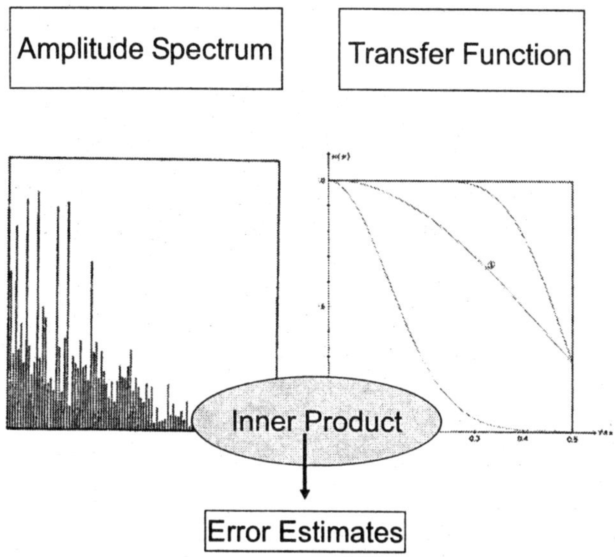

Implementation

- Fast Fourier Transform
- look-up tables for transfer function
- inner product calculation

Sampling & Interpolation is a Linear System

behaviour of S&I can be studied on a single sinewave

Transfer Function of S&I

$$s(x) = \sin(2\pi\nu x)$$
$$\Delta x \rightarrow \hat{s}(x)$$
$$e(x) = s(x) - \hat{s}(x)$$
$$E = \sqrt{\frac{1}{N}\sum_{i}^{N} e(i\delta)^2}$$
$$H = 1 - \sqrt{2}E$$
$$u = \nu\Delta x \quad \in (0, \tfrac{1}{2})$$
$$u_k = \nu\Delta x_k$$

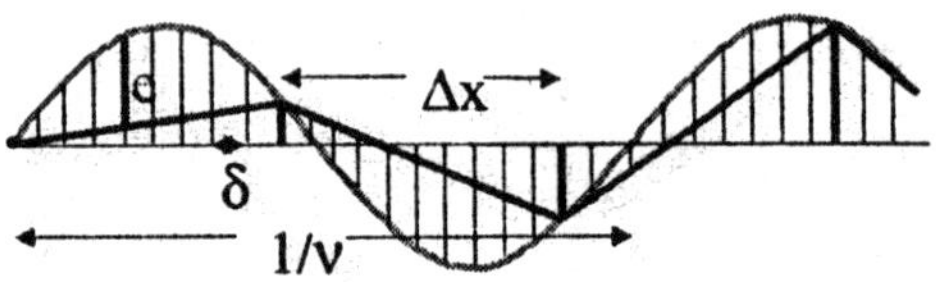

Estimation Procedure

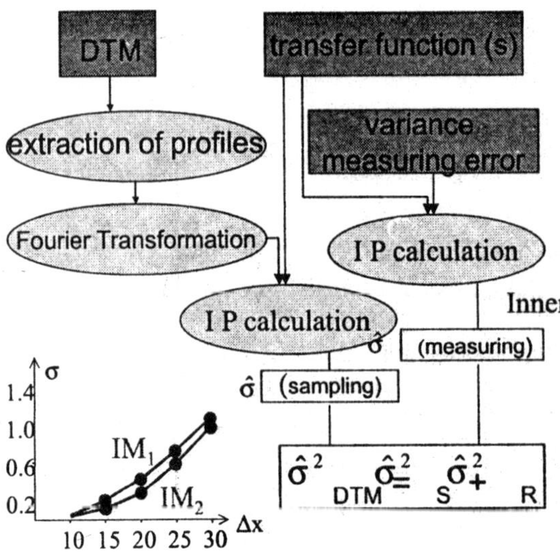

Image matching
Manual:
$\sigma_M \sim 0.01\text{-}0.02\%$ H
Automatic:
$\Delta_Z \sim \Delta_p$ H/B
pixel size <60μm

DTM Design

Example 1: DTM for orthophoto

$$\sigma_{ortho} \leq 0.2mm \text{ at map scale}$$

$$\sigma_{ortho} = \tan\alpha \quad \sigma_{DTM}$$

Example 2: DTM for flood prediction

$$Q = R^{2/3} \cdot S^{1/2} \cdot A \cdot n^{-1}$$

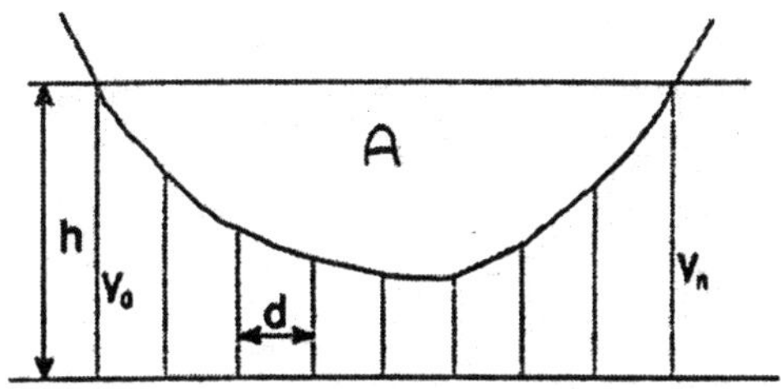

Q:=discharge
R:=hydraulic radius
S:=longitudinal slope of river bed
A:=area of cross section
n:=Manning's roughness coefficient

Analytical Approach

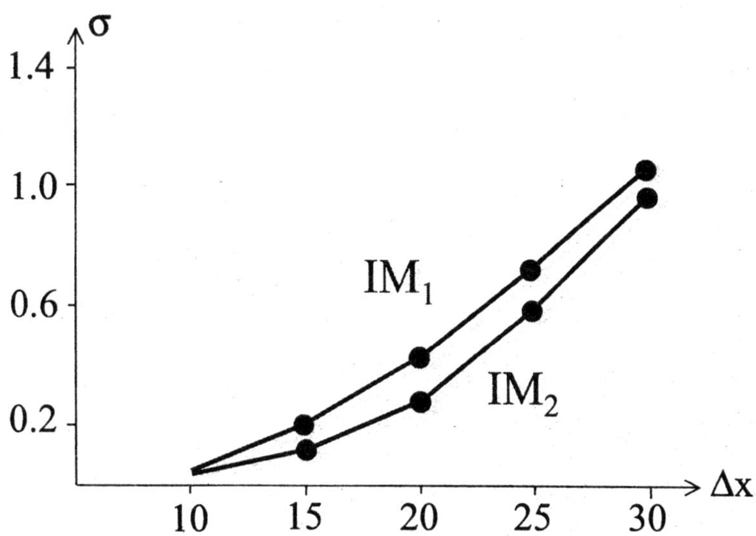

Empirical Validation

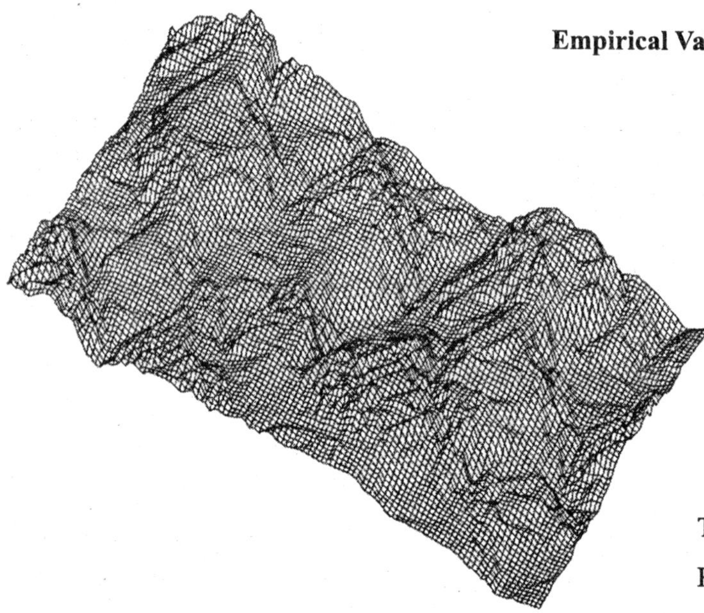

D
σ_M
IM

T6, $8*12\ km^2$
RMSE 2-20m

Conclusions

- Fair accuracy estimates for grid DTM by FFT, σ_M, TF
- no additional measurements for existing DTM
- no exorbitant analysis fidelity of DTM
- DTM design
- comparison of interpolation methods

? Skeleton

? Irregular point pattern

DTM

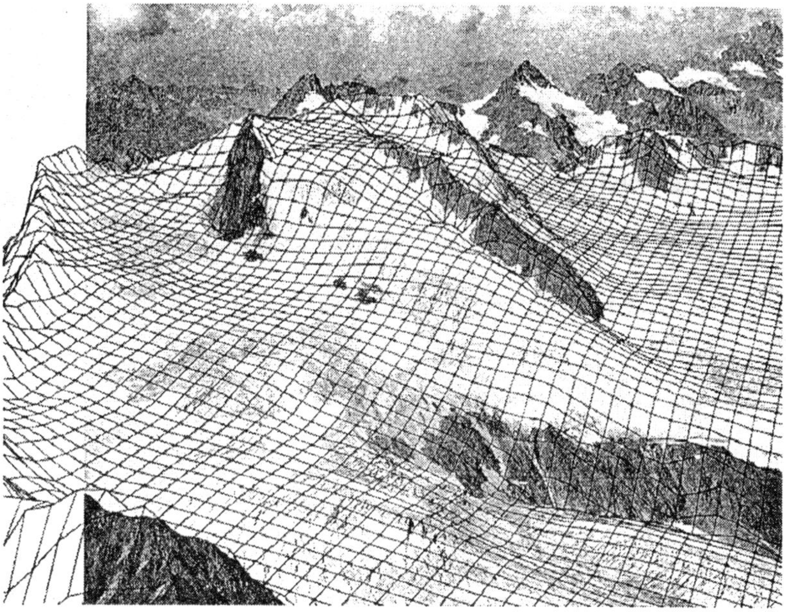

DI

Scanning Photographs

(a) for geometric operations

correcting image geometry (transformation)
image orientation
elevation determination (image matching)

location

(b) for pictorial (semantic) operations

image enhancement
classification
interpretation

variation of light intensity

pixel size?

Pictorial Considerations

sensor response

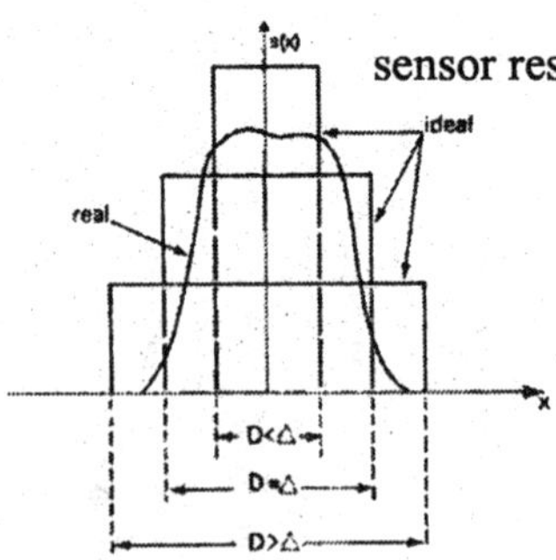

sampling interval Δ
sensor 'aperture' D

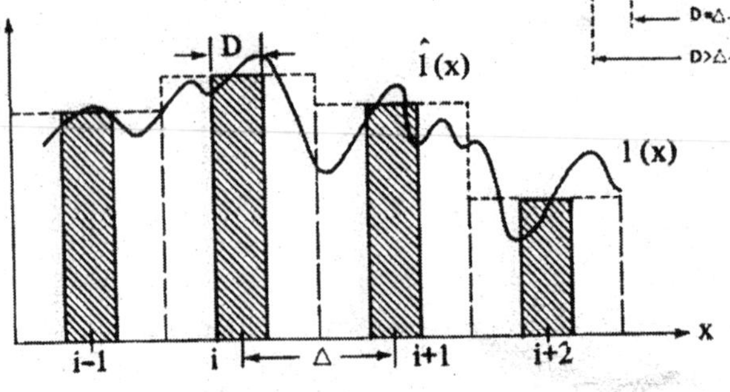

Transfer Function

$$s(x) = \sin(2\pi\nu x)$$
$$\Delta x \rightarrow \hat{s}(x)$$
$$e(x) = s(x) - \hat{s}(x)$$
$$E = \sqrt{\frac{1}{N}\sum_{i}^{N} e(i\delta)^2}$$
$$H = 1 - \sqrt{2}E$$
$$u = \nu\Delta x \quad \in (0, \frac{1}{2})$$
$$u_k = \nu\Delta x_k$$

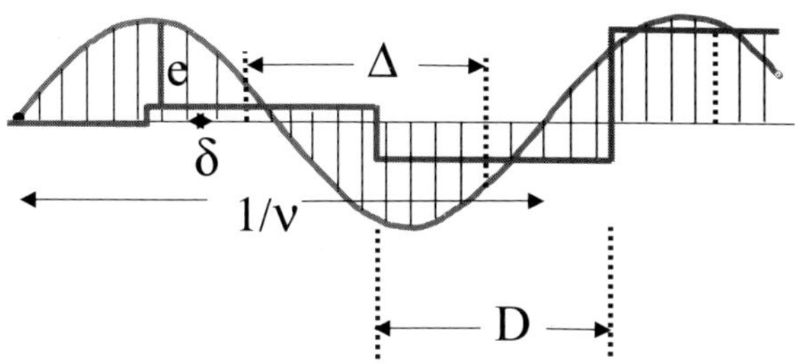

TF of Scanning D=Δ

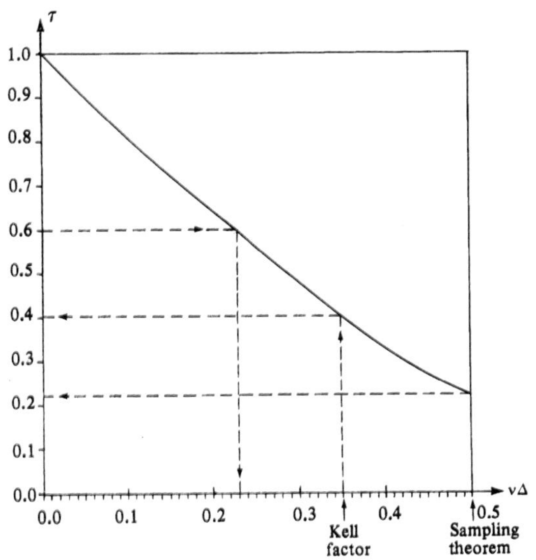

Example:
film resolution of 50 lp/mm:

maintaining smallest detail at 60%
$\nu\Delta = 2.3$
$\Delta =$ **4.6 μm**

'Sloppy' Sensors

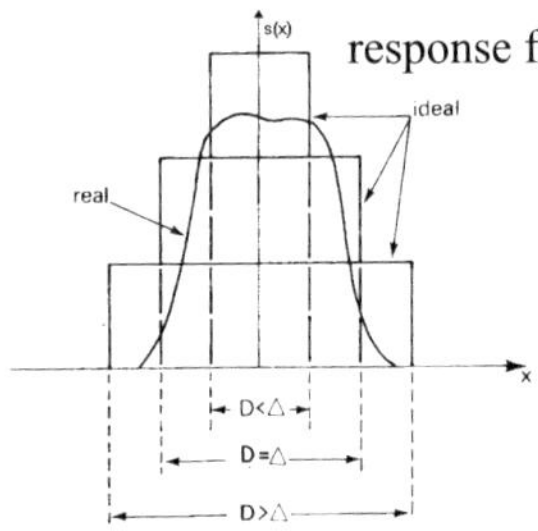

response function

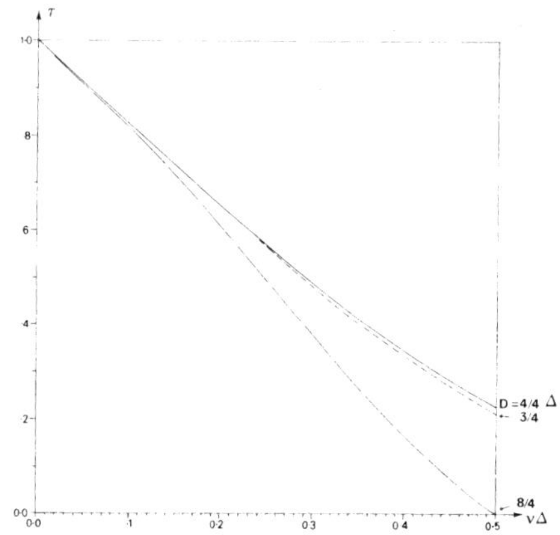

$$\hat{l}_i = {}_{x_i-D/2}^{x_i+D/2} s(t)l(x_i-t)dt$$

Fields and objects

Keep on extracting info from sensor data !

30

ACQUISITION AND EVALUATION OF MAP AND CARTOGRAPHIC DATA FROM DATA SHEET

MILAP PUNIA

Contents and Structure

- **Lecture 1:** ***General Principles and Applications***
  - concept, end-product characterization involved processes
- **Lecture 2:** ***Data Sources and Acquisition Techniques***
  - specifications
  - characterisation of various techniques
- **Lecture 3:** ***Photogrammetric Data Collection***
  - preparation and manual acquisition
  - principles and procedure of image matching
  - checking and editing/supplementing
- **Lecture 4:** ***Systems, Data Processing***
  - hardware and software for DTM
  - pre-processing, interpolation, derived information
- **Lecture 5:** ***Conversion, Visualisation, Accuracy***
  - data restructuring, archiving, graphic presentation
  - empirical accuracy assessment

Terrain Relief

Topographic maps

- *terrain features* (roads, rivers, buildings, trees, etc.)
- *terrain relief* (contour lines, hill shading, etc.)

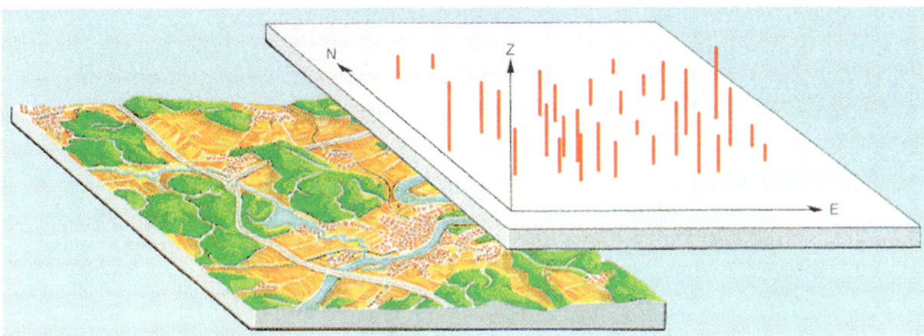

terrain relief:
change of elevation of the ground surface (of the earth) with position

Digital Terrain (Relief) Model: DTM

Description of the geometry of the ground surface in terms of:

- an elevation data set,
- the elements of data structuring,
- an interpolation algorithm.

$Z_i = f(X_i, Y_i)$ *elevation as single-valued function of horizontal position*

regular distribution of points $(X,Y)_i$ (grid)

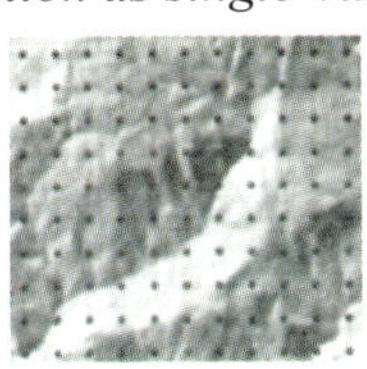

irregular distribution of points $(X,Y)_i$ (*e.g.*, TIN)

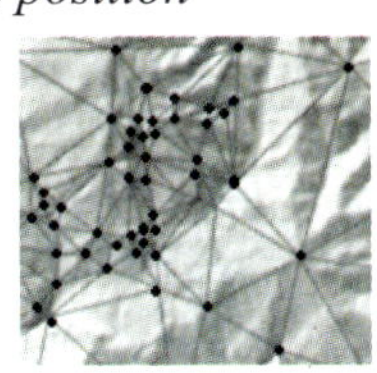

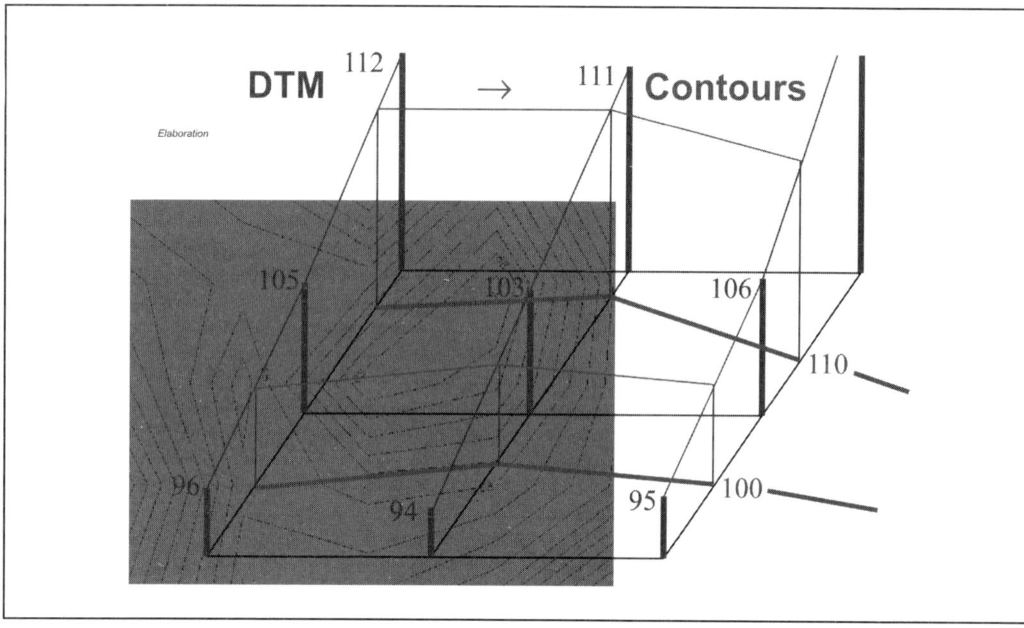

Field View → Raster/Tessellation

DTM has no scale but a resolution!
Resolution of the digital model:
spacing between points

→ database specifications

Example:

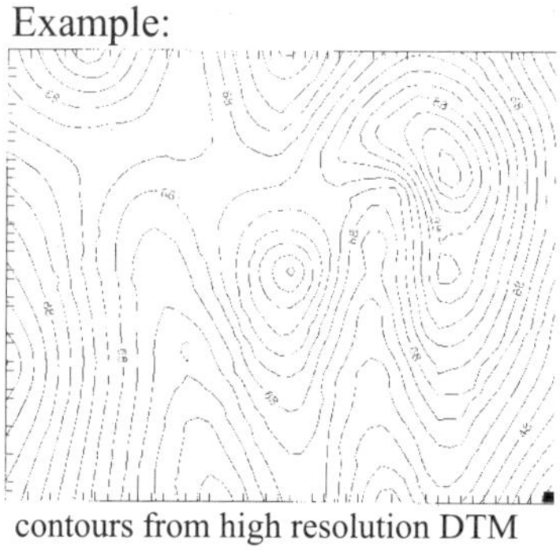

contours from high resolution DTM

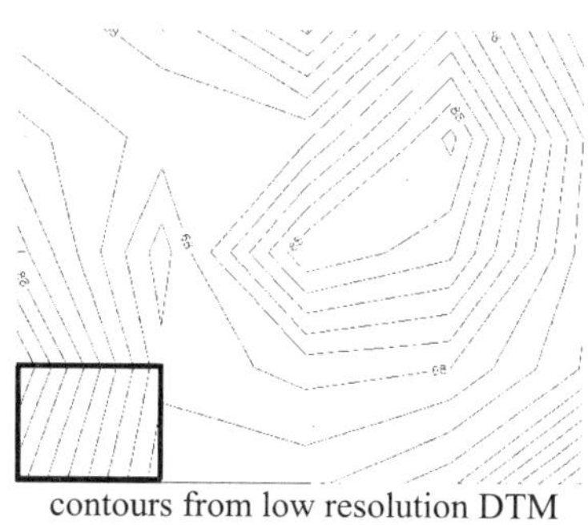

contours from low resolution DTM

Basic Notions (1)

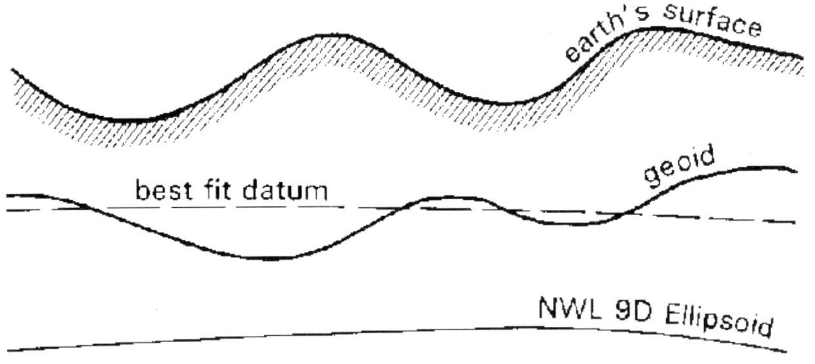

Basic Notions (2)

Horizontal position: defined in a geographic, geodetic or map coordinate system (error free)

Elevation: height of the ground above a defined datum (geoid or ellipsoid) (DTM error = error of a given/computed elevation value)

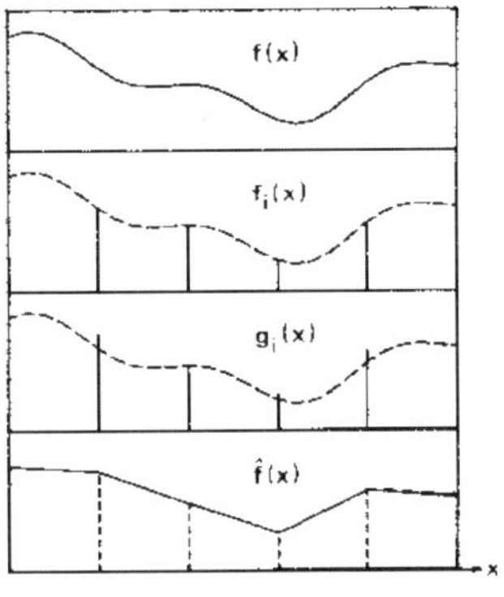

- Sampling discretising
- Measuring determining coordinates
- Interpolation computing elevation at
- an arbitrary point

DTM: not a 3D description of the *topography* but a "2.5D" model of the ground

DEM: use if you do not know what you are modelling

Applications: Disciplines

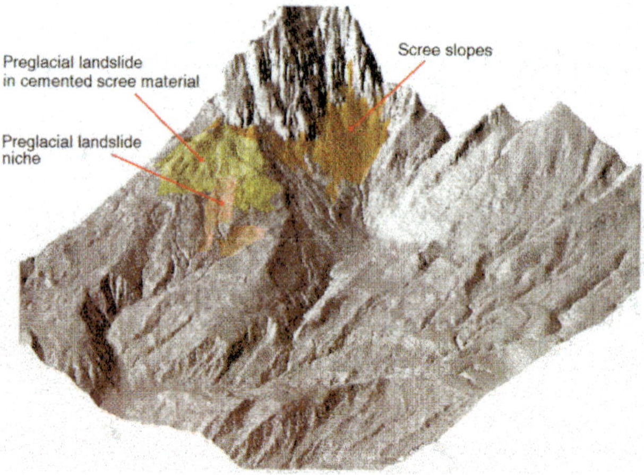

- Civil Engineering
- Mining/Dumping
- Mapping
- Surveying
- Geo-Sciences
- Land Management
- Military

Applications: Prime Examples (1)

Road design

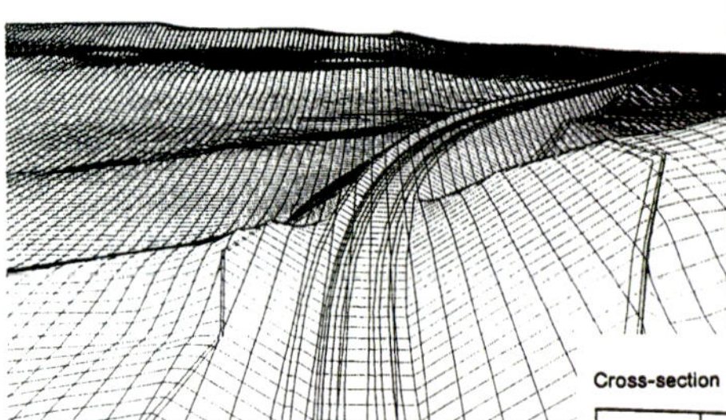

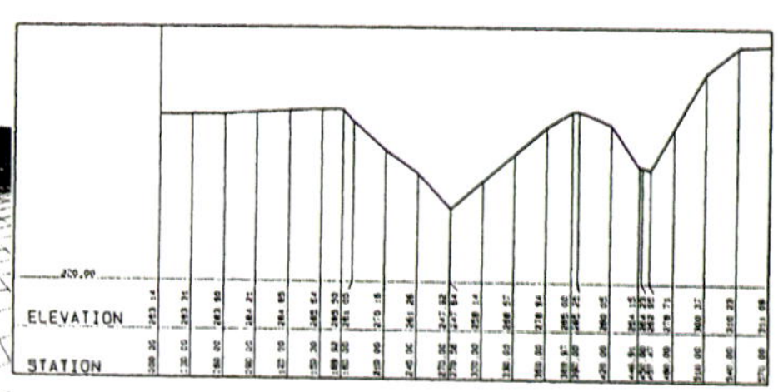

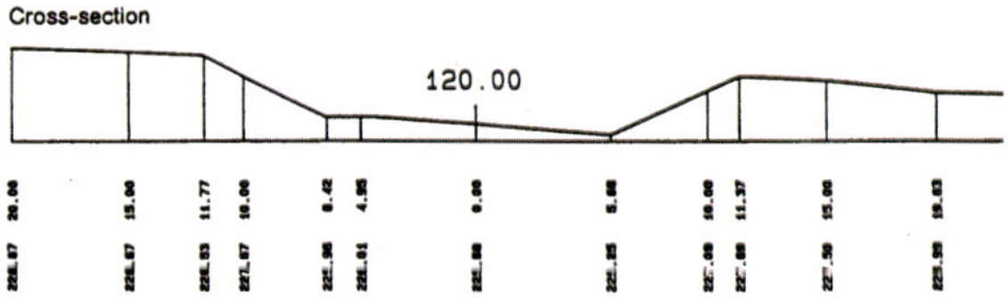

	FILL	CUT
AREA m^2	6053	4846
VOLUME m^3	156.1	157.3

Applications: Prime Examples (2)

Orthophoto production

Topographic mapping

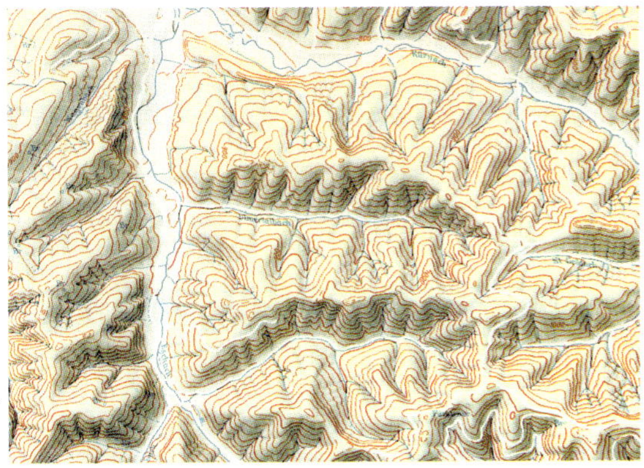

Applications: Prime Examples (3)

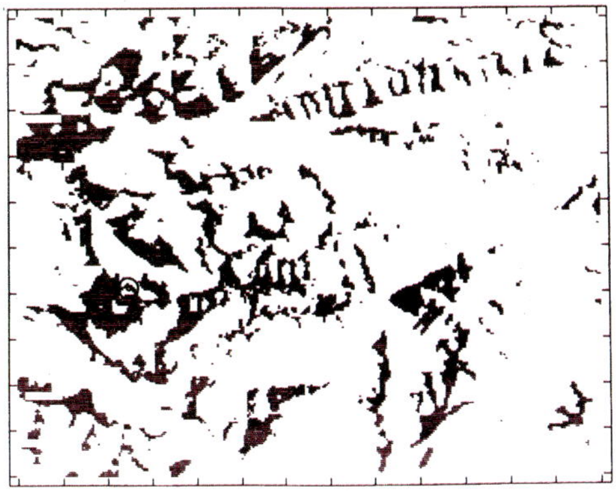

Cellular planning

Water studies

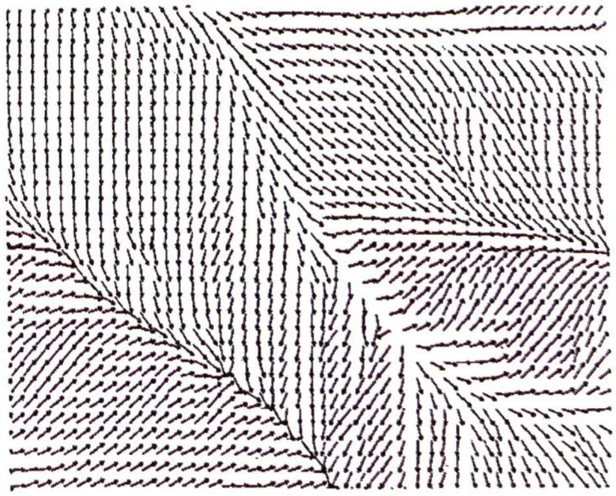

Production

- resolution dependent
- project or program
- elevation database or (integrated) GIS

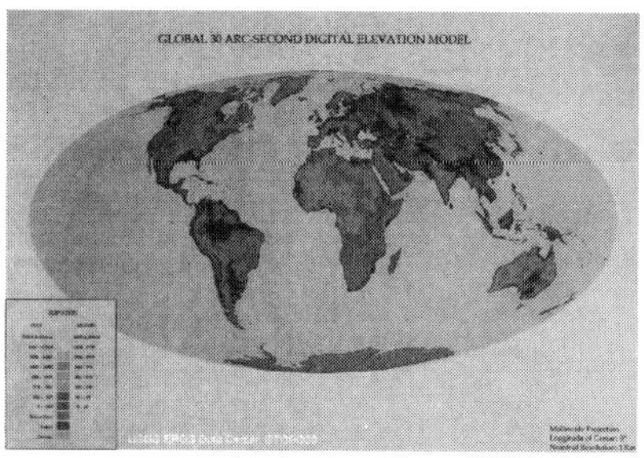

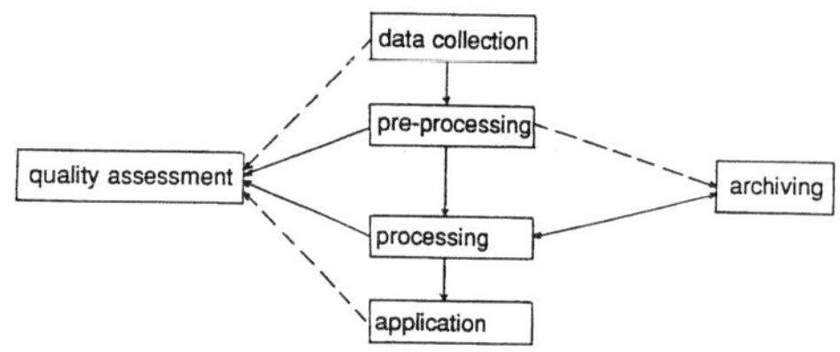

Problems of Terrain Modelling (1)

- specifications and quality assessment
- selection of best data source and system
- efficiency of data collection (large data amount)
- data from images:
  - — unfavourable scenes (little texture, repeating image patterns, obscured ground)
  - — bad image quality
- selecting best sampling strategy/parameters
- detection of blunders and systematic errors
- selecting best interpolation
- efficient data storage

Problems of Terrain Modelling (2)

Photogrammetric Process

- provide specifications
- provide flight planning/check of images
- provide ground control
- provide photogramm. plotters & DTM software
- preparation
- specify sampling parameters, define digitizing rules for morphologic features, prepare project files, ingest and orient images
- digitizing/matching:
  - — interpret, code, digitize, monitor, check, edit
- data conversion
  - — from collection to modelling system

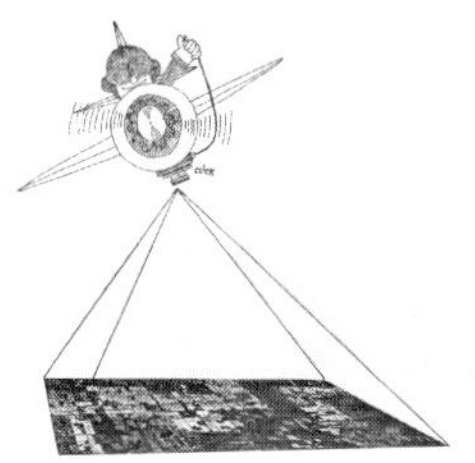

500001	840233.13	174372.45	362.87
500001	840225.81	174379.67	360.49
500001	840230.19	174386.23	360.19
500131	841438.63	176026.78	329.17
500131	841432.38	176025.81	328.95
500131	841426.00	176026.63	328.40
500131	841419.81	176026.09	327.92

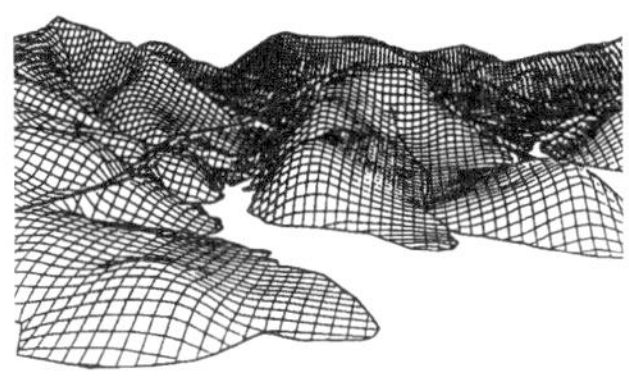

Data Acquisition: Specifications (1)

(i) verbal description

1. What exactly is the ground surface

(ii) specifying resolution

Specifications (2)

2. Type of Relief Information

(a) Elevation at arbitrary points: 'bulk/filling data'
(b) Elevation at distinct features: 'skeleton data'

Skeleton:
ridge lines, drainage lines, breaklines
peaks, pits, passes

Relevance:

minimising data amount
improving interpolation

Specifications (3)

3. Resolution and Precision
   Sampling density:
   sampling spacing relative to **terrain variability**
   Precision:
   standard deviation of measuring error
   Accuracy of the DTM:
   model error
4. Sampling pattern
   - grid
   - profiles
   - contours
   - morphologic

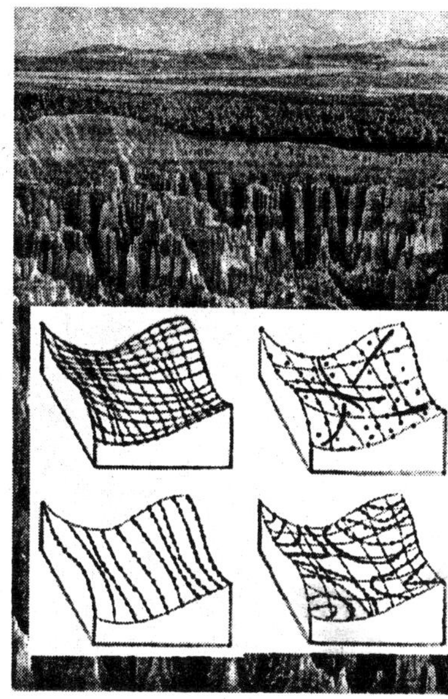

Example of DTM Requirements

Orthophoto:
σ_{OP} 0.2mm at map scale

Eg. wide angle photograph:
$\tan\alpha_{max} = \frac{230}{2}\sqrt{2} \ / \ 150 = 1$

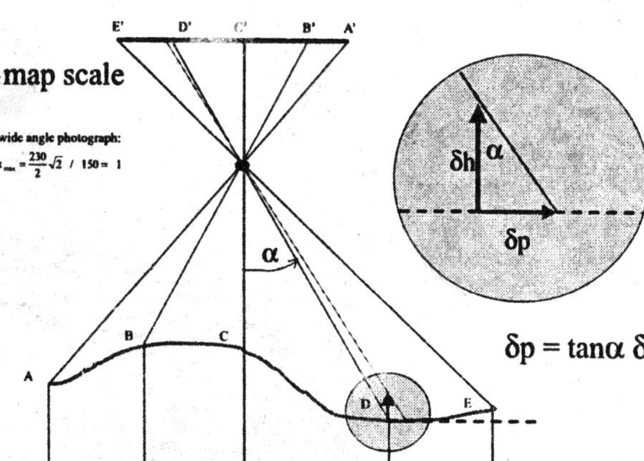

$\delta p = \tan\alpha \ \delta h$

Data Acquisition Techniques

Sampling pattern	Properties	Data collection	Terrain type	Specifications
Grid	+ objective + sample + automation – low adaptivity	Photogrammetry levelling	Regular	Small scale (e.g. for image mapping) or small area and high accuracy (e.g. airport maintenance) data volume and production time non-critical
Profiles	As grid but lower accuracy	Photogrammetry laser, radar, sonar	Flat-undulated	High speed and moderate accuracy (e.g. for orthophoto or real time product (special projects, e.g. open pit mining)
Morphometric points and lines	+ adaptivity + min. volumes – subjective – expertise – no automation – unfavourable data structure	Tachometry, GPS photogrammetry	Irregular, man-made surfaces	Large scale (e.g. civil engineering) small areas if ground survey accuracy and production time non-critical
Contours	+ objective + adaptivity – skill (unless map scanning) – hetero-geneous point distribution – unfavourable data structure	Scanning/digitizing of maps photogrammetry	Hilly-mountainous	Low cost low quality short time (excluding photogrammetry) data volume non-critical
Composite patterns	Mixed	From maps and photographs	Any	High quality production time non-critical

Alternative Data Acquisition Methods

- Ground Survey
  - field surveying
  - laser
  - radar
- Photogrammetry
  - (semi-automatic) digitizing
  - image matching
- Digitizing Maps
  - (semi-automatic) line following
  - scanning

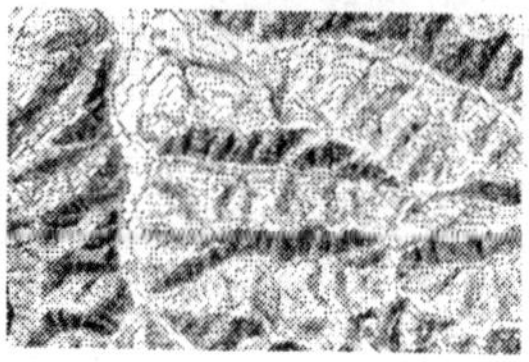

Field Surveying

PROPERTIES and ACCURACY
economic for small area (revision, project work)
requires accessible terrain
highest accuracy

SYSTEMS

• GPS	
• kinematic mode	a few dm
• all weather, 24 hours, obstruction by trees and buildings	ellipsoidal height
• TACHEOMETRY	
• total station	a few cm
• urban ground surfaces, public utilities	
• LEVELLING	
• levels	a few mm
• engineering construction and maintenance	

Laser profiling/Scanning

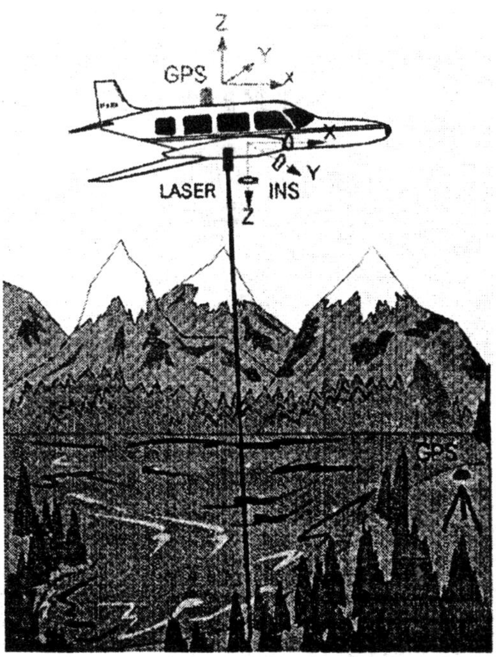

SYSTEM

- GPS
  - position of the aircraft
- INS
  - attitude of the aircraft
- LASER
  - range aircraft-ground
- VIDEO CAMERA
  - visual control

Laser Profiling/Scanning

PRODUCTION

calibration (tuning of 3 different sub-systems)

flight with GPS navigation

interpolation and transformation (of 3 data sets)

filtering (of terrain profiles)

resampling (of the DTM)

Laser profiling/Scanning

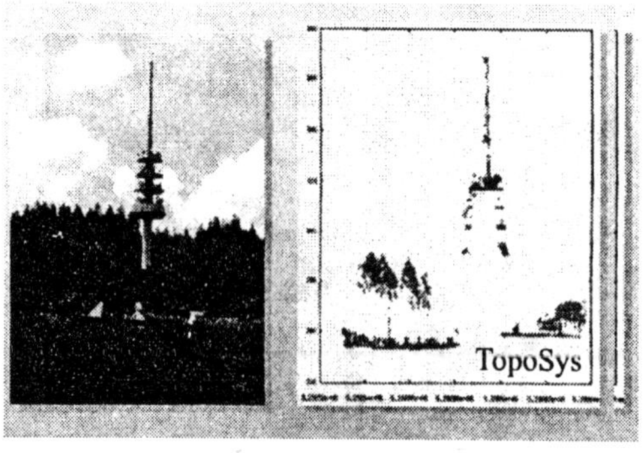

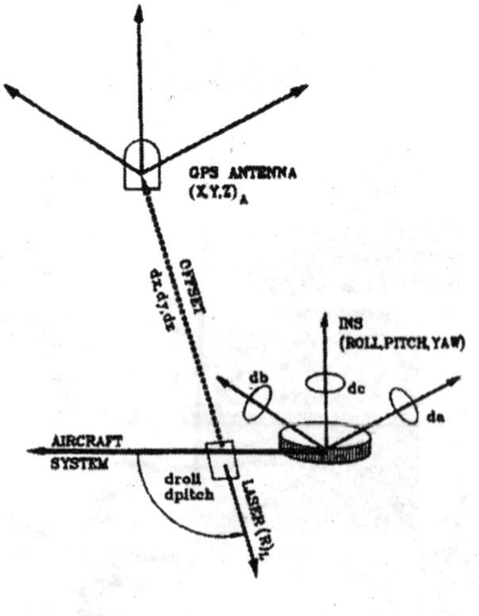

Laser Profiling/Scanning (1)

PRODUCTION

- **calibration** (tuning of 3 different sub-systems)
- **flight with GPS navigation**
- **interpolation and transformation** (of 3 data sets)
- **filtering** (of terrain profiles)
- **resampling** (of the DTM)

Laser Profiling/Scanning (2)

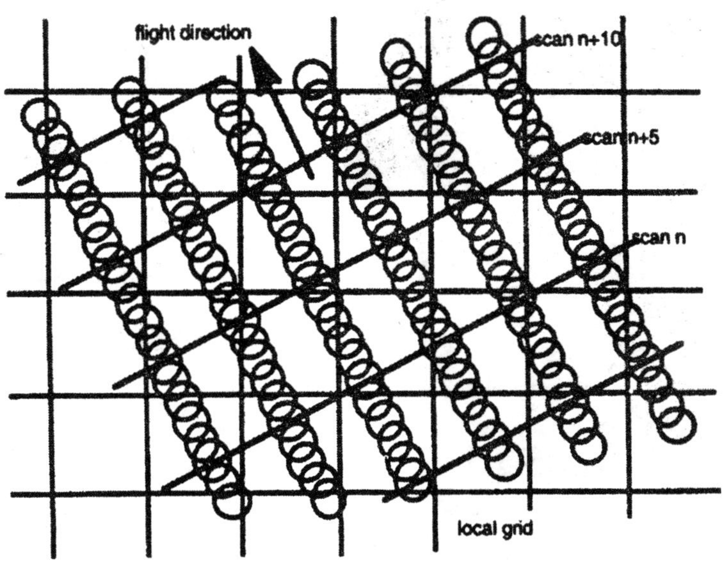

Laser Profiling/Scanning

PROPERTIES and ACCURACY

- high accuracy ⇒ < 2 dm

  dependent on hardware, ground type, weather

- capability to penetrate tree foliage
- attractive where photogrammetric data collection is difficult or expensive

  — forest areas, coastal monitoring, urban surface models

- offers near real-time production

  — open pit mining, waste dumping, special hydrological and geological applications

Radar

SYSTEMS

- airborne and spaceborne
- SAR

  (Synthetic Aperture Radar)

- INSAR

  (Interferometric SAR)

- RS Internet services
- radar processing software

PROPERTIES and ACCURACY

- SAR is an active instrument, producing images under all weather conditions
- Satellite SAR covers large areas with high
  - location accuracy
  - *e.g.,* ERS SAR scene: 100 km by 100 km
  - absolute localisation accuracy < 50m
- Radargrammetry produces a DTM from a stereopair of radar images
- INSAR is based on the combination of two SAR images from slightly different positions/orbits, exploiting the phase difference; high accuracy possible in open terrain

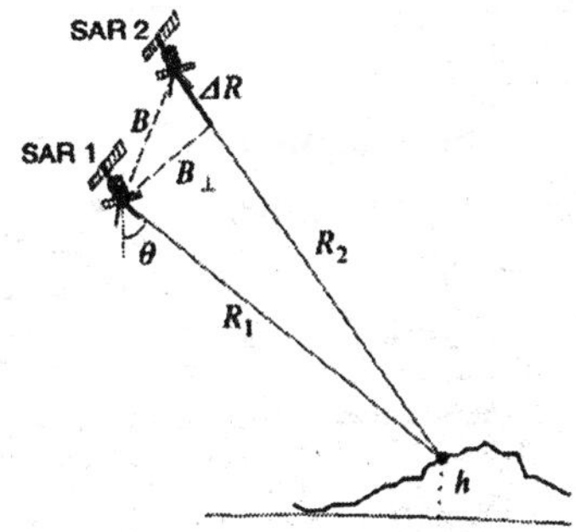

⇒ a few metres dependent on system, surface stability, vegetation, baseline, atmospheric artefacts

Digitizing Maps

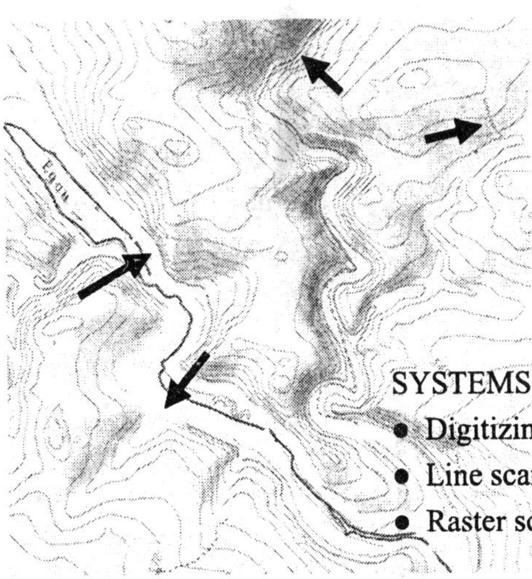

PROPERTIES

- economic, especially if contour separates available
- offers automation
- limited quality

SYSTEMS

- Digitizing tablets & SW (manual line following)
- Line scanners (semi-automatic line following)
- Raster scanners & SW (possibility of automation)

Digitizing Maps

PRODUCTION

Manual operation

- orientation of map
- specifying sampling parameters
  - — data types and codes, digitizing increment, threshold for data compression
- digitizing
  - — contours, spot heights, drainage
- checking and editing
  - — geometry and codes

Raster scanning

- scanning
- thinning
- vectorizing
- compression
- completion
- orientation
- coding
  - — interactive or automatic
- interactive editing

Digitizing Maps

ACCURACY

- dependent on:
  - — map quality
  - — contour interval
  - — operator/system performance
- best expected:
  - — 1/3 of the contour interval

- **Major cost factors**:
  - personnel and equipment rates
  - production time (~ 1 week per sheet)

⇒ a few m
for topographic maps
1:25,000 and smaller

Photogrammetry

PROPERTIES

- economic for large area
- offers automation
- high quality potential
- versatile
  on sampling pattern
  and accuracy

SYSTEMS

- Stereoplotters
  - analogue with encoders and PC
  - analytical
  - digital
- Digitizing software for manual d.c.
  - from aerial photographs & SPOT
- (a) Software for semi-automated d.c.
  - profiling
  - progressive sampling
- (b) Software for automated data col.
  - signal/area based matching
  - feature based matching

Preparation

Manual operation

- provide product spec's, validated images, camera calibration, ground control
- test/calibrate stereoplotter (analogue/analytical)
- define output file (*e.g.*, using Microstation: DGN)
- specify sampling parameters
  - (a) skeleton: feature types and codes, symbols for checkplot, digitizing rules (strategy, mode)
  - (b) grid: spacing in X and Y
- orientation of each stereo pair
- define/digitize boundary polygon per model

Automated operation

- provide product spec's, camera calibration, ground control; validated images
- scan images
- define output file, select images
- specify sampling parameters
  - region of interest, (grid) spacing
  - strategy and respective parameters
  - precision
- orientation of each stereo pair

Data Collection Procedure

Manual operation
(exclusive or supplementary)

- (a) interpret, code, digitize skeleton and excluded regions
- (b) set floating mark on grid intersections
- detect blunders and check visually (superimposed skeleton, contours or grid) edit interactively

Automated operation

- press 'start' button
- monitor and intervene when necessary
- check (quality indicators and visually for blunders)
- edit (interactive manual correction, automatic editing tools, manual supplement)

- Convert data from collection system (*e.g.,* DGN, LH) to surface modelling system (*e.g.,* SCOP, ARC/INFO)

Check by Superimposition

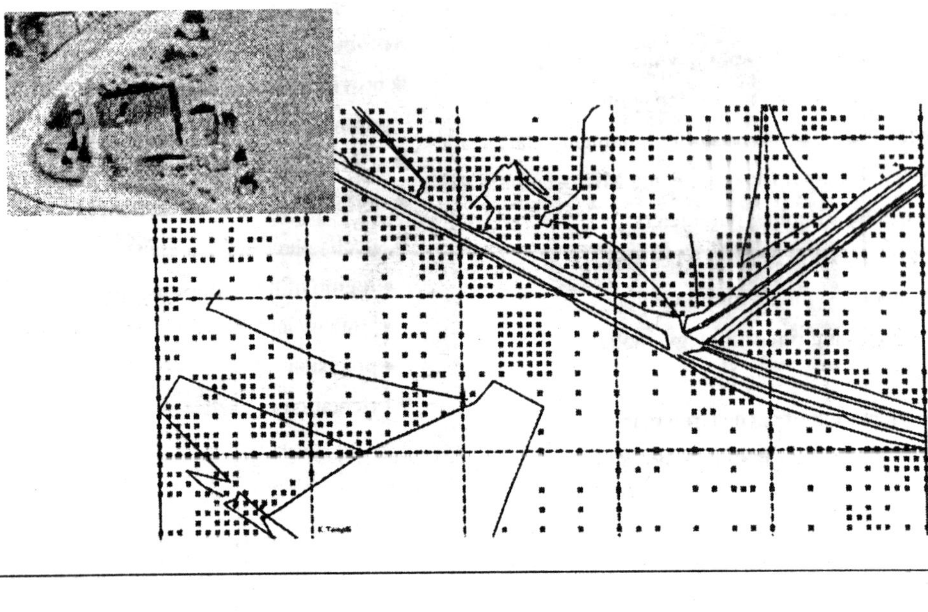

Photogrammetry

ACCURACY

- dependent on:
  - image scale and quality
  - scene properties (texture, etc)
  - sampling spacing
  - operator/system performance
- best possible:
  - 0.2% of the flying height
  - 8m from SPOT

given favourable conditions, automated image matching is as precise as manual, but 5 times faster

- Major Cost Factors:
  - image acquisition
  - personnel and equipment rates
  - production time

⇒ a few dm

Simple 'manual error' model: $\sigma^2_{zmeasured} = \sigma^2_{AT} + \sigma^2_{o} + \sigma^2_{ident.} + \sigma^2_{digit}$

Principles of Image Matching

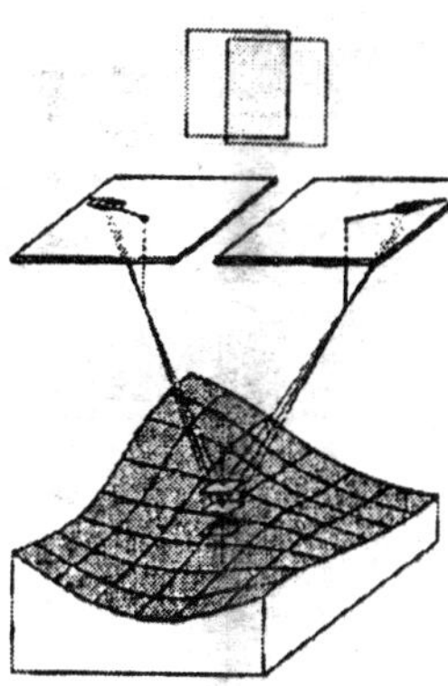

- find conjugate points
- orient stereo pair

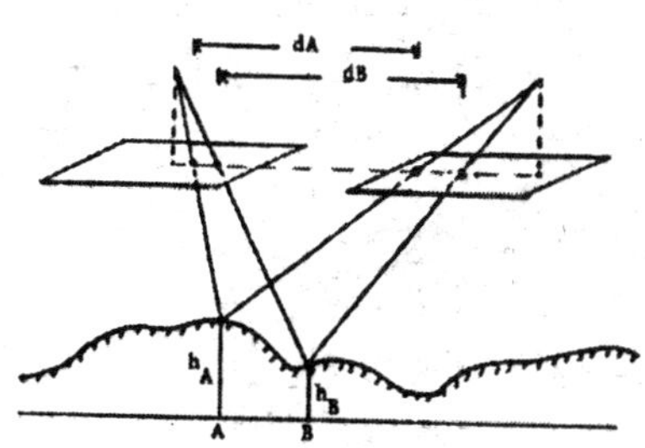

- compute parallax
- compute elevation

Finding Conjugate Points

(A) Signal Based Matching

Similarity Measures:
(a1) correlation, etc.
(search for maximum)
(a2) least squares
(direct solution by equations)

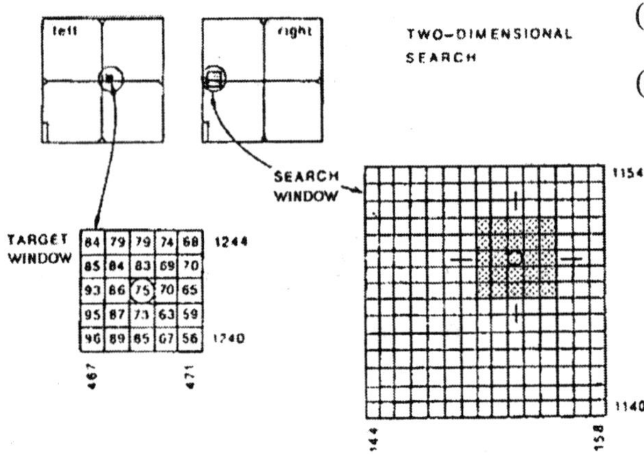

Search for Max. Correlation

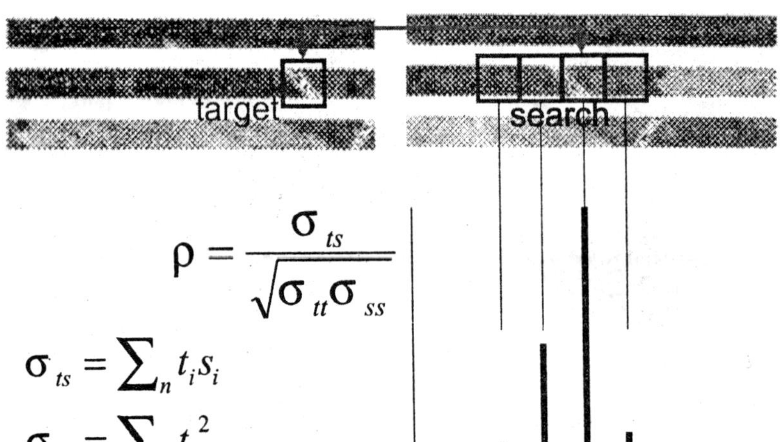

$$\rho = \frac{\sigma_{ts}}{\sqrt{\sigma_{tt}\sigma_{ss}}}$$

$$\sigma_{ts} = \sum_n t_i s_i$$

$$\sigma_{tt} = \sum_n t_i^2$$

$$\sigma_{ss} = \sum_n s_i^2$$

Finding Conjugate Points

(B) Feature Based Matching

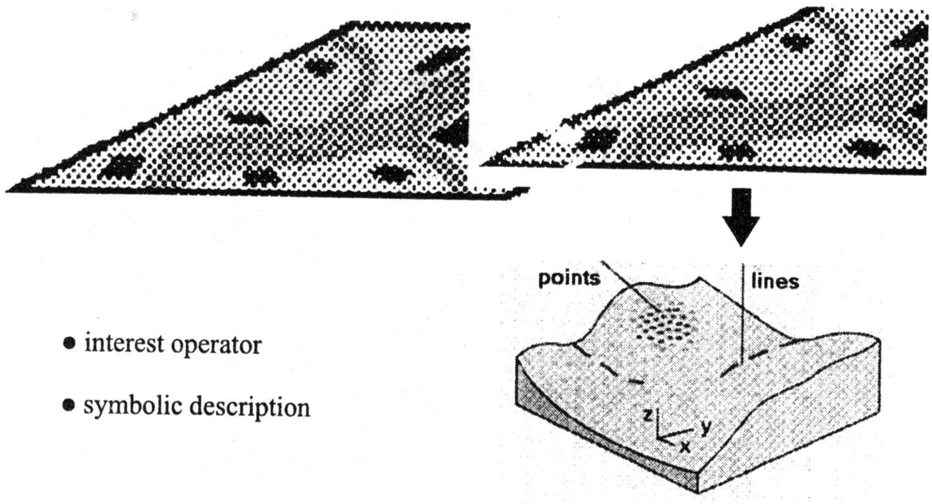

- interest operator
- symbolic description

Reducing Complexity (1)

- pre-processing (radiometric and geometric corrections)
- resampling to epipolar lines
- matching in image pyramids

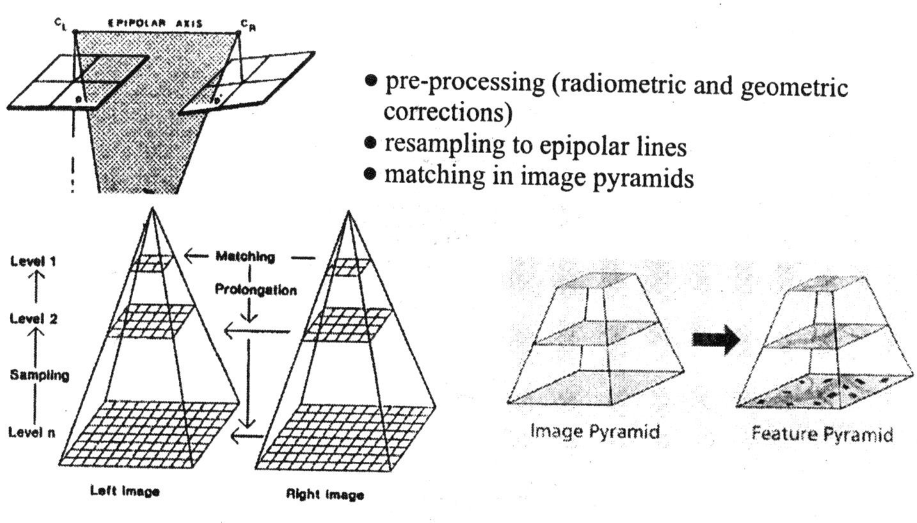

Reducing Complexity (2)

Remaining Problems:

- occlusions
- shadows
- poor texture
- repeating patterns
- multivalued surfaces

Parallax - Elevation

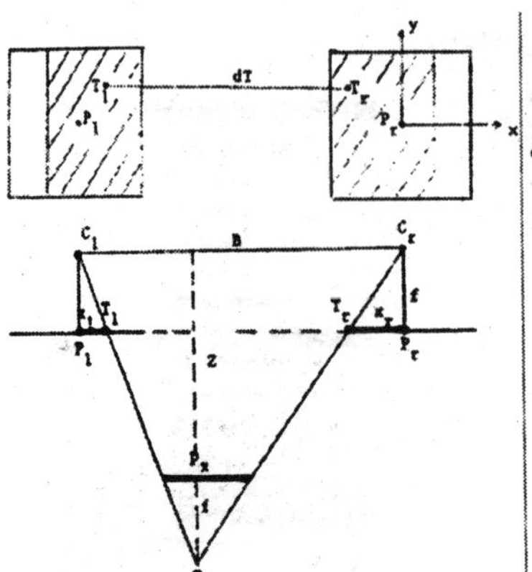

B	base of the stereo pair
f	focal length (principal distance)
C_l, Cr	projection centres of left and right photograph
P_l, P_r	principal points
T_l, T_r	images of point T
T	terrain point
Z	flying height above ground
h_T	elevation of point T
h_c	elevation of the projection center

$x_l = x_{T_l} - x_{P_l} > 0$ displacement in x of T in the left image

$x_r = x_{T_r} - x_{P_r} < 0$ displacement in x of T in the right image

$p_x = x_l - x_r$ x - parallax of point T

$dT = B - p_x$ shift of the images of point T

From the triangle TC_lC_r we get:

$$\frac{B}{p_x} = \frac{Z}{f}$$

Thus $Z = \frac{fB}{p_x}$

and $h_T = h_c - Z$

an error in determining the parallax causes on error in height

$$\delta h_T = \delta p_x \frac{Z}{b}$$

Pre-processing

Pre-processing (Tayloring)

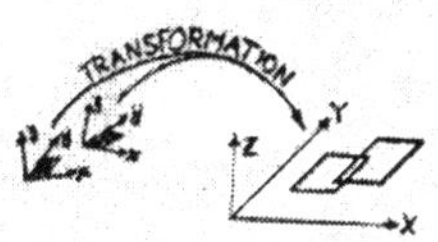

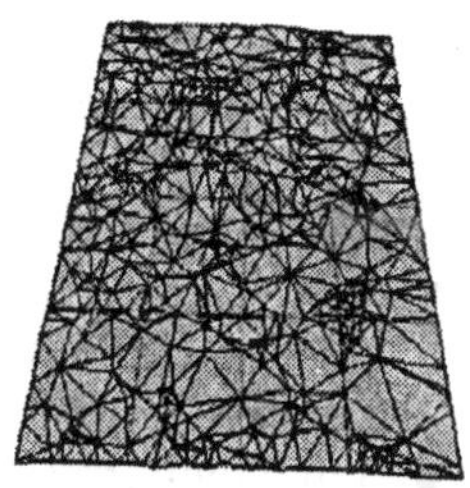

2. Tayloring:

- coordinate transformation
  from measuring system(s) to common/client system

- data restructuring

  eg, generate TIN, indexing to tiles

- data compression
  if wanted

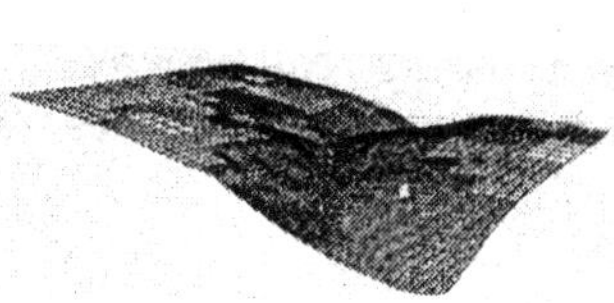

HOW?

By data collection software or DTM package (modelling software)

Numerical Quality Control

Accuracy:
spec's ↔ statistics from checkpoints

e.g., RMS error:

$$\sigma_Z = \sqrt{\frac{\sum (Z_i - {}_cZ_i)^2}{n}}$$

Field Completion and Check

- hardly done, unless in the context of topographic mapping
- survey gaps and check points

Photogrammetric Systems

Hardware:

- analogue plotters (+ encoders + interface + PC)
  no positioning of floating mark (unless profiling), no superimposition
  manufactures: Qasco, upgrading Kern, Wild, Zeiss, etc.
- analytical plotters
  computer support (positioning of floating mark, data analysis), mono superimposition
  manufactures: Leica, Zeiss, etc.
- digital plotters (softcopy plotters, digital photogrammetric workstations, etc.)
  (multiple) image matching, data analysis, digital editing tools, stereo superimposition, combination with orthoimage production
  manufactures: legio new ones (LH, IZ, ISM, DAT/EM, R-Wel, ERDAS, Autometric, etc.)

Data Processing

Function	*Hardware and Software*
Processing	UNIX-WS, PC
Pattern conversion (interpolation)	DTM package (interpolation software, contouring program)
contours to grid	
grid to contours	
irregular points to grid	
irregular points to contours	
Deriving new information	
elevation difference (DTM intersection)	DTM package, special modules or application packages
slope, aspect	
fall lines	
relief shades	
drainage pattern	
watershed delineation	*Examples of DTM packages:*
breaklines	*SURFER, SCOP, HIFI, SPANS, ILWIS,*
areas of visibility	*ARC/INFO TIN, MOSS, GENASYS,………*
volumes	
Accuracy assessment	
interpolation and statistics	DTM package, (plus statistical package)

Interpolation

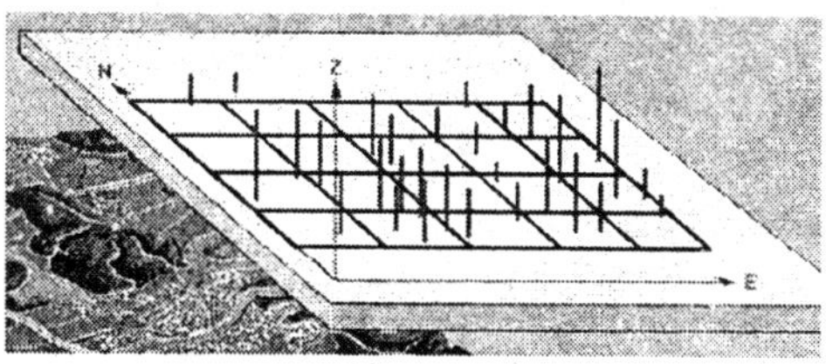

given: $Z_I = f(E_I, N_I)$

Compute value at intermediate point from values given at neighbouring points

E.g., linear/cubic interpolation, moving average, finite elements, kriging

Filtering

given: $Z_i' = g_i = f_i + e_i$

Modify values at given points
using values at neighbouring points

$$\hat{Z}(E,N) = a_1 g_1 + a_2 g_2 + \ldots\ldots + a_n g_n$$

E.g., moving average, finite elements, kriging /linear prediction

Interpolation

$$\hat{Z} = a_1 Z_1 + a_2 Z_2 + \ldots\ldots + a_n Z_n$$

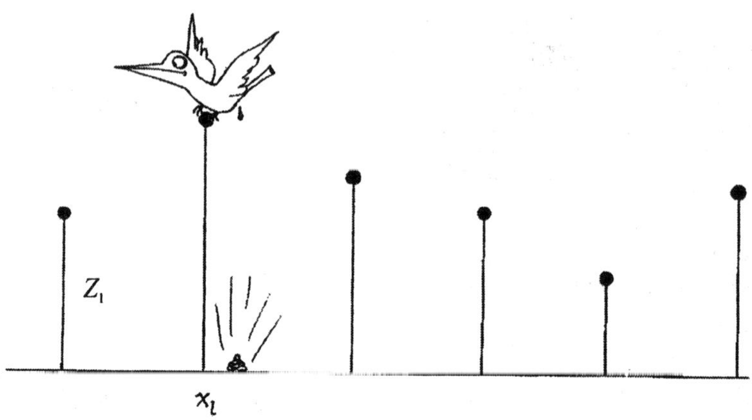

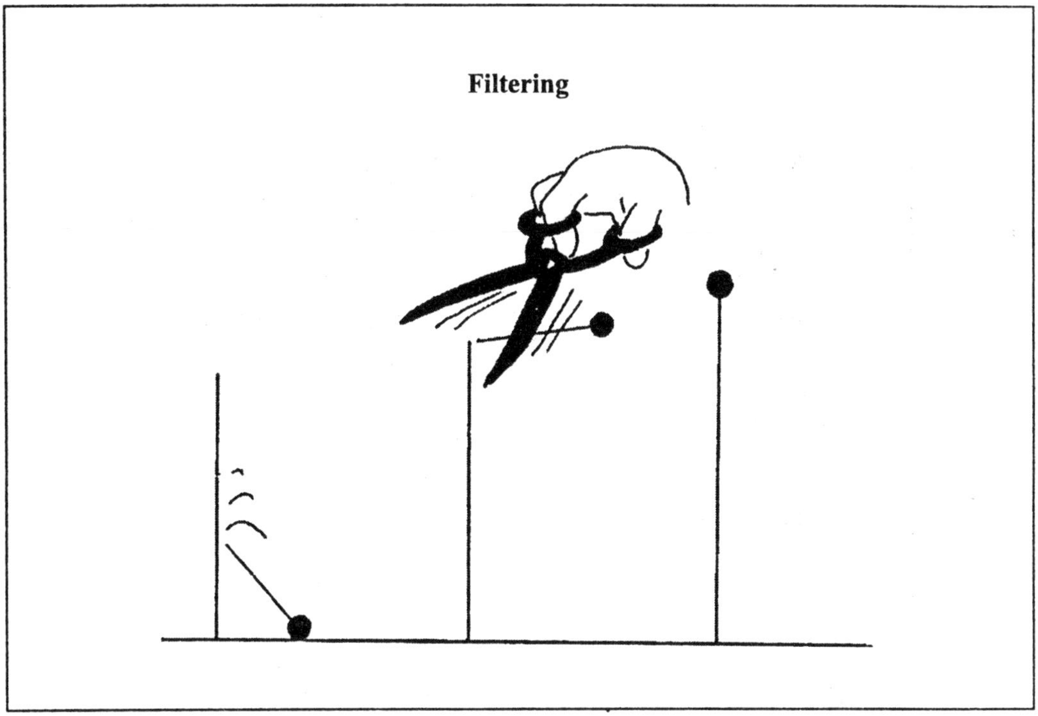
Filtering

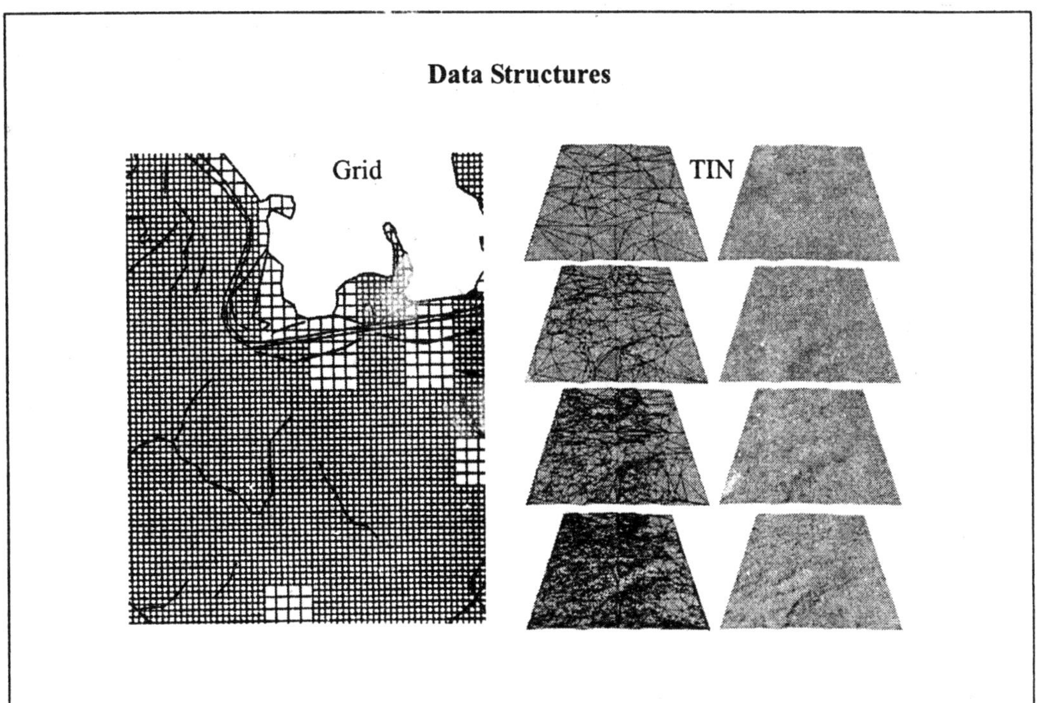
Data Structures
Grid
TIN

Presentation

Function	*Hardware and Software*
Presentation	UNIZ-WS, PC
projection type	Colour monitor
orthogonal	Harcopy graphics device
oblique	DTM package
Graphics type	GIS software
lines	VR browsers
layer tints	
grey value raster	
pictures and animation	

Orthographic Contour Lines

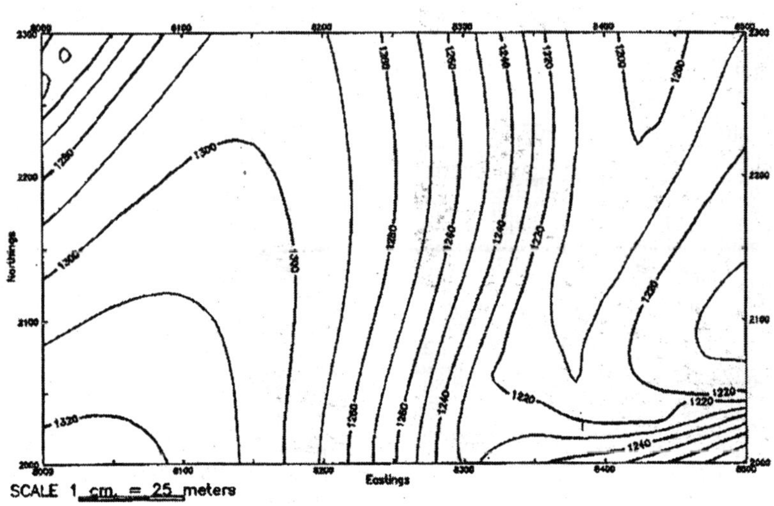

Layer Tints

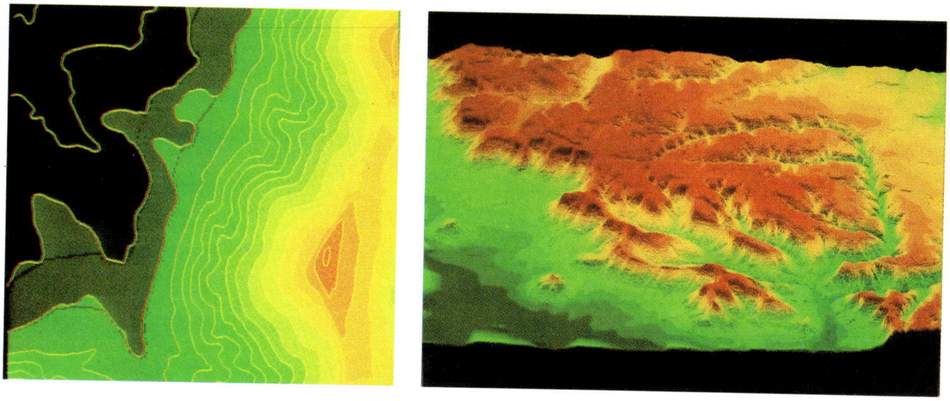

Relief Shades

Wireframe/Draping

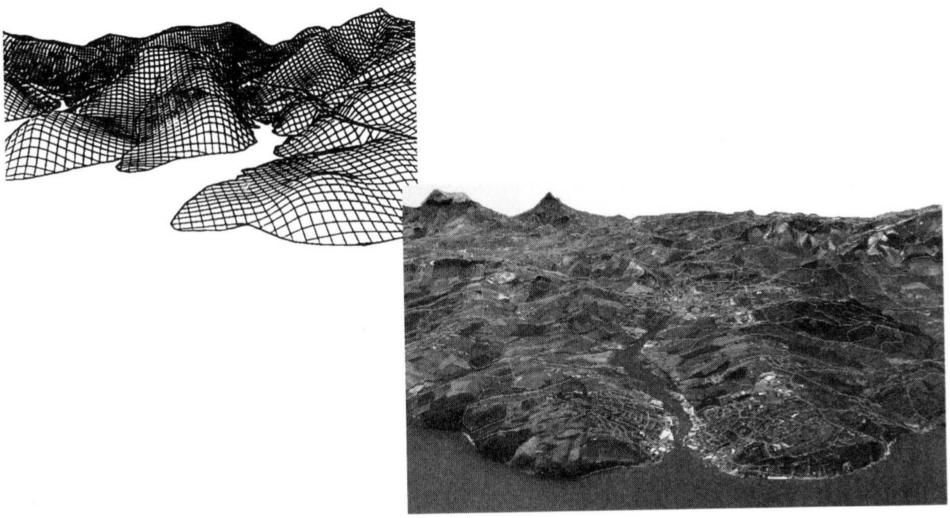

Topomap Production

- derive (labelled) contours
- cartographically touch up (morphologic generalization)
- match with plotted terrain features (clip, displace, etc.)
- supplement relief symbols (embankments, cliffs, rock drawings)
- produce contour separate

Quality and Accuracy

PERFORMANCE of DTM process

accuracy (precision and comprehen)
appearance (of derived products)
efficiency (time and costs)
reliability (of operations)
simplicity (of operations)
versatility (of DTM system to I/O)

Accuracy Influencing Factors

Precision of data:
– quality f source
  (terrain type, image/map quality, scale, B/H)
– equipment, operator
– mode of scanning, digitizing

Comprehensiveness of data:
– sampling density
– sampling pattern

quality of derived product:
– accuracy of data
– interpolation method

Quality and Accuracy

Empirical Approach

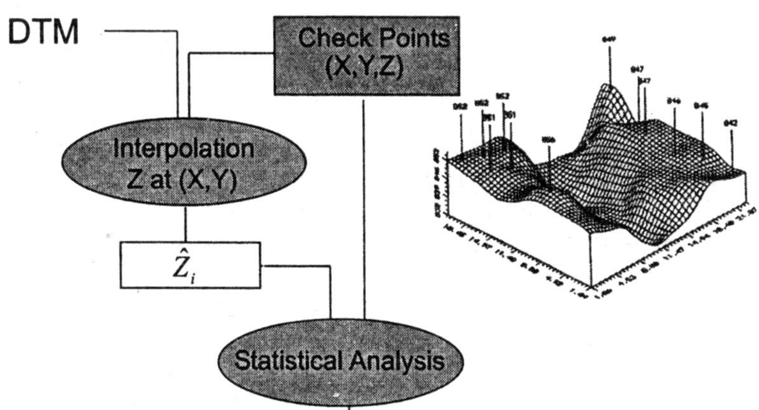

$$\hat{\sigma} = \sqrt{\frac{\sum (Z_i - \hat{Z}_i)^2}{n}}$$

31

BASIC CONCEPTS OF GPS

SHAILENDRA SINGH SOLANKI

GEOGRAPHICAL INFORMATION SYSTEM—BACKGROUND

Geographic Information System (GIS) today is an indispensable tool in the planning for tomorrow's requirements, such as rail network, dam site location, urban development etc. Until computers were available, geographic data were drawn on a piece of paper, as points, lines and areas. With the advent of computers, accurate and rapid analyses of diverse geographic data sets are perceived as a requirement for more effective planning. The information about the resources, thus collected from the real world, converted into digital form and is integrated and analysed in a GIS to make quick decisions as well as their implementation on the real world.

Work on GIS began in late 1950s, but first GIS software came only in late 1970s from the lab of the ESRI. Canada was the pioneer in the development of GIS as a result of innovations dating back to early 1960s. Evolution of GIS has transformed and revolutionized the ways in which planners, engineers, managers etc. conduct the database management and analysis.

BASICS OF GIS

Early definitions of GIS treat it as a spatial database, but the recent trend is to consider it in following several ways (ESRI, 1996):

- Information technology which stores, analyses and displays spatial and non-spatial data.
- Decision support system involving the integration of spatially referenced data in a problem solving environment.
- Tool for management, analysis and manipulation of spatially referenced data information in a problem solving environment.

From above definitions, it is clear that the common factor between all is the notion of spatial data processing. In this context, spatial data may be defined as objects that have physical dimensions on the Earth. "Every object present on the Earth can be geo-referenced", is the fundamental key of associating any database to GIS. Here, term 'database' is a collection of information about things and their relationship to each other and 'geo-referencing' refers to the location of a layer or coverage in space defined by the co-ordinate referencing system.

COMPONENTS OF GIS

GIS refers to three integrated elements:

(i) Geography which is the real world,
(ii) Information which is data, their meaning and uses, and
(iii) Systems which is computer technology and its support infrastructure.

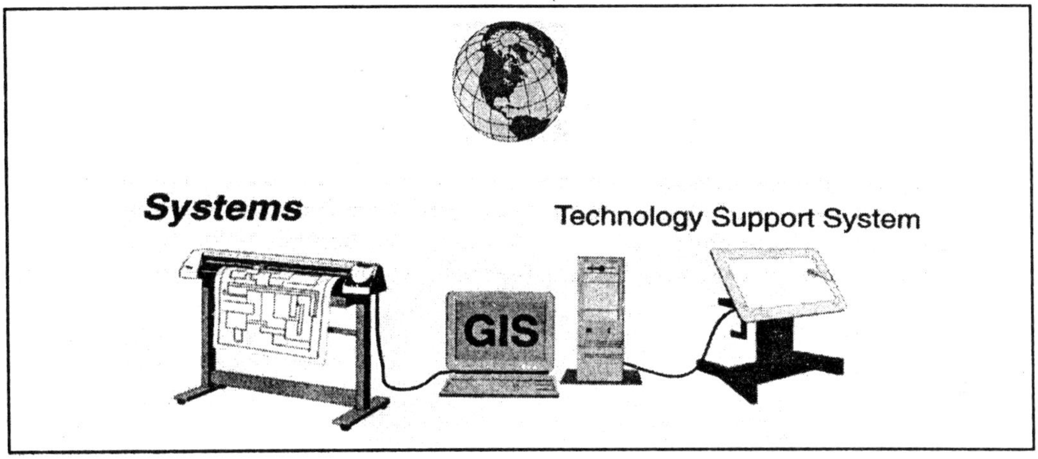

COMPUTER HARDWARE

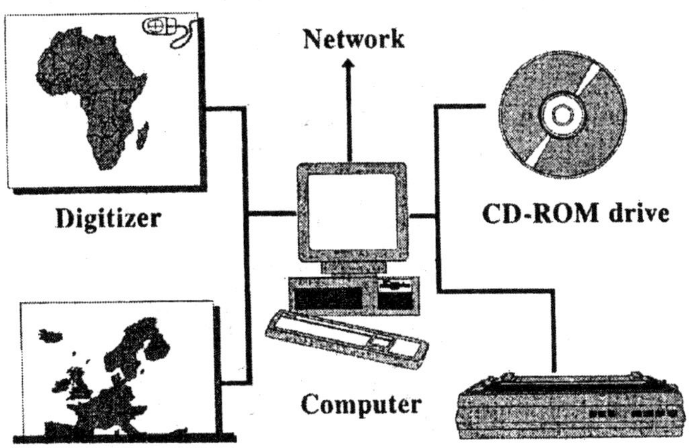

Major Hardware Components of a GIS

GIS SOFTWARE

The software for a geographical information system may be split into following functional groups Main Software Components of a GIS

- Data input and verification,
- Data storage and database management,
- Data output and presentation,
- Interaction with the user.

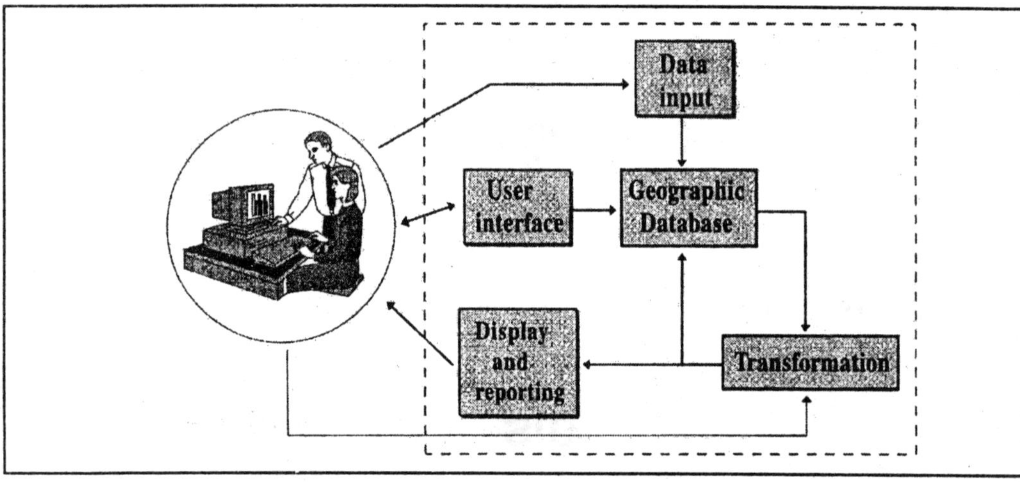

FUNCTIONAL ELEMENTS OF GIS

Essential elements of a GIS:

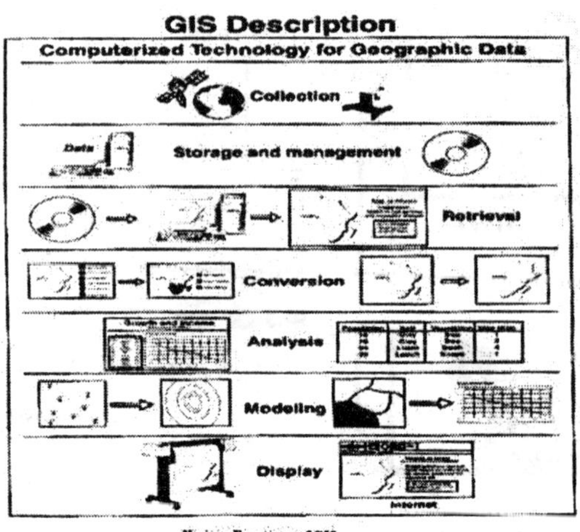

Various Functions of GIS

- Data acquisition,
- Preprocessing,
- Data management,
- Manipulation and analysis,
- Product generation.

DATA ACQUISITION

- Main methods of data collection are:
    - Data capture and
    - Data transfer.
- Data Sources are:
    - Primary (direct measurement) and
    - Secondary (indirect derivation from other sources)

Classification of Geographic Data for Data Collection Purposes

	Raster	*Vector*
Primary	Digital remote-sensing images	GPS measurements
	Digital aerial photographs	Survey measurements
Secondary	Scanned maps or photographs	Topographic maps
	Digital elevation models from maps	Toponoymy (place name) databases

VECTOR DATA CAPTURE

Primary vector data capture is a major source of geographic data. The two main branches of vector data capture are:

- Ground surveying and
- GPS.

ABOUT GPS

- GPS stands for Global Positioning System and it allows users to determine their location on land, sea, and in the air around the Earth. It does this using satellites and receivers. There are currently 24 satellites in orbit operated by the US Department of Defence that provide worldwide 3 dimensional coverage 24 hours a day, 7 days a week, in all weather.
- The Global Positioning System (GPS) is a satellite-based system that can be used to locate positions anywhere on the earth. Operated by the U.S. Department of Defence (DoD), NAVSTAR (NAVigation Satellite Timing and Ranging) GPS provides continuous (24 hours/day), real-time, 3-dimensional positioning, navigation and timing worldwide.

- The GPS is a network of satellite, monitoring stations, and inexpensive receivers used for primary GIS data capture.
- The Global Positioning System is a collection of 27 NAVSTAR satellite orbiting the Earth at a height of 12500 miles, five monitoring stations, and individual receivers.
- The GPS NAVSTAR (Navigation Satellite timing and Ranging Global Positioning System) is a satellite-based navigation, timing and positioning system.

A BIT OF HISTORY

- The Global Positioning System was conceived, designed, and deployed by the US Department of Defense.
- Primarily, if not solely, for military purposes.
- NAVSTAR (Navigation Satellite Timing and Ranging), the official DoD name for GPS.

A BIT OF HISTORY

- 1964: US Navy built a low earth orbiting satellite-based 2-dimensional navigation system called *Timation.*
- Mid 1960s: US Air Force began developing a 3-dimensional satellite based navigational system called *621B.*
- The Air Force was designated as the executive agent to consolidate the *Timation* and *621B* concepts into a comprehensive system; the result was the *Global Positioning System (GPS).*

A BIT OF HISTORY

- 1973 – 1977: Proof of concept.
- 1976: US Congress approved development.
- 1978: First satellites launched.
- First real operational test for GPS technology was the Gult war in 1990.
- Full constellation achieved in 1995.
- Selective Availability turned off by President Clinton in 2000.

A BIT OF HISTORY

- National Defense Authorization Act for Fiscal Year 1994 (Public Law 103-160)
    - National Academy of Public Administration
    - National Research Council
    - National Academies of Sciences and Engineering

"It is clear that GPS offers the potential to revolutionize the movement of goods and people the world over. Civil and commercial exploitation of GPS could soon dwarf that of the Department of Defense and lead to large productivity gains and increased safety in all transportation sectors."

WHAT IS GPS?

- 24 (+ spare) earth-orbiting satellites
- 5 ground-based monitoring stations
- Individual GPS receivers

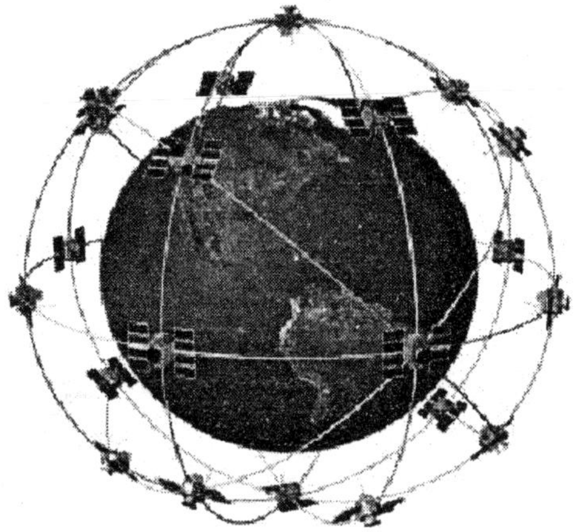

All these make up the Global Positioning System.

COMPONENTS OF A GPS

- *Space segment:* GPS satellites continuously emit signals down to earth.
- *User segment*: Any appropriately configured receiver can interpret these signals at no cost.
- *Control segment*: Ground stations provide corrections and updates to satellites and receivers.

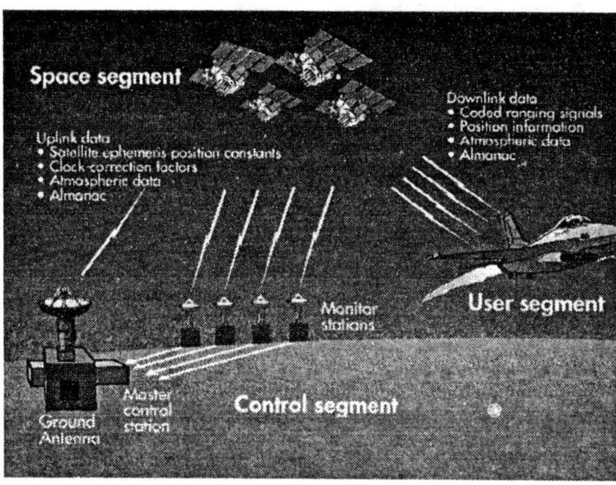

THE CONTROL SEGMENT

- The Control Segment monitors and tracks the satellites, synchronises their operation, performs satellite positional computations and transmits orbital and corrected time data to the satellites.
- To perform these operations, the Control Segment consists of five (5) Monitor Stations, one (1) Master Control Station, and three (3) Data Upload Stations.
- The Monitor Stations are located in the Ascension Island, Hawaii, Diego Garcia, Kwajelei and Colorado.
- The Master Control Station and one Data Upload Station are located in Colorado and the remaining Data Upload Stations are located in Hawaii and Ascension Island.
- The Control Segment is the sole responsibility of the DoD who undertakes construction, launching, maintenance, and virtually constant performance monitoring of all GPS satellites.

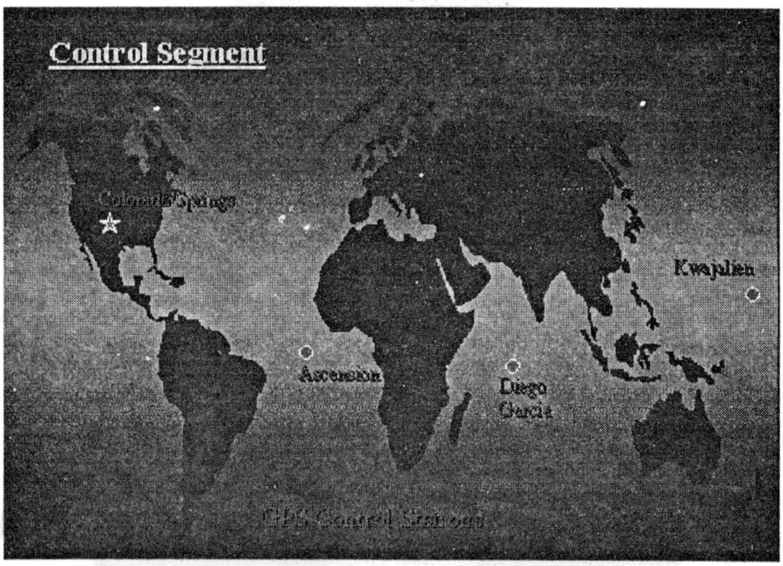

THE SPACE SEGMENT

- The Space Segment consists of the Constellation of NAVASTAR earth orbiting satellites. The current system consist of full constellation of 24 Block II satellites (21 operational and 3 in-orbit spares).
- The satellites are arrayed in 6 orbital planes, inclined 55 degrees to the equator. They orbit at altitudes of about 12000 miles each, with orbital periods of 12 sidereal hours (i.e., determined by or from the stars), or approximately one half of the earth's periods, approximately 12 hours of 3-D position fixes.
- Each satellite contains four precise atomic clocks (Rubidium and Cesium standards) and has a microprocessor on board for limited self-monitoring and data processing.

- The satellites are equipped with thrusters which can be used to maintain or modify their orbits.

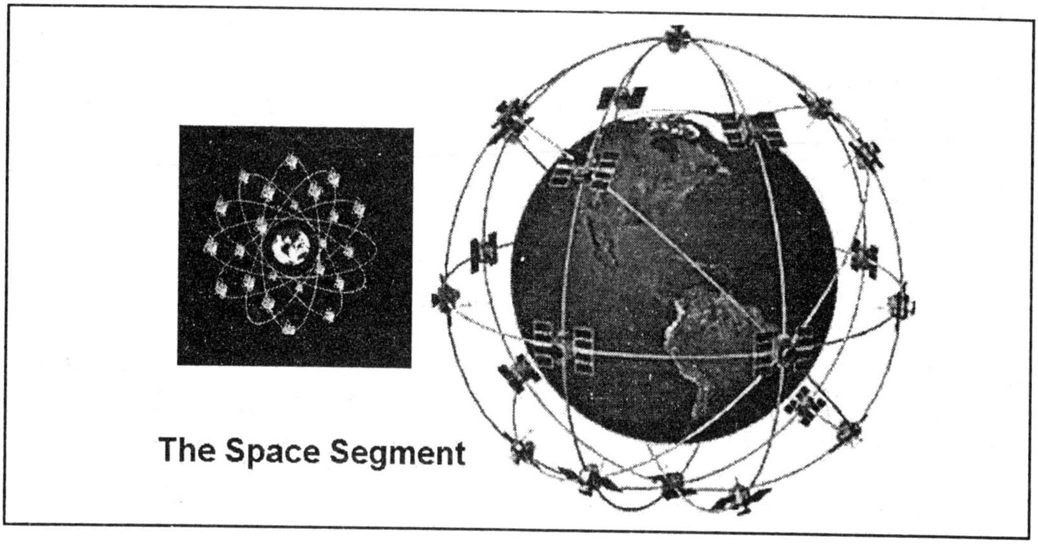

THE USER SEGMENT

- The user segment is a total user and supplier community, both civilian and military.
- The User Segment consists of all earth-based GPS receivers.
- The typical receiver is composed of an antenna and preamplifier, radio signal microprocessor, control and display device, data recording unit, and power supply.
- The GPS receiver decodes the timing signals from the 'visible' satellites (four or more) and, having calculated their distances, computes its own latitude, longitude, elevation, and time.
- The position is updated on a second-by-second basis.

WHAT IS GPS?

Intended use is determined
ONLY
by the receiver's interpretation of the
satellite and station signals!

The system is ever-present and
anyone can use it!

WHO USES GPS?

- Land, sea, and air vehicle navigation
- Defence and combat systems

- Public safety, e-911
- Fishing industry
- Surveying
- Scientific research
- Precision agriculture
- Endangered species and other animal tracking
- Hiking, biking, etc.
- Cell phones
- Utility companies
- Automobile routing

GPS APPLICATIONS

The applications of the Global Positioning System fall into five categories: location, navigation, timing, mapping, and tracking. Each category contains uses for the military, industry, transportation, recreation and science.

- *Location*: This category is for position determination and is the most obvious use of the Global Positioning System.
- *Navigation*: Navigation is the process of getting from one location to another. This was the what the Global Positioning System was designed for.
- *Timing*: GPS brings precise timing as each satellite is equipped with an extremely precise atomic clock.
- *Mapping:* This is used for creating maps by recording a series of locations.
- *Tracking*: The applications in this category are ways of monitoring people and things such as packages.

US POSITIONING SYSTEMS

Navstar-GPS

- Navstar-GPS system managed by the Navstar GPS Joint Program Office at Air Force Materiel Command's Space and Missile Systems Center at Los Angeles AFB, CA.
- The GPS master control station, operated by the 2nd Space Operations Squadron at Falcon AFB, CO, is responsible for monitoring and controlling the GPS satellite constellation.

NAVSTAR GPS

- Dedicated system
    - The satellites cannot be used for any other purpose
    - Minimize disturbances in times of national crisis
- Communication

- Currently, satellites only communicate with ground control stations
- In the future, satellites will be able to "talk" to each other (increase speed of repositioning)

NON-US POSITIONING SYSTEMS

- GLONASS
  - *Glo*bal *Na*vigation *S*atellite *S*ystem
  - Russian Federation Ministry of Defense
  - Has deteriorated over time; not complete
- GALILEO
  - European Union
  - Still in debate over cost/benefit
  - 2008?

GNSS

- Global Navigation Satellite System
  - GPS is a subset of GNSS
- GNSS receivers would utilize:
  - NAVSTAR, GLONASS, GALILEO
  - ...or all three...
- Currently no GNSS systems in operation, but several are planned.

GPS INDUSTRY

- Even though GPS technology predates personal computers, applications using GPS exploded in the mid-1990s.
- It continues to grow at a rapid pace, accelerated by consumer applications and terroristic/military events.

GPS INDUSTRY

- In March of 1996, President Clinton finally signed the Presidential Decision Directive (PDD), a document formalizing GPS policy and the government's commitment to the commercialization of GPS:
  - removal of selective availability within a four-to ten-year time frame,
  - continued maintenance and funding of the system by the Department of Defense,
  - encouragement of private sector investment in and use of U.S. GPS technologies and services.

COMMERCIAL APPLICATIONS DOMINATE THE GPS MARKET

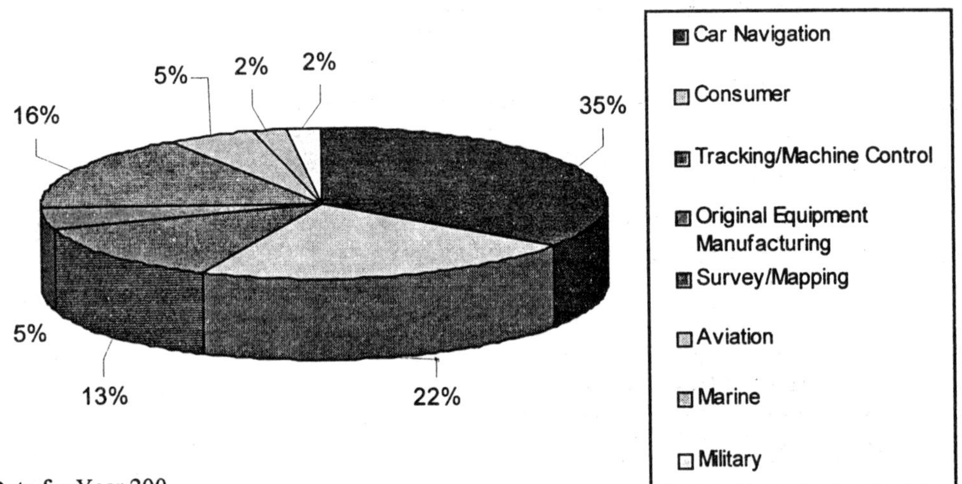

Data for Year 200
Source: DOC, 1998

GPS INDUSTRY

- Worldwide sales revenue of GPS receivers:
  - 1991: $300 million
  - 1995: $700 million
  - 1997: < $2 billion
- GPS products and services is predicted to be at least $14 billion by 2005.

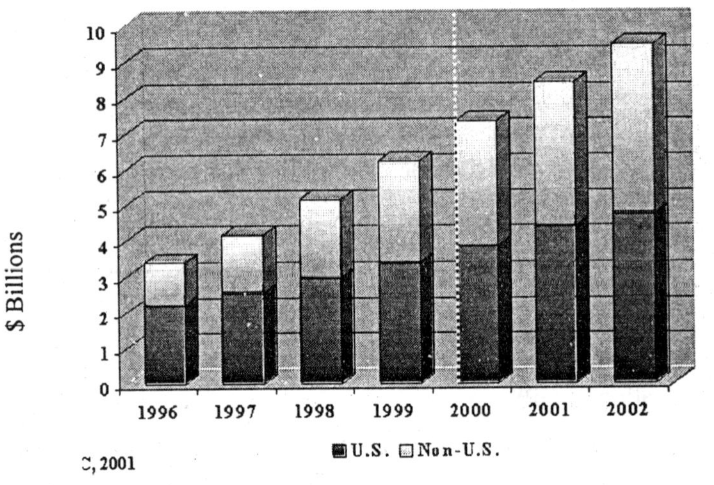

Source: DOC, 2001

UNIT COST OF RECEIVERS IS FALLING AT ~ 30% PER YEAR

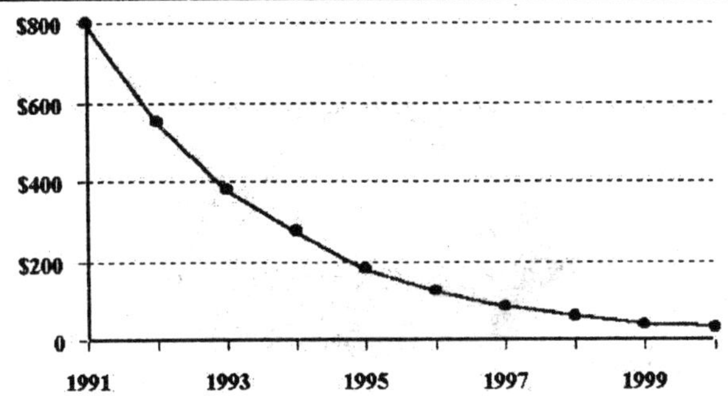

Source: U.S. GPS Industry Council, 1995

US NAVSTAR GLOBAL POSITIONING SYSTEM

How Does it Work?

SATELLITE ORBITS

- 24 operational satellites (+ 3 or 4 spare)
- 6 orbital planes
- 4 equally spaced per orbit
- 20,200 km (10,900 n. mi.) altitude
- 12 hour orbital period
- Solar powered, backup batteries, rocket boosters

FYI: The 24 satellites of GLONASS maintain 19,100 km altitude in 3 orbital planes.

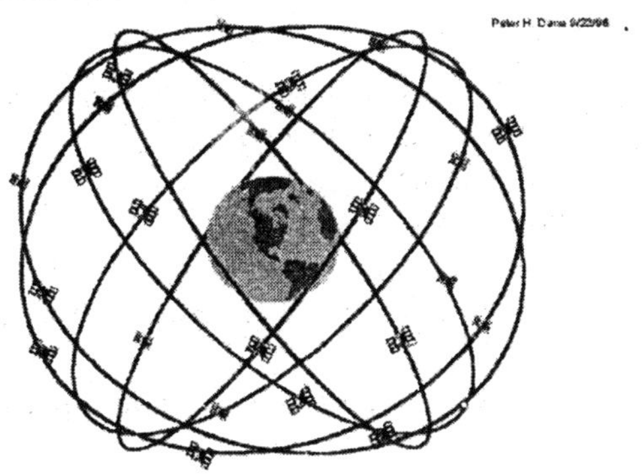

GPS Nominal Constellation
24 Satellites in 6 Orbital Planes
4 Satellites in each Plane
20,200 km Altitudes, 55 Degree Inclination

SATELLITE ORBITS

- Usually 6 (at least 4) satellites are within view from any location on earth.

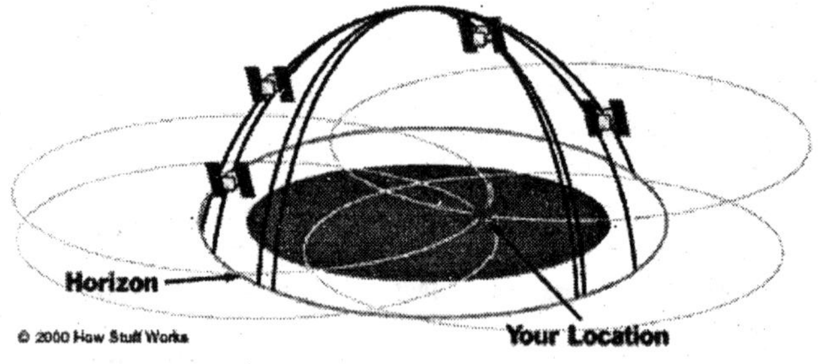

SATELLITE CHARACTERISTICS

- Block IIA
  - 5′ × 17.5′ (w/ panels)
  - 2200 lbs
  - 7.5 years design life
- Block IIR
  - 5′ × 6.33′ diam × 6.25′ high (38.025′ w/panels)
  - 2370 lbs
  - 10 years design life
- Block IIF
  - 8′ × 70.42′ (w/panels) × 12′
  - 3758 lbs
  - 15 years design life

TIME IS OF THE ESSENCE

- Each GPS satellite is equipped with a cesium atomic clock
  - accurate to 3 nanoseconds (0.000000003 s)
  - +/-1 second every 360,000 years
- Ground control stations correct satellites
- Satellites correct GPS receivers
- A discrepancy between satellite and receiver of 1/100 sec could result in an error of 1860 miles
- Each satellite continuously emits low power radio signals in several channels
- Only 20-50 Watts
- Timing and identity signal
- Not much else within signal
- Simplicity keeps receiver's cost down

HOW IT WORKS

- Each satellite continuously emits low power radio signals in several channels
- Only 20-50 Watts
- Timing and identity signal
    - Not much else within signal
    - Simplicity keeps receiver's cost down

HOW IT WORKS

- GPS receiver "receives" and decodes signals from 4 (or 3) satellites
- Signal knows the exact time it was sent and from which satellite
- Based upon receiver's own time keeping, it calculates length of time the signal traveled from satellite
- Receiver computes distance to each satellite (D = V × T)

HOW IT WORKS

- There are 4 variables unknown:
    - Longitude (X)
    - Latitude (Y)
    - Elevation (Z)
    - Time (T)
- Therefore, you need info from four satellites!
- If only 3 satellites are present, the receiver will guess at elevation and compute latitude and longitude.

HOW IT WORKS

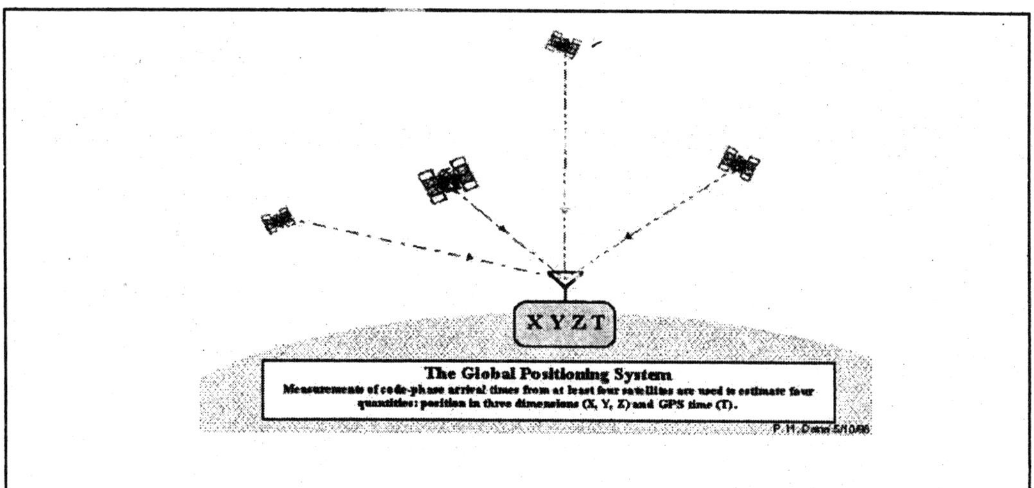

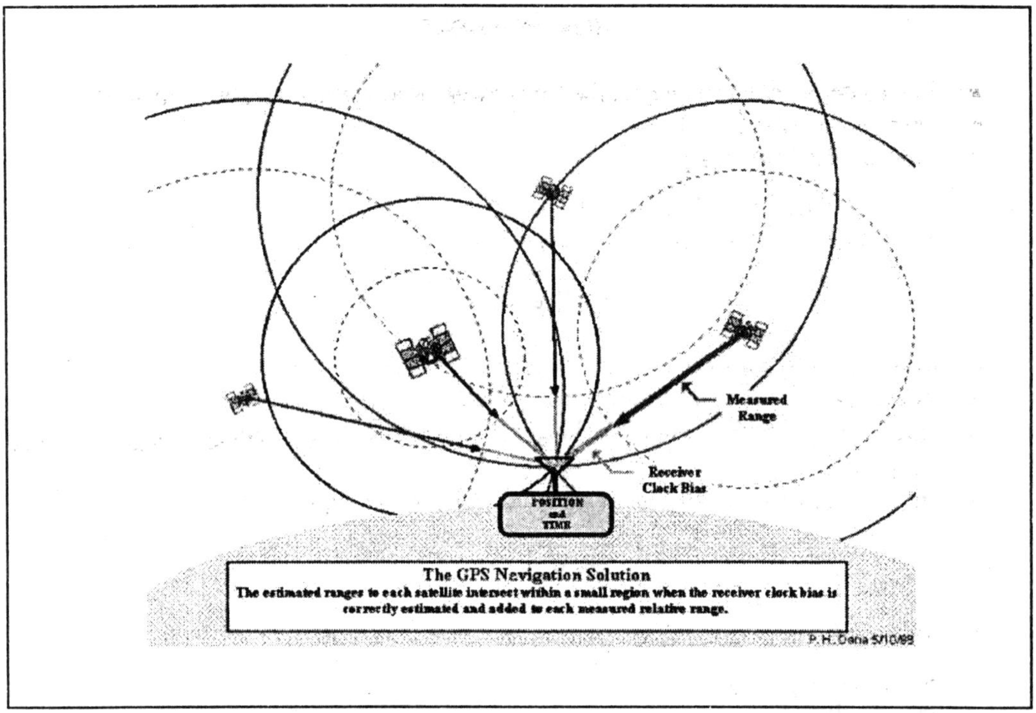
Measured
Range
Receiver
Clock Bias
POSITION
and
TIME
The GPS Nevigation Solution
The estimated ranges to each satellite intersect within a small region when the receiver clock bias is correctly estimated and added to each measured relative range.
P. H. Dana 5/10/98

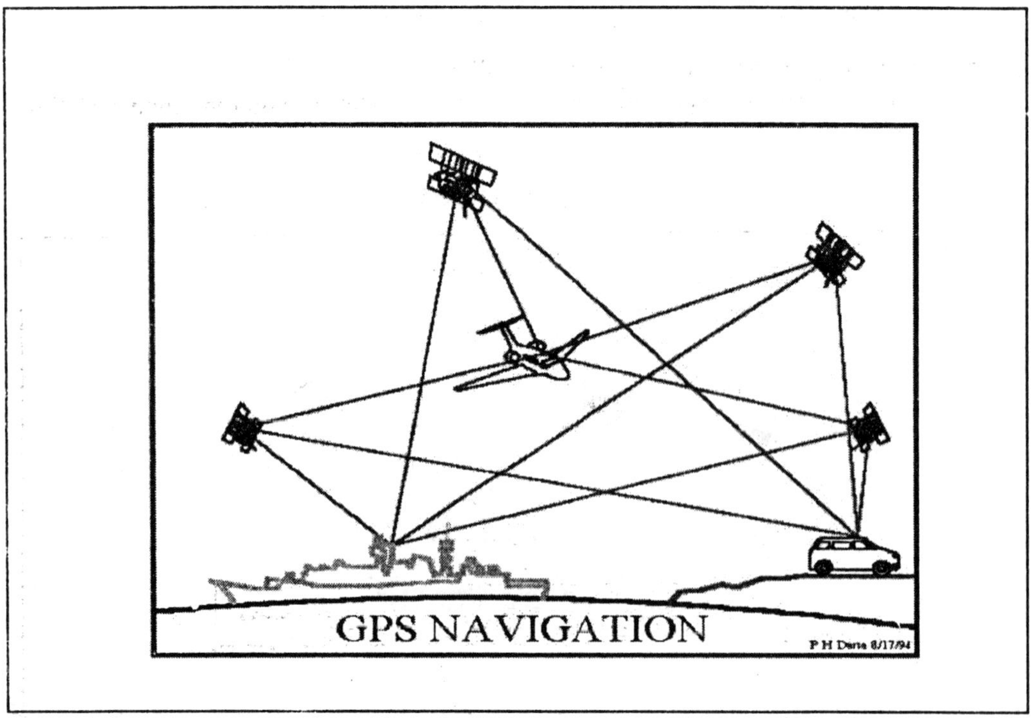
GPS NAVIGATION
P H Dana 8/17/94

SATELLITE SIGNALS

- Easily blocked and reflected by solid surfaces and water bodies
- Can penetrate standard atmosphere, water vapor, clouds for travel over long distances
- Primary purpose is signal transit time

SATELLITE SIGNALS

- Microwave region of EM spectrum; makes for all weather communication
  - L1 Band: 1585.42 MHz
  - L2 Band: 1227.60 MHz
  - 'L3 Band: 1381.04 MHz
- L1 and L2 bands used for positioning
- Pseudorandom code
  - Extremely complicated patterns of 0s and 1s repeating every 1023 bits

SATELLITE SIGNALS

- Atomic clocks on satellites generate a pure sine wave
  - Fundamental frequency, f_o 10.23 MHz
- L1 and L2 carrier waves
  - $f_1 = f_o \times 154 = 1575.42$ MHz (19 cm)
  - $f_2 = f_o \times 120 = 1227.60$ MHz (24 cm)
- These signals must be modified, or *modulated*, to contain information

SATELLITE SIGNALS

- C/A (Course/Acquisition) Code
  - Modulates on L1 carrier
  - 1.023 MHz Pseudo Random Noise (PRN) code
  - Entire message 1023 bits
  - Repeats every millisecond
  - "Wavelength" of 300 m
  - Total sequence is 300 km
  - Unique from each satellite

SATELLITE SIGNALS

- (Precise or Private) Code
  - Much more complicated

- Modulates on L1 and L2
- 10.23 MHz PRN code
- "Wavelength" of 30 m
- 266.4 days long code
- Anti-Spoofing mode includes an encrypted W code to create the Y code
- Classified decryption

SATELLITE SIGNALS

- Navigation Message
  - Modulates on L1 and L2
  - 50 Hz
  - Entire message is 1500 bits; 30 sec
  - Includes information such as:
    - Predicted **satellite ephemerides.**
    - Predicted **satellite clock correction model** coefficients.
    - GPS **system status** information.
    - The GPS system **ionospheric model**

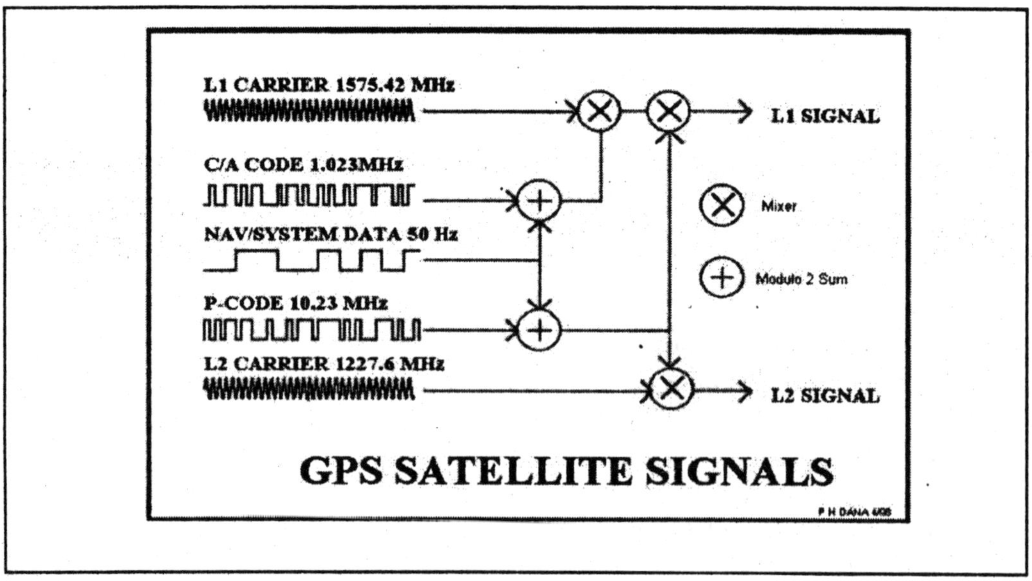

SATELLITE SIGNALS

- Listen-only system
- Multiples bands allow for calculation of ionospheric delay
- Simultaneous measurements from multiple satellites
- Two codes enables civilian use of one, military use of the other

- Anti-jamming system
- Real-time corrections to system

GPS MONITORING STATION NETWORK

Global Positioning System (GPS) Master Control and Monitor Station Network

GPS CONTROL STATIONS

- Each satellite passes over a ground monitoring station every 12 hours
- Stations receive satellite data
- Four unmanned stations send information to master control station
- Master control station determines any corrections that may be necessary
- All stations "uplink" the corrections to satellites

GPS RECEIVERS

- Receivers are usually configured with quartz clocks (< accurate than atomic)
  - Therefore four satellites are needed
  - Can be corrected to GPST from information encoded in Navigation Message
- When reading satellite signals, receivers try to duplicate C/A and P codes, then use offset to extract info
- Type of receiver uniquely determines the type of decryption possible, and therefore the accuracy of GPS measurement

GPS RECEIVERS

- Standard Positioning Service
  - Single frequency (L1) C/A code ranging
  - Civilian uses, typical receivers
- Precise Positioning Service

- Dual frequency (L1 and L2) P code
- P code contains higher resolution info
- Two frequency channels enables atmospheric corrections
- Commonly used for national "interests"

GPS SOURCES OF ERROR

- Minor disturbances in satellite orbits (called *ephemeris*)
    - Gravitational pulls of sun and moon; solar wind
- Atmospheric signal delay
    - Ionosphere, water vapor
- Slight fluctuations in atomic clocks
- Receiver quality
    - Faulty clocks or internal noise
- Multi-path
    - Multiple signal reflections

GPS SOURCES OF ERROR

- Standard *vs.* Precise Positioning Service
    - Classified *vs.* unclassified
- Selective Availability (SA)
    - SA is the intentional degradation of the signals by a time varying bias. SA is controlled by the DOD to limit accuracy for non-U.S. military and government users.

GPS SOURCES OF ERROR

- System Noise: 1-2 metres
- Ephemeris: 1 metre
- Satellite clock: 1 metre
- Multi-path: 0.5 metres
- Tropospheric delay: 1 metre
- Ionospheric delay: 10 metres
- S/A: ~ 70 metres
- Human and software errors: 1 – 100s metres

GPS MEASUREMENT IMPROVEMENT

- SA is turned off by DoD in 2000
    - Increased standard accuracy from 100 to 30 m
    - Can be turned on anytime, anywhere

- Use of Precise Positioning Service
  - Classified decryption receivers
- Real-time averaging of measurements
- High quality receivers
- Differential GPS

DIFFERENTIAL GPS (DGPS)

- Use two receivers!
  - Base station with known location and elevation
  - Mobile receiver
- Velocity of signal from satellite to receiver is dependent on atmospheric conditions
- Known error at base station can be applied to measurements by mobile receiver
- Works best for measurements within 300 miles of base station
  - Same atmospheric conditions over both receivers

GPS ACCURACY SUMMARY

Selective Availability

- For an off-the-shelf, inexpensive GPS receiver
- SA on – 100 metres
- SA off – 15 metre

Differential GPS

- Higher quality receiver with DGPS
  - Accuracy increases to 1 meter or less

WORLD GEODETIC SYSTEM 1984

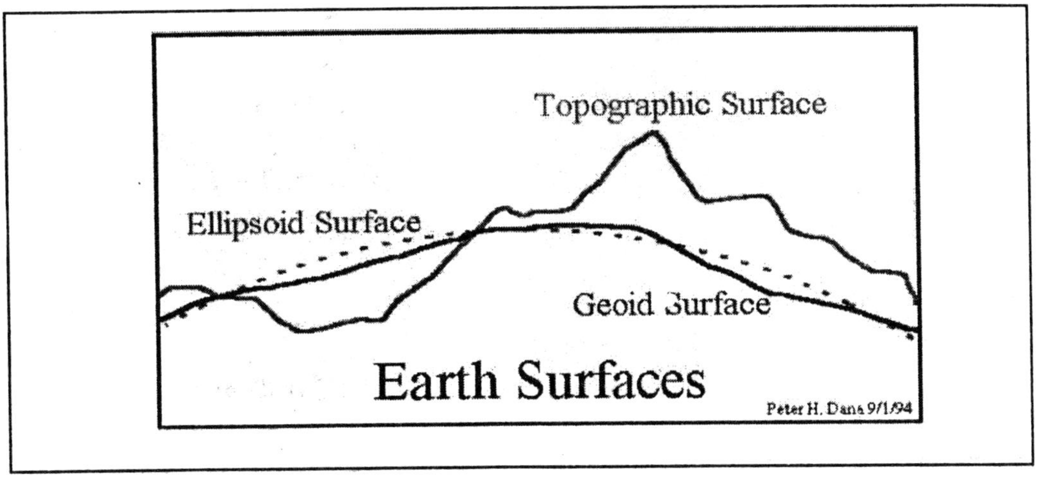

MOBILE MAPPING USING GPS

- The GPS 12XL features a 12 parallel channel receiver for amazingly fast satellite aquisition and tight satellite lock—even under tree cover.
- The GPS 12XL is loaded with navigation information including 106 map datums and 7 grid formats, plus advanced orienteering features like user datum/UTM/and user grid formats to define your own map parameters.
- Store position as one of 500 waypoints or add it to any of 20 routes.
- The proximity waypoint feature gives advanced warning of nearby hazards, while the position averaging function helps get the best fix possible.
- Packed in a rugged case to handle tough conditions, the GPS 12XL uses an ergonomic rocker keypad for easy one-hand operation and features a user-friendly operating system to keep you on the move instead of pushing buttons.
- Four AA batteries provide up to 24 hours of use and a variety of accessories can be added like an external antenna and mounting bracket.

BRIEF INTRODUCTION TO GPS – 12 XL

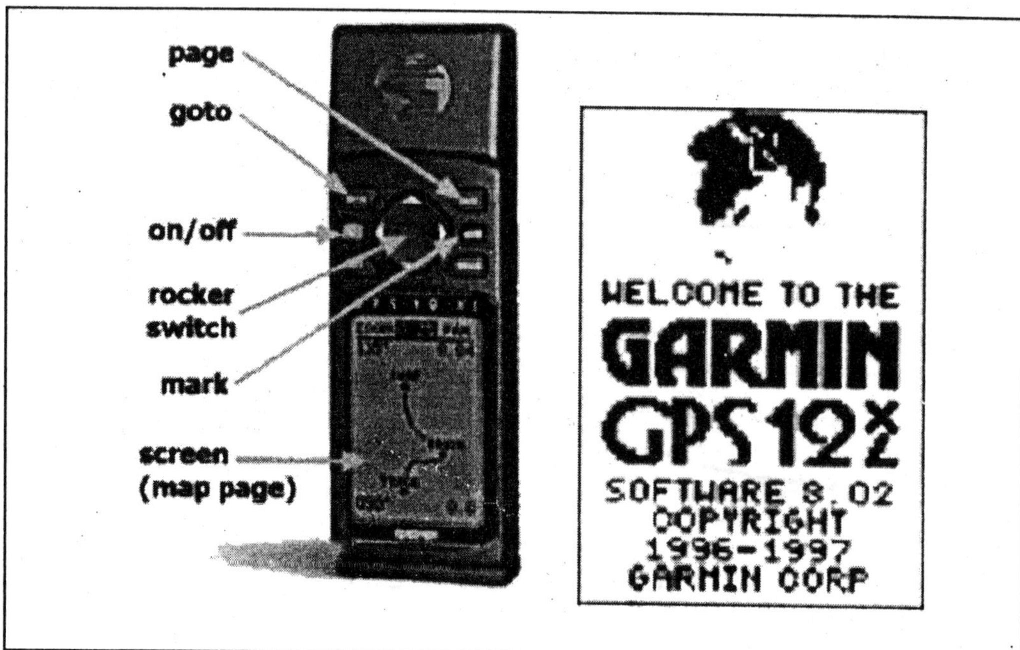

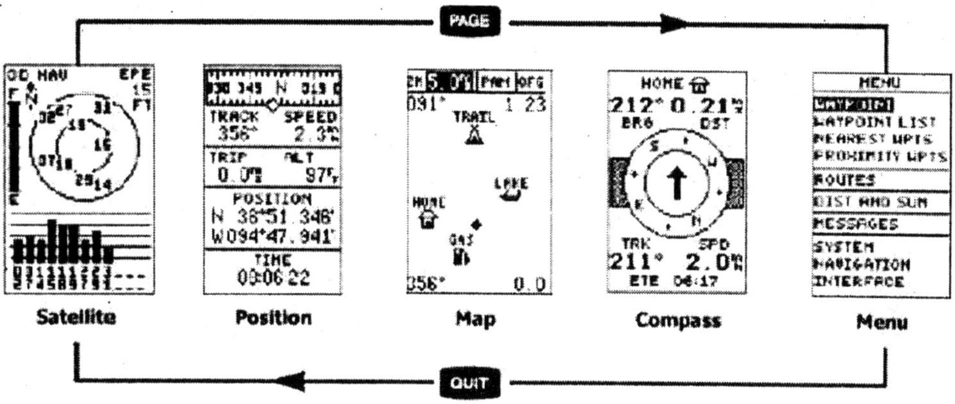
PAGE
Satellite
Position
Map
Compass
Menu
QUIT

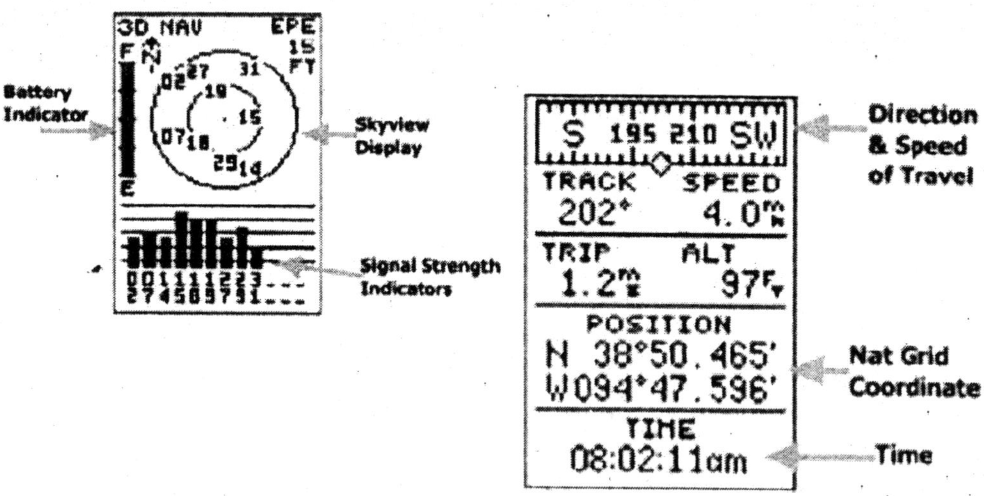
Battery Indicator
Skyview Display
Signal Strength Indicators
Direction & Speed of Travel
Nat Grid Coordinate
Time
3D NAV
EPE 15 FT
TRACK SPEED
202° 4.0
TRIP ALT
1.2 97
POSITION
N 38°50.465'
W094°47.596'
TIME
08:02:11am

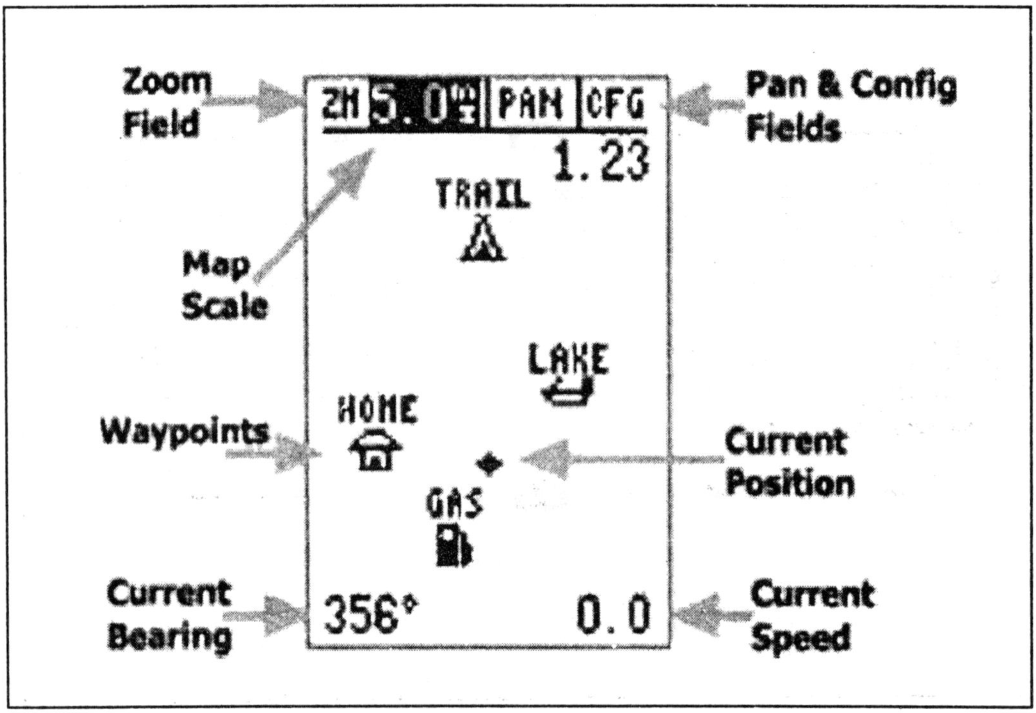
Zoom Field
ZM 5.0
PAN
CFG
Pan & Config Fields
1.23
TRAIL
Map Scale
LAKE
HOME
Waypoints
Current Position
GAS
Current Bearing
356°
0.0
Current Speed

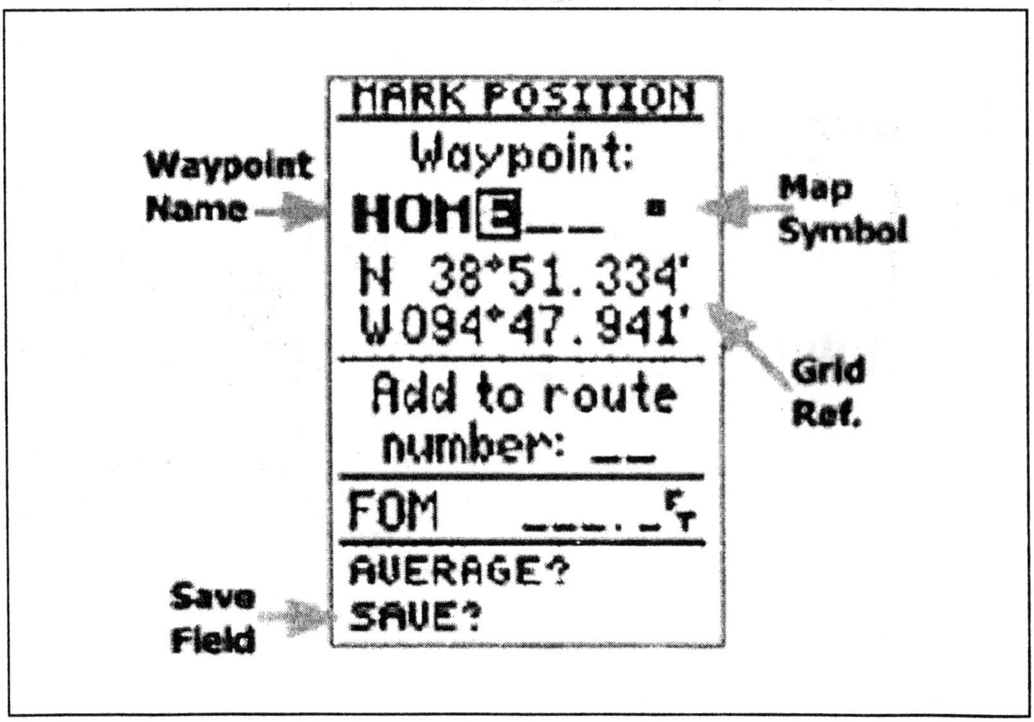
MARK POSITION
Waypoint:
Waypoint Name
HOME
Map Symbol
N 38°51.334'
W094°47.941'
Grid Ref.
Add to route number:
FOM
AVERAGE?
SAVE?
Save Field

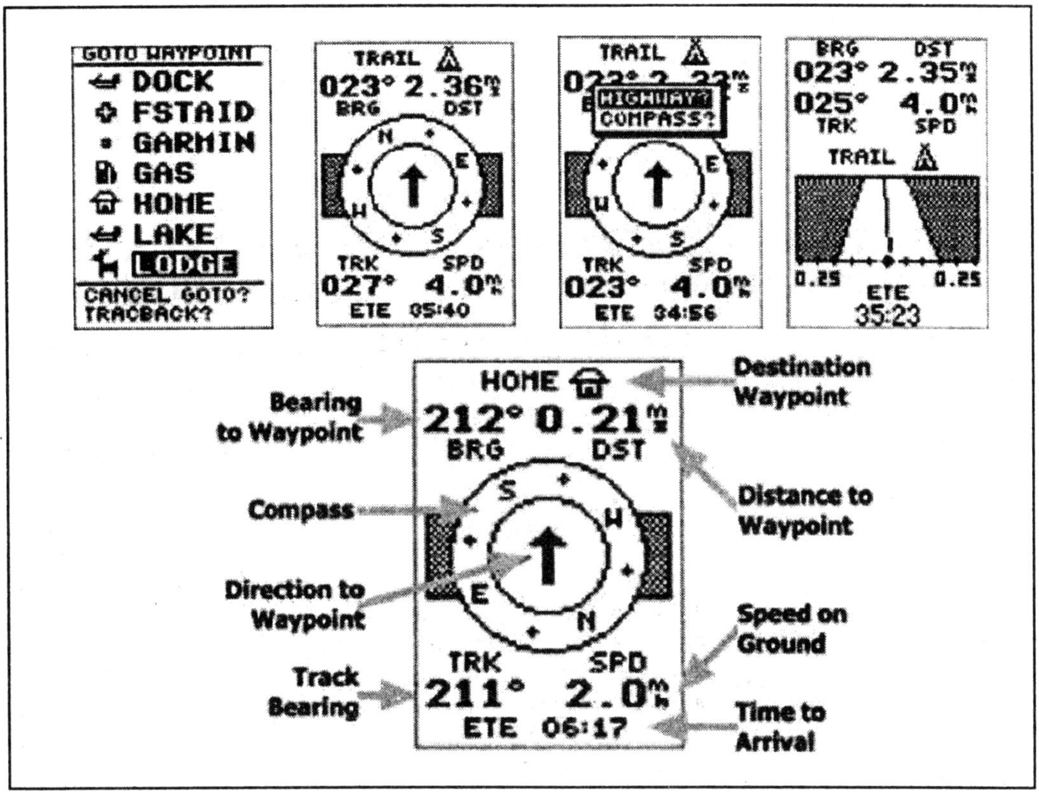

WWW REFERENCES

Navstar GPS Joint Program
http://gps.losangeles.af.mil/
GLONASS
http://www.rssi.ru/SFCSIC/SFCSIC_main.html
Galileo Program
http://europa.eu.int/comm/dgs/energy_transport/galileo/
GPS Satellite Tracking
http://sirius.chinalake.navy.mil/satpred/
Principles and Practices of GPS Surveying
http://www.gmat.unsw.edu.au/snap/gps/gps_survey/principles_gps.htm

32

MODELLING GENERATION, MANAGEMENT AND TRANSFORMATION OF RASTER DATA FOR LANDUSE STUDY

V.S.P. SINHA

Topics

- The Spatial Analyst extension
- Display and explore data
- Surface analysis
- Distance analysis
- Reclassification
- Raster Calculator
- Conversion
- Statistics
- Density
- Interpolations

Loading extension and toolbars

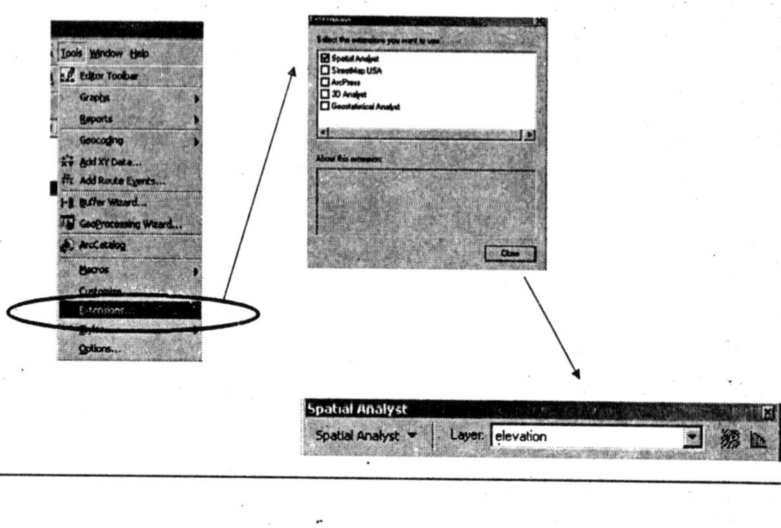

Topics

Display and explore data
- Grid data
- Temporary outputs (convert to permanent)
- Create a histogram
- Setting analysis properties
  - extent
  - cell size
  - apply analysis mask
- applying transparency
- open effect toolbar
- properties

Grid data

- A raster dataset consists of cells (pixels)
- Cells are grouped in rows and columns
- Each cell has a value
- Any two or more cells with the same value belong to the same *zone*
- Each group of connected cells in a zone is considered a *region*
- Cells can have the value NoData
- Floating point girds (continuous) and integer grids (discrete)
- Integer grids have attribute table including count

  (total number of cells in the dataset that have the same value)
- Resolution

Temporary outputs

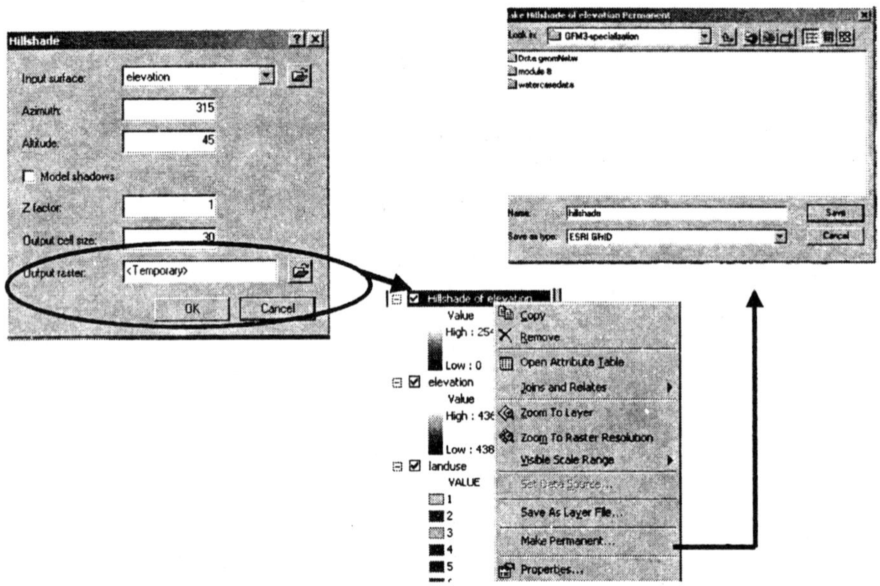

Histogram

- All values included
- only selected
- only pixels that intersect ....

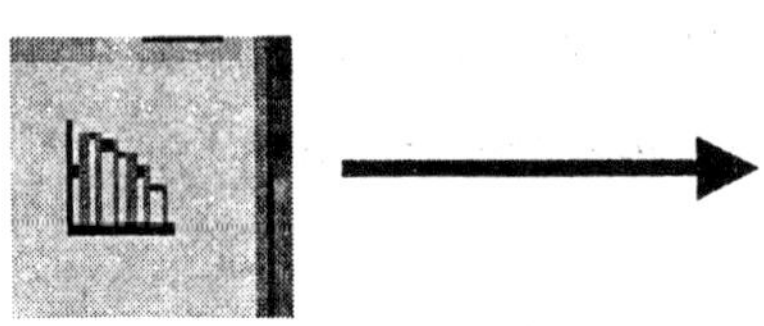

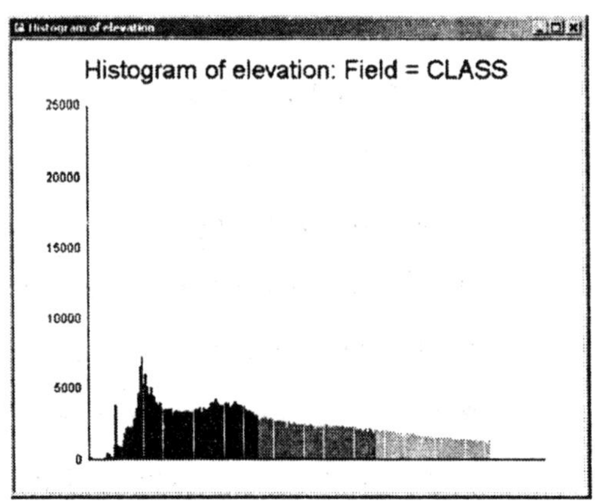

Histogram

Histogram for the features within the selected graphic

Analysis properties

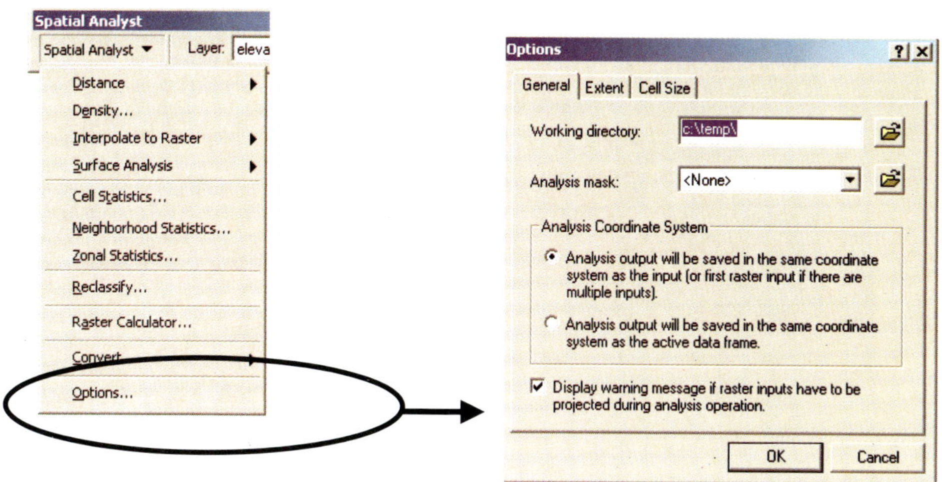

Analysis mask

Analysis mask: is a grid identifying cells within the analysis window that will be considered when and expression is evaluated. Masks are useful to exclude certain cells from evaluation (example: water bodies, everything outside study area)

Mask Input grid Output grid

Analysis Mask Applied

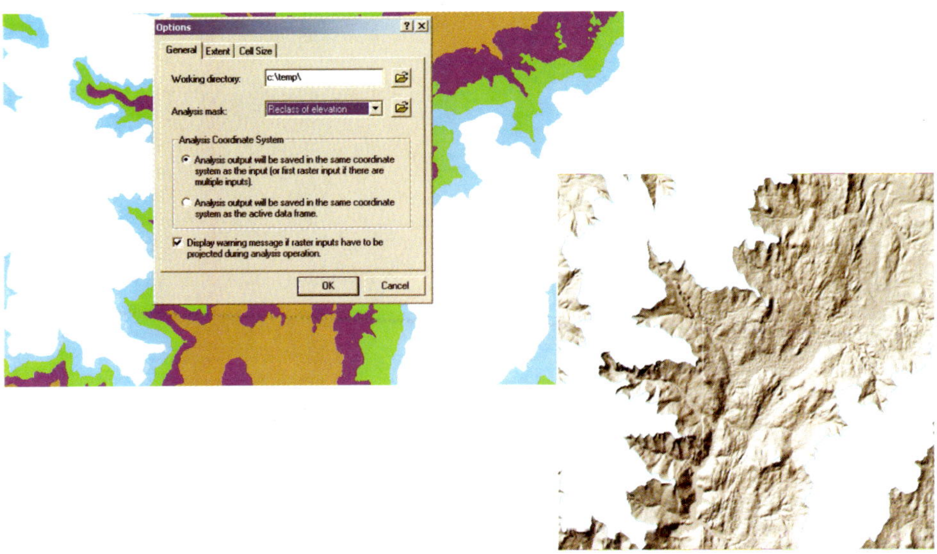

Analysis Extent

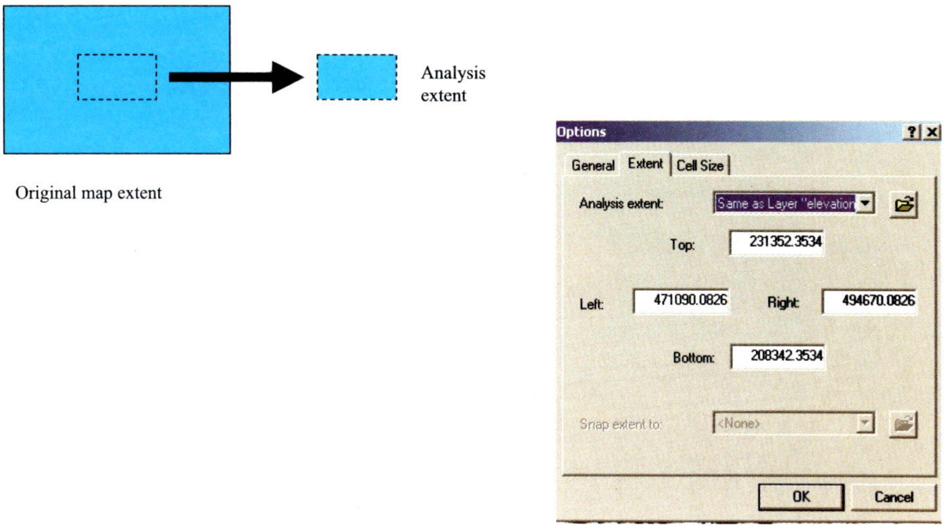

Output Cell size

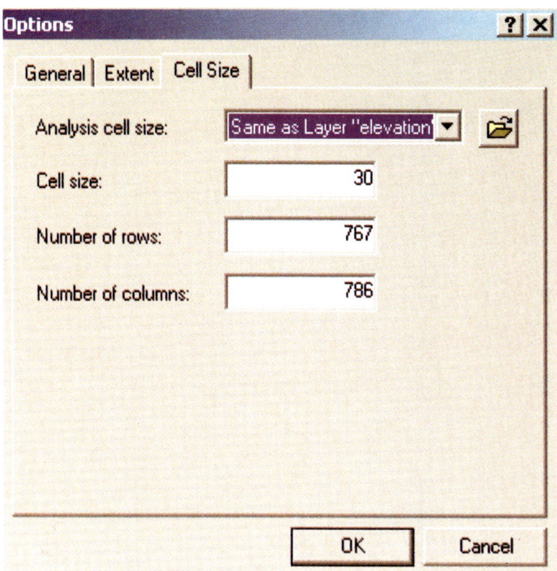

Transparency

Open the effect toolbar

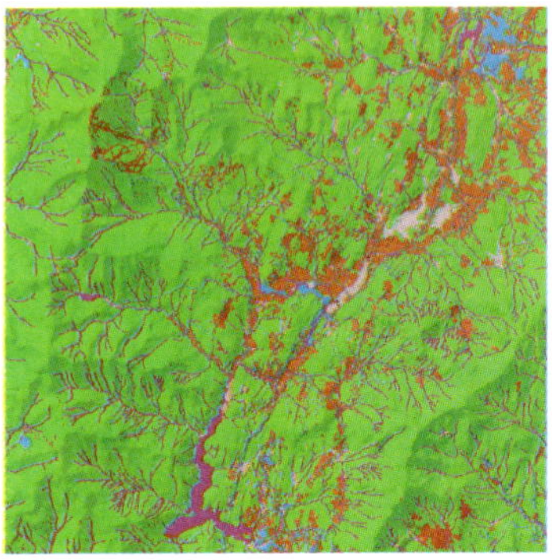

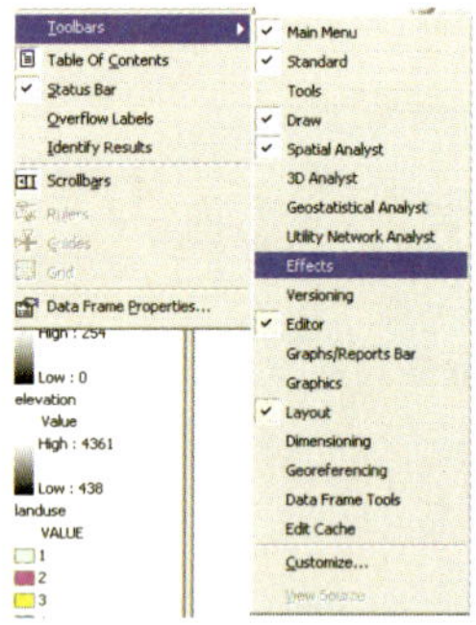

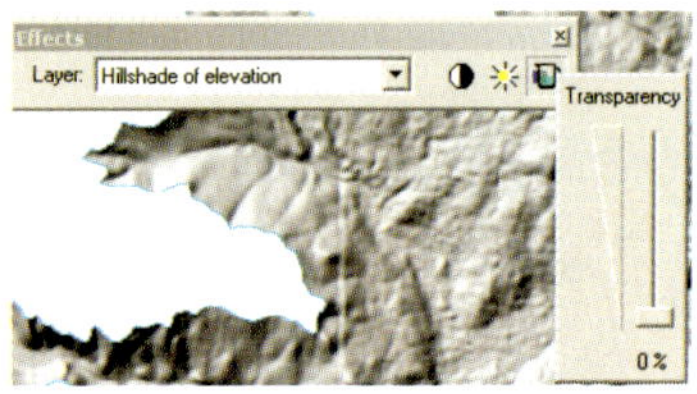

Topics

Surface analysis

- Contour
- Slope
- Aspect
- Hillshade
- Viewshed

Contour

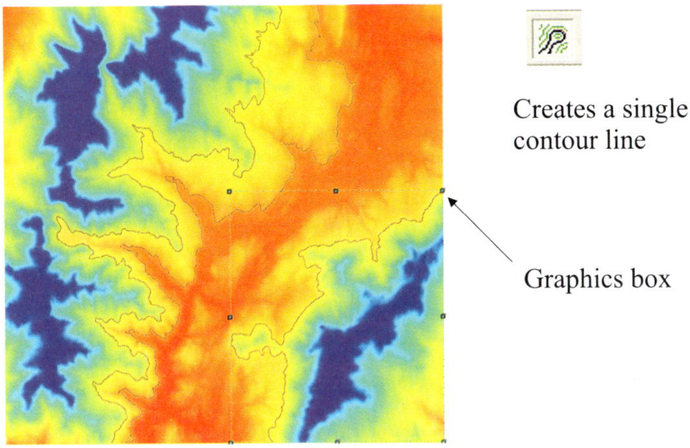

Contour

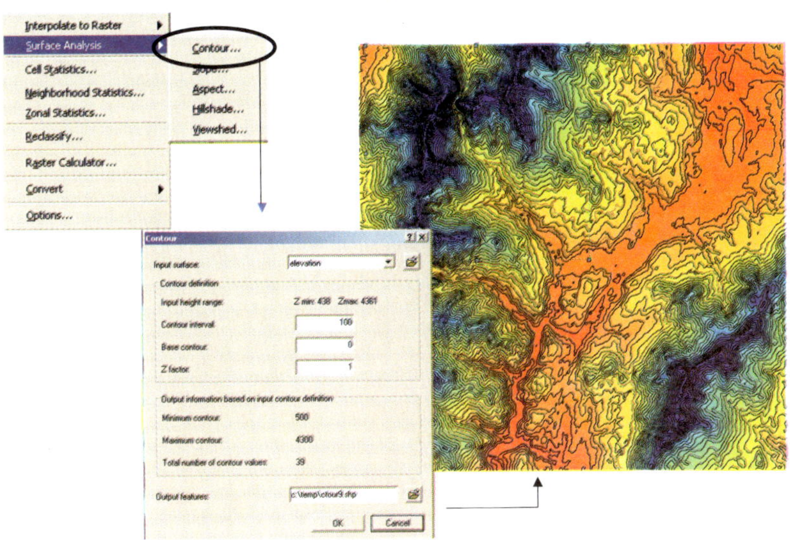

Slope

In degrees
in percentage

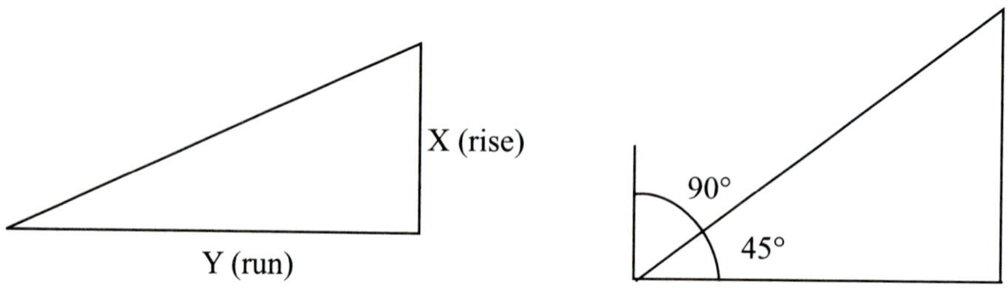

slope in percentage is X/Y *100

Slope

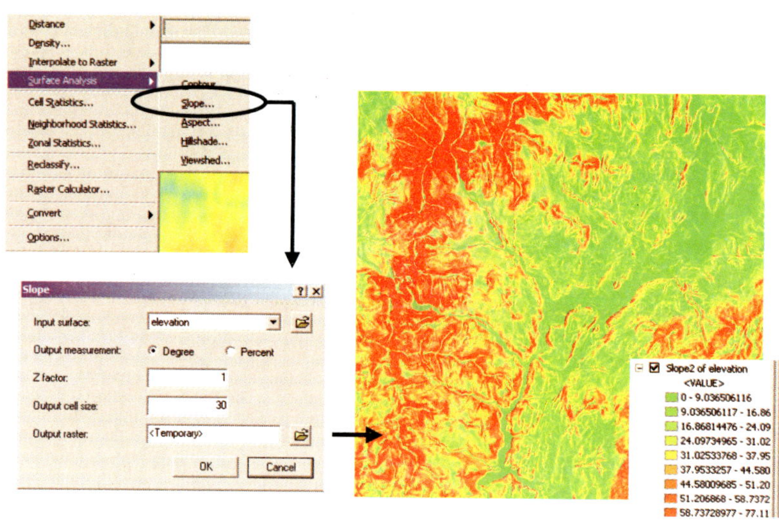

Aspect

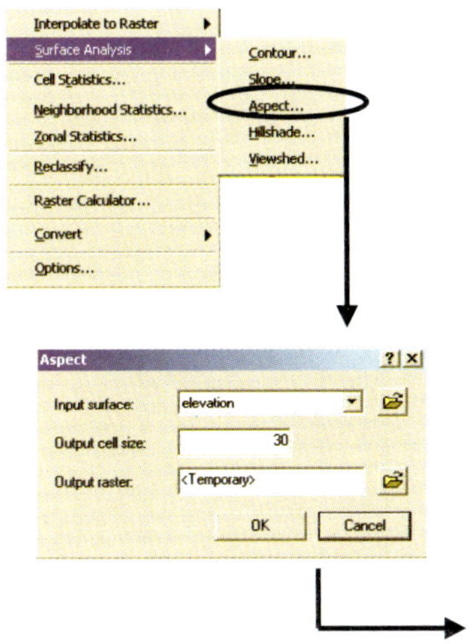

Aspect identifies the steepest down slope direction from each cell to its neighbours.

Hillshade

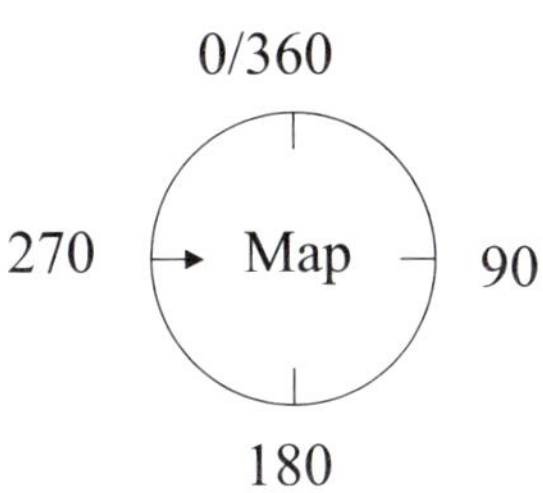

Azimuth: Orientation of the map towards the sun (in degrees)

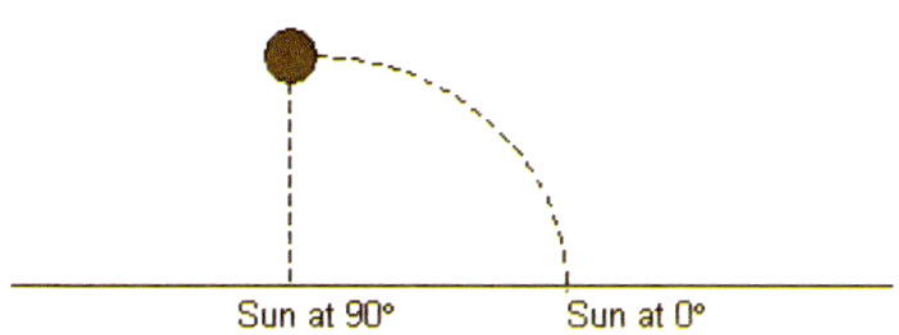

Altitude: height of the sun with 90° being straight above, and 0° being on the earth surface

Hillshade

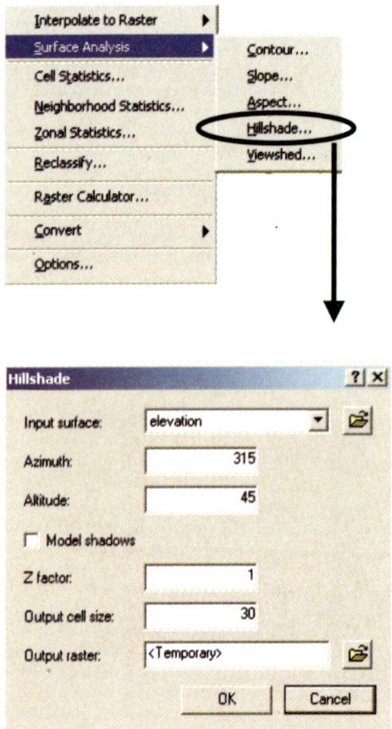

Hillshading is a visualization tool to display hills and valleys in relief

Viewshed

- Viewshed identifies the observation points that can be seen from each cell in the input raster.

- Each cell in the output raster receives a value that indicates how many observer points can be seen from each location. If you have only one observer point, each cell that can see that observer point is given a value of 1. All cells that cannot see the observer point are given a value of 0.

Viewshed

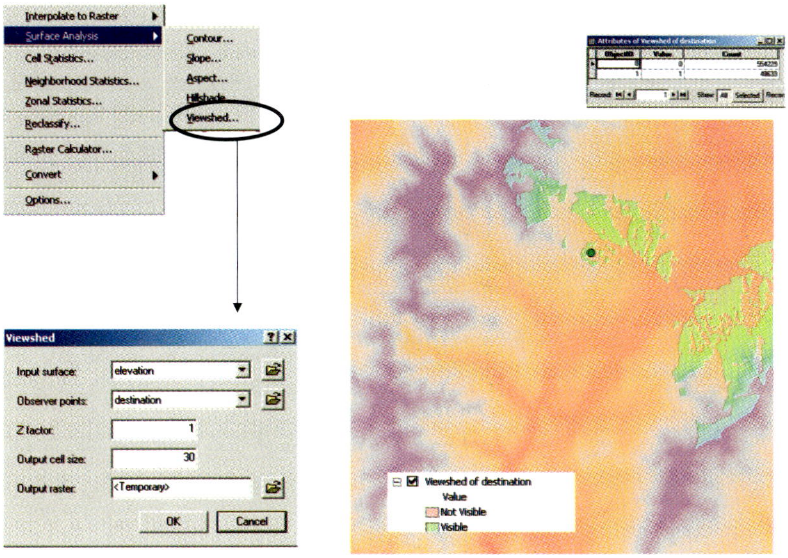

Topics

Distance analysis

- straight line distance (Euclidean Distance)
- Allocation (Thiessen polygons)
- Cost weighted
- Shortest Path

Euclidean Distance

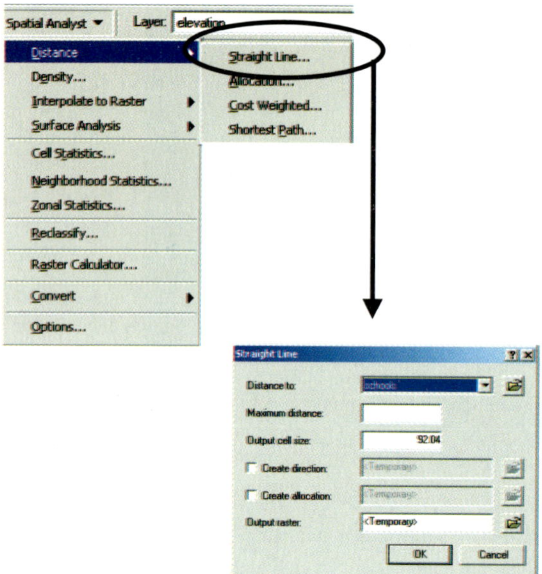

- The straight line distance function measures the straight line distance from each cell to the closest source (the source identifies the objects of interest such as wells, roads, or a school).
- The distance is measured from cell center to cell centre.

Euclidean Distance

- The source identifies the locations of the objects of interest. If the source is a raster, it must contain only the values of the source cells – all other cells must be NoData
- If the source is a feature, it will internally be transformed into a grid

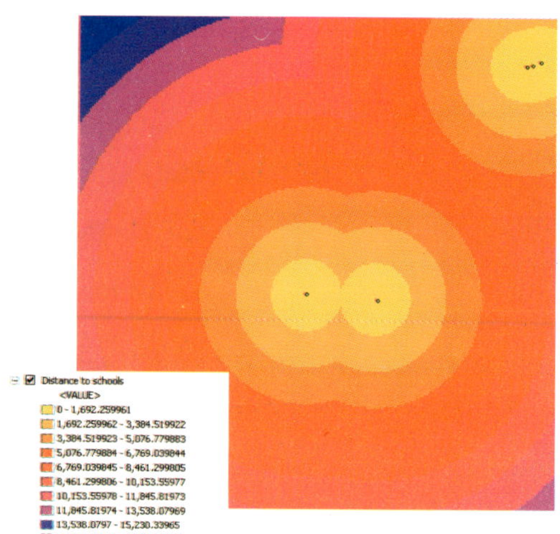

Allocation - Direction

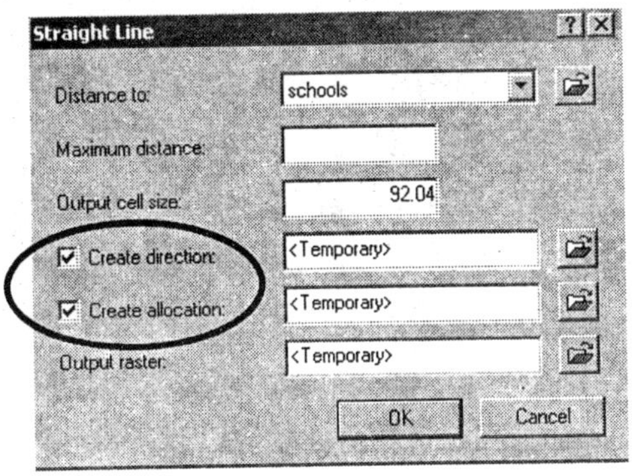

- Straight line allocation identifies the cells that are to be allocated to which source, based on proximity
- Straight Line direction gives the direction for each cell to the closest source

Direction grid

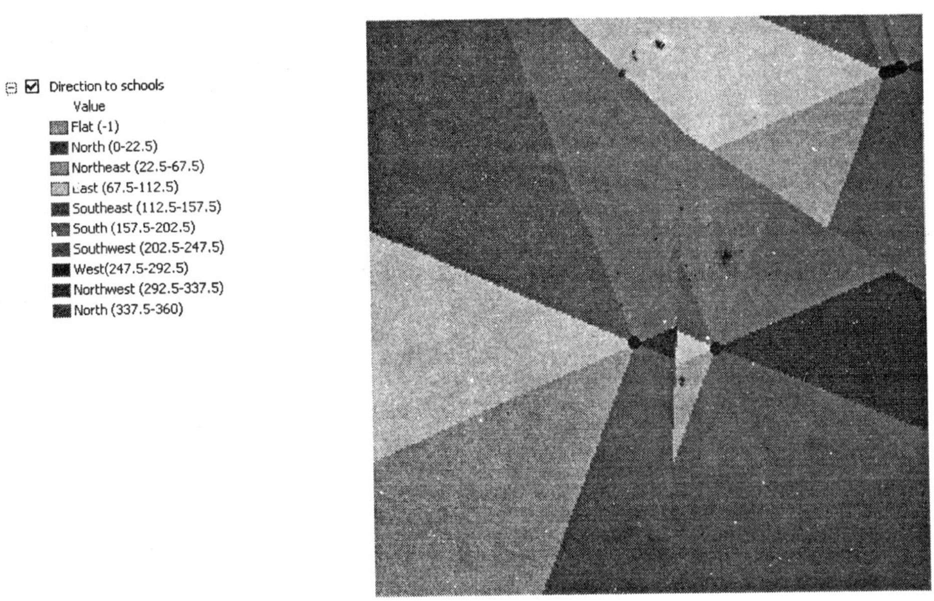

Allocation Grid

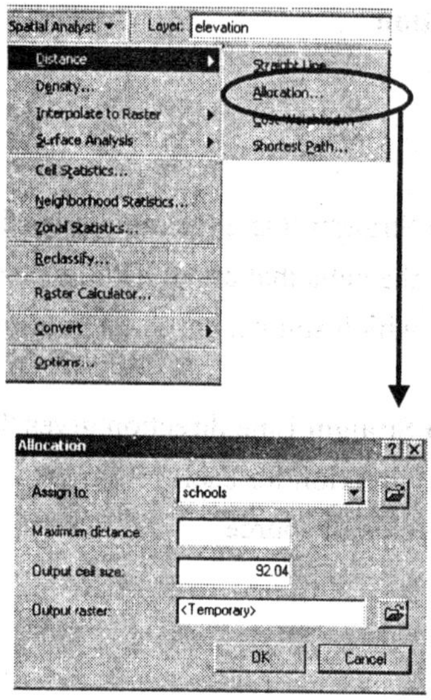

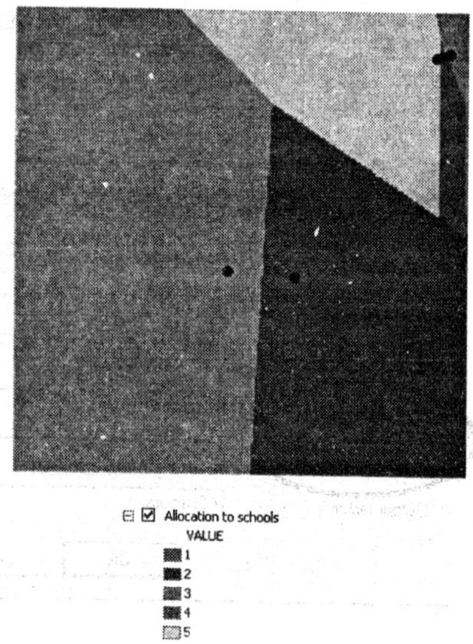

Cost Weighted

Cost weighted distance mapping finds the least accumulative cost from each cell to the nearest (cheapest) source.

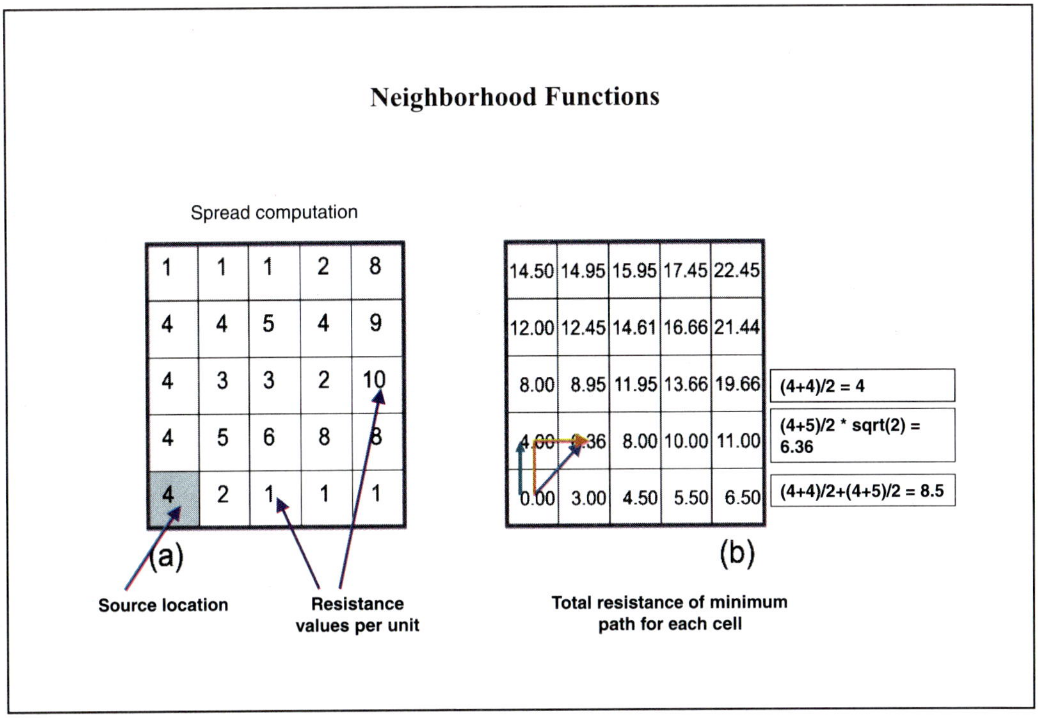
Neighborhood Functions
Spread computation
1 1 1 2 8
4 4 5 4 9
4 3 3 2 10
4 5 6 8 8
4 2 1 1 1
(a)
Source location
Resistance values per unit
14.50 14.95 15.95 17.45 22.45
12.00 12.45 14.61 16.66 21.44
8.00 8.95 11.95 13.66 19.66
4.00 6.36 8.00 10.00 11.00
0.00 3.00 4.50 5.50 6.50
(b)
Total resistance of minimum path for each cell
(4+4)/2 = 4
(4+5)/2 * sqrt(2) = 6.36
(4+4)/2+(4+5)/2 = 8.5

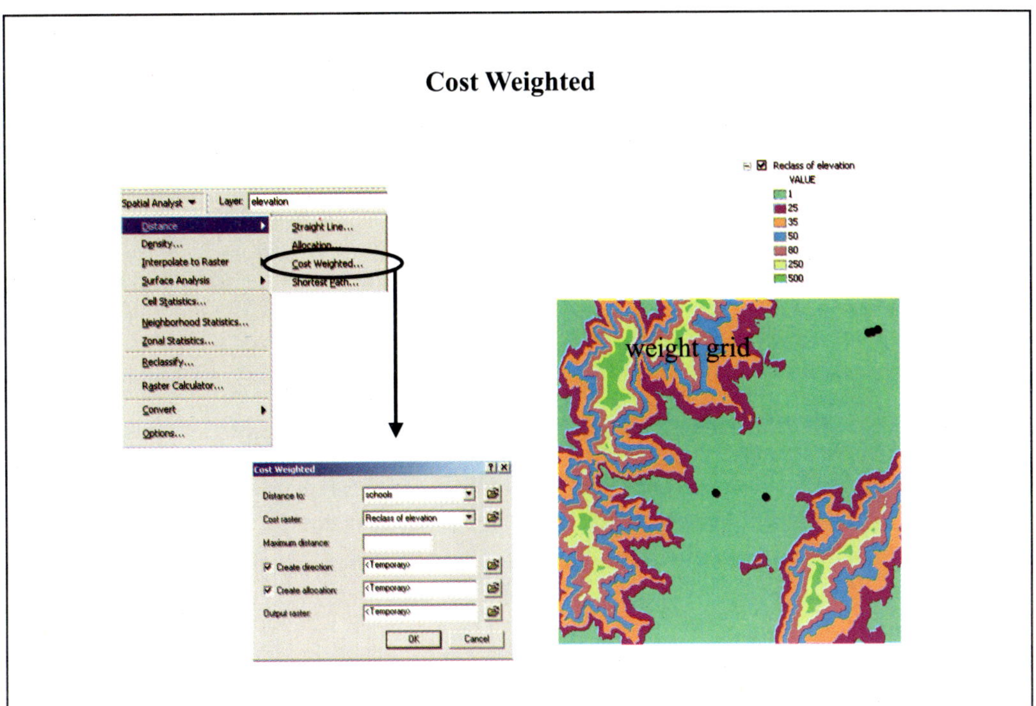
Cost Weighted
Spatial Analyst
Layer: elevation
Distance
Density...
Interpolate to Raster
Surface Analysis
Cell Statistics...
Neighborhood Statistics...
Zonal Statistics...
Reclassify...
Raster Calculator...
Convert
Options...
Straight Line...
Allocation...
Cost Weighted...
Shortest Path...
Cost Weighted
Distance to: schools
Cost raster: Reclass of elevation
Maximum distance:
Create direction: <Temporary>
Create allocation: <Temporary>
Output raster: <Temporary>
OK
Cancel
Reclass of elevation
VALUE
1
25
35
50
80
250
500
weight grid

Cost Weighted

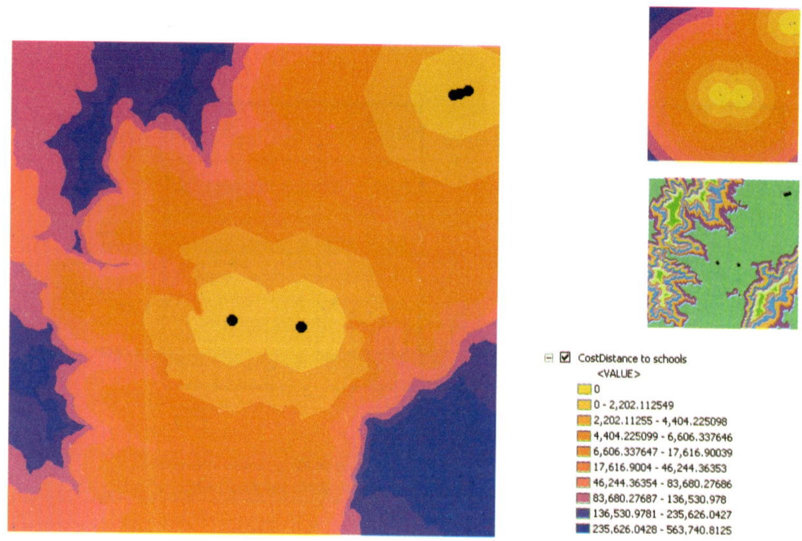

Cost Weighted Direction Grid

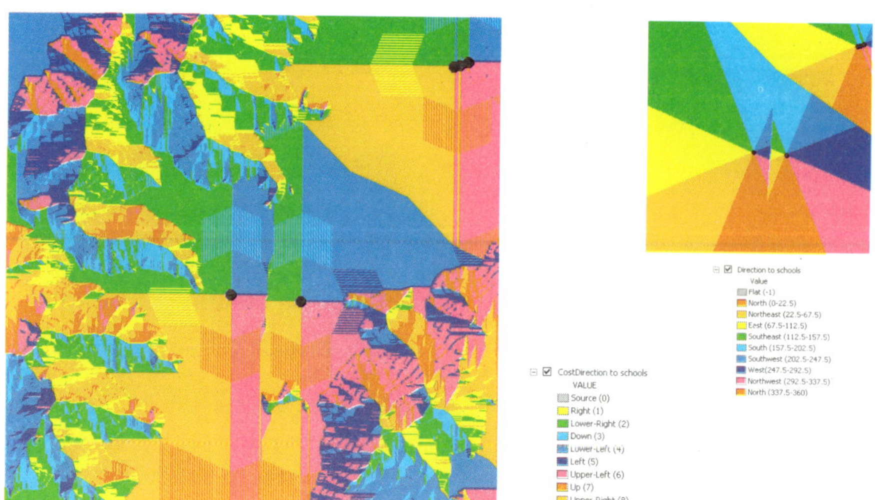

Cost Weighted Allocation Grid

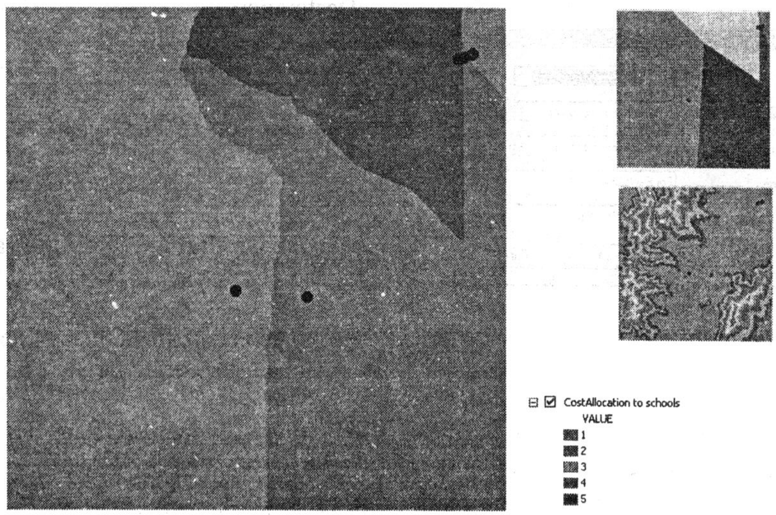

Shortest Path

- The shortest Path function determines the path from a destination point to a source.

  Before you can create a Shortest Path you first have to perform the Cost Weighted distance function (including direction and distance grid)

Shortest Path

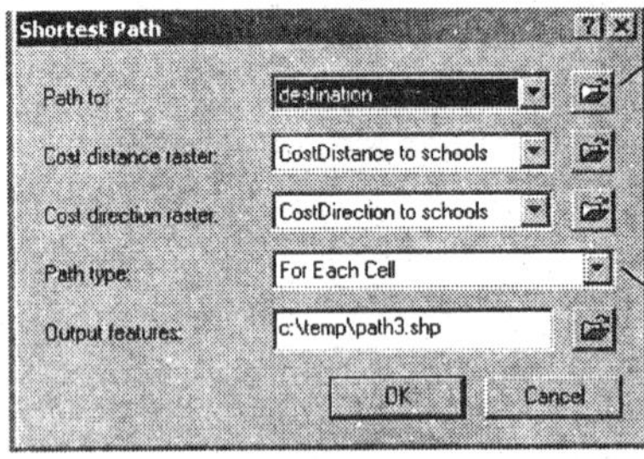

Destination

For your source

- For each cell : finds a path to each cell in each zone
- For each zone: finds the one, least-cost path for each zone
- The best Single path: finds the least-cost path for all zones

Shortest Path

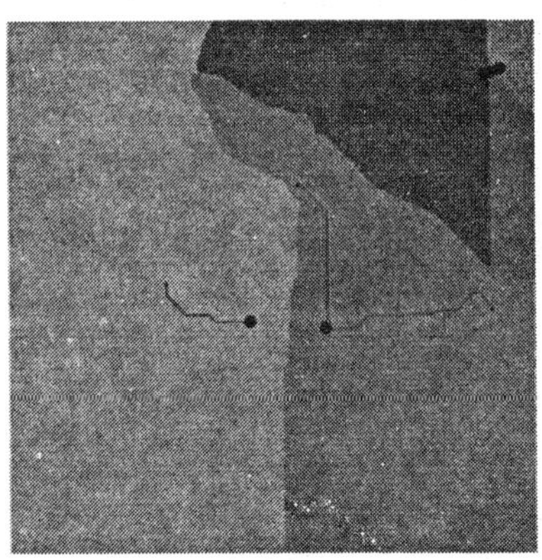

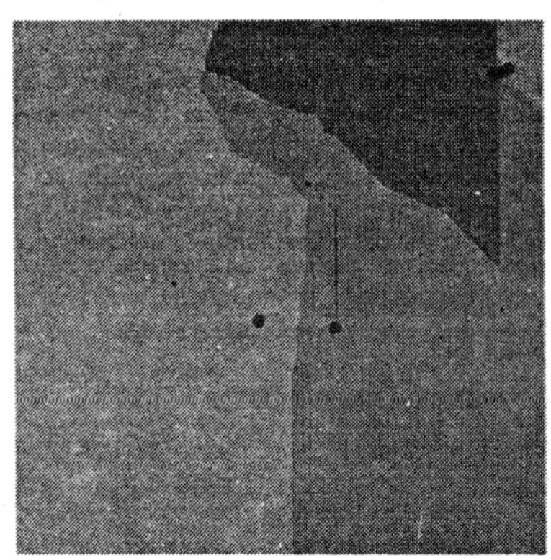

For Each Cell

Best Single

Topics

- Reclassification
- Raster Calculator
- Conversion
- Statistics
- Density
- Interpolations

Reclassify

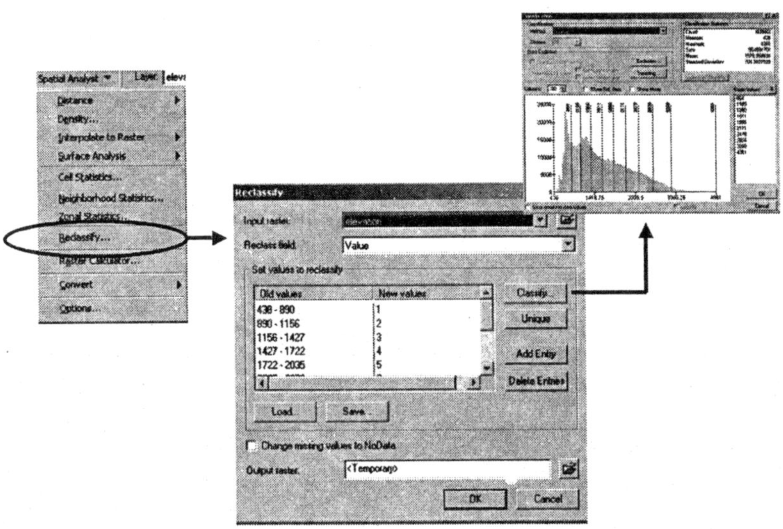

Reclassify

- Change the new values
- Delete classes
- Group entries (select more, right click, select group)
- Change classification method, number of classes
- Set values to NoData

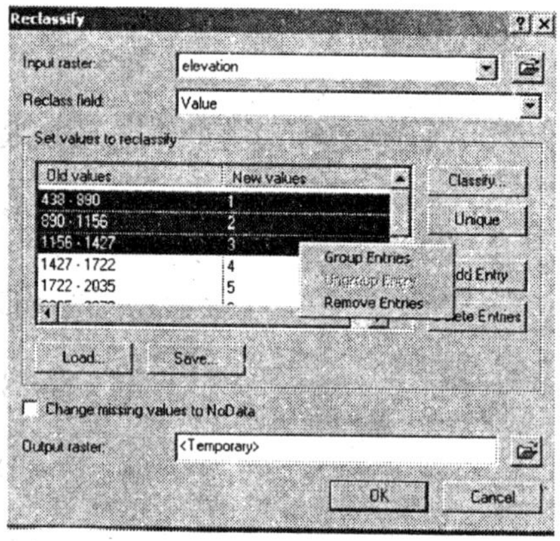

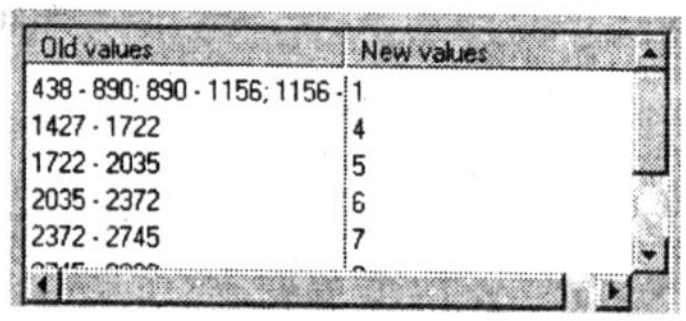

33

GIS MODELLING FOR LANDUSE STUDY

SANDEEP GOYAL

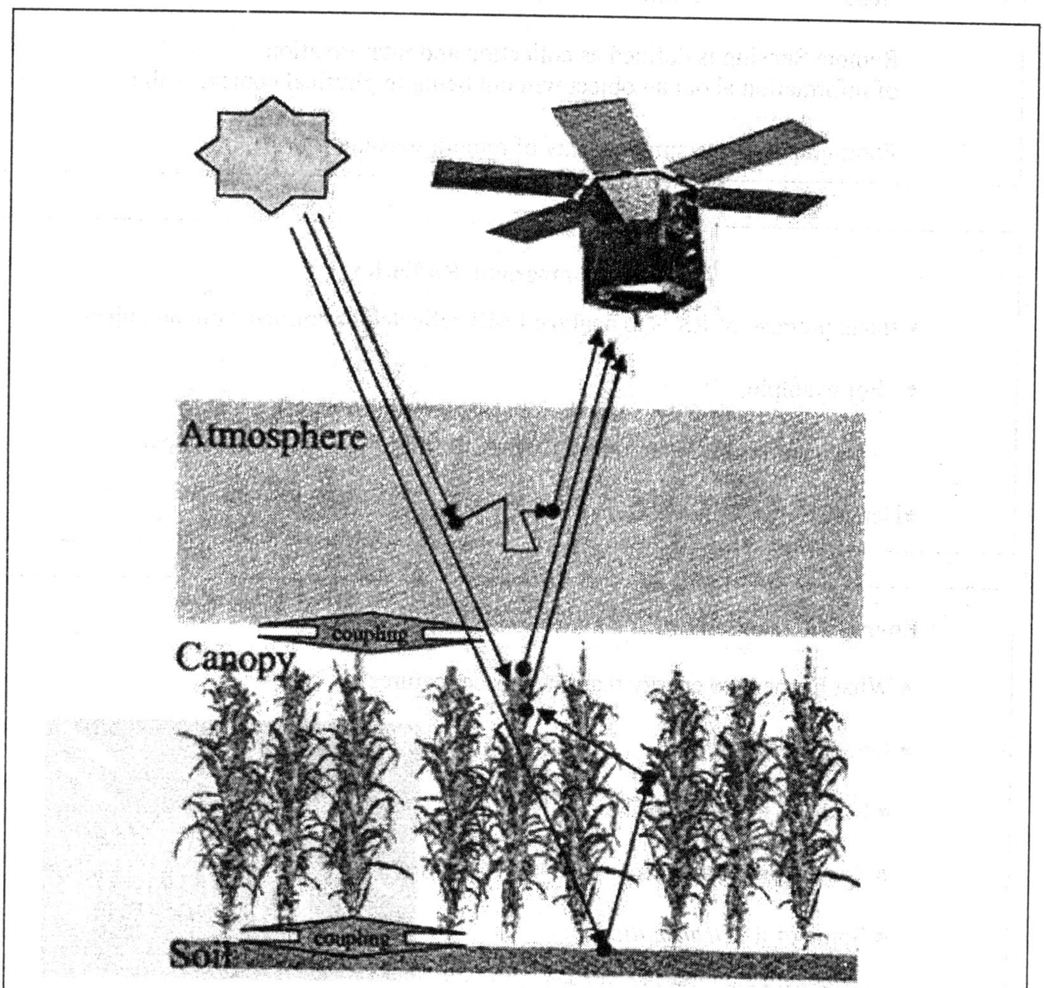

Sensing Can be Carried Out by

Smell
Hear
Vision
Taste
Touch

VISION will qualify for Remote Sensing

VISION	REMOTE SENSING
Light	EMR
Eye	Sensor
Head	Platform

Remote Sensing is defined as collection and interpretation of information about an object without being in physical contact with it.

Photography is the simplest way of remote sensing.

Electromagnetic Radiation

- Basic premise of RS is to analyze EMR reflected or emitted from an object.
- For example,
  - because visible light is being reflected from the screen into our eyes.
- Hence visible light is a form of electromagnetic energy.

Energy Interactions with Earth Surface Features

- What happens to energy that hit surface features?
- One of three things
  - Some of it is *absorbed*
  - Some of it is *reflected*
  - Some of it is *transmitted*
- This is known as *radiative transfer*

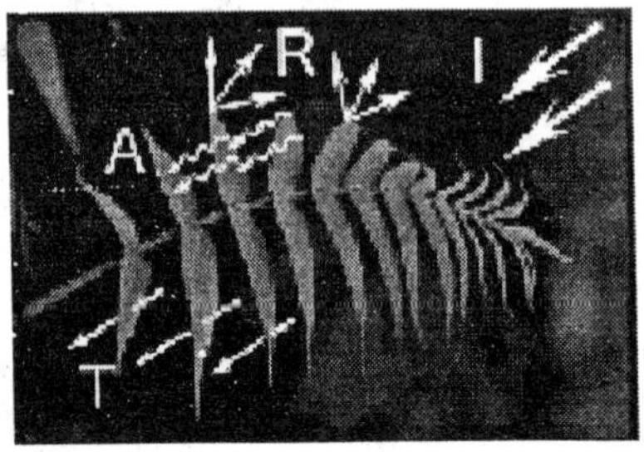

Active Remote Sensing: It generates its own source of energy, e.g. RADAR

Passive Remote Sensing: Use external energy source (Solar) for data capturing, e.g. Satellite remote sensing, Aerial photography

Satellites
Geo-stationary: Altitude 36,000 Km, views a portion of globe (INSAT series, NOAA series)

Sun-Synchronous: Altitude 700-900 km, covers entire earth (IRS, LANDSAT, SPOT, JERS, ERS, IKONOS etc.)

Stages
Design, Launch, Operate and Manage

Satellite Data

- Available in
  - Digital form
  - B&W photographs
  - Colour Photographs

What is an Image?

- Picture
- Photograph
- Visual data
  - Usually two or three dimensional

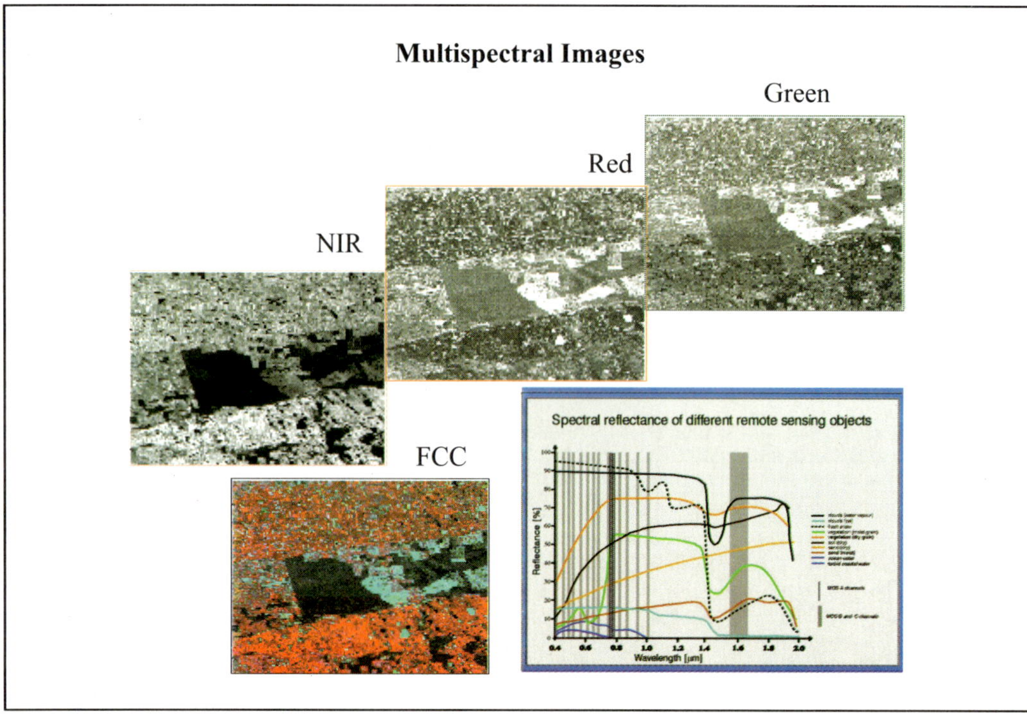
Multispectral Images
Green
Red
NIR
FCC
Spectral reflectance of different remote sensing objects
Reflectance [%]
Wavelength [μm]

143/47
TM L5
19.11.85

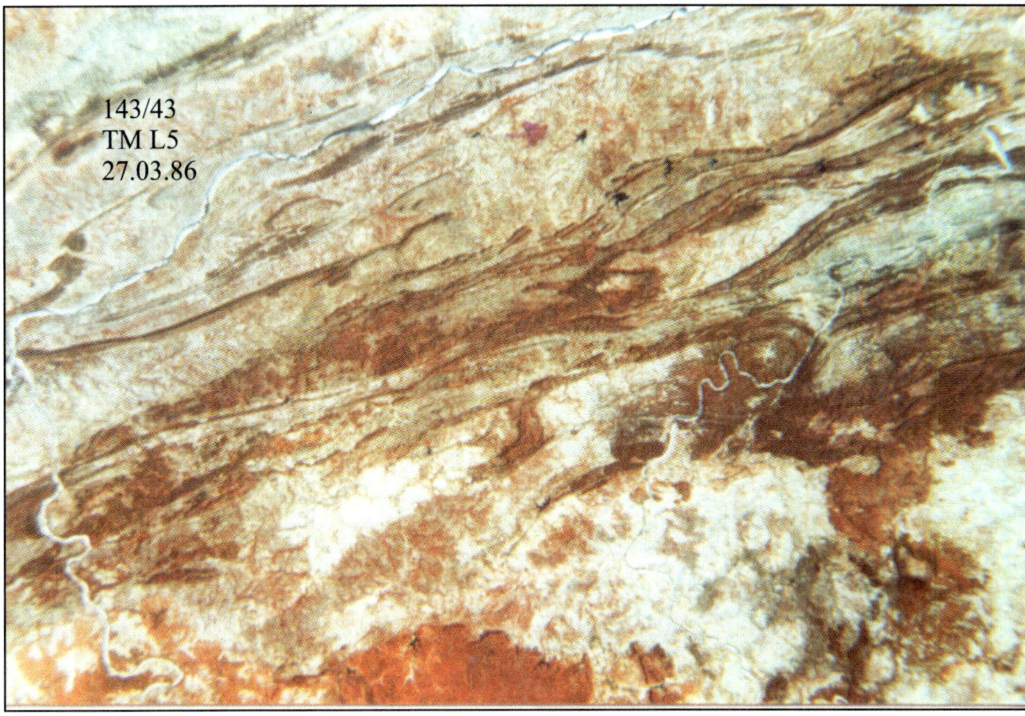
143/43
TM L5
27.03.86

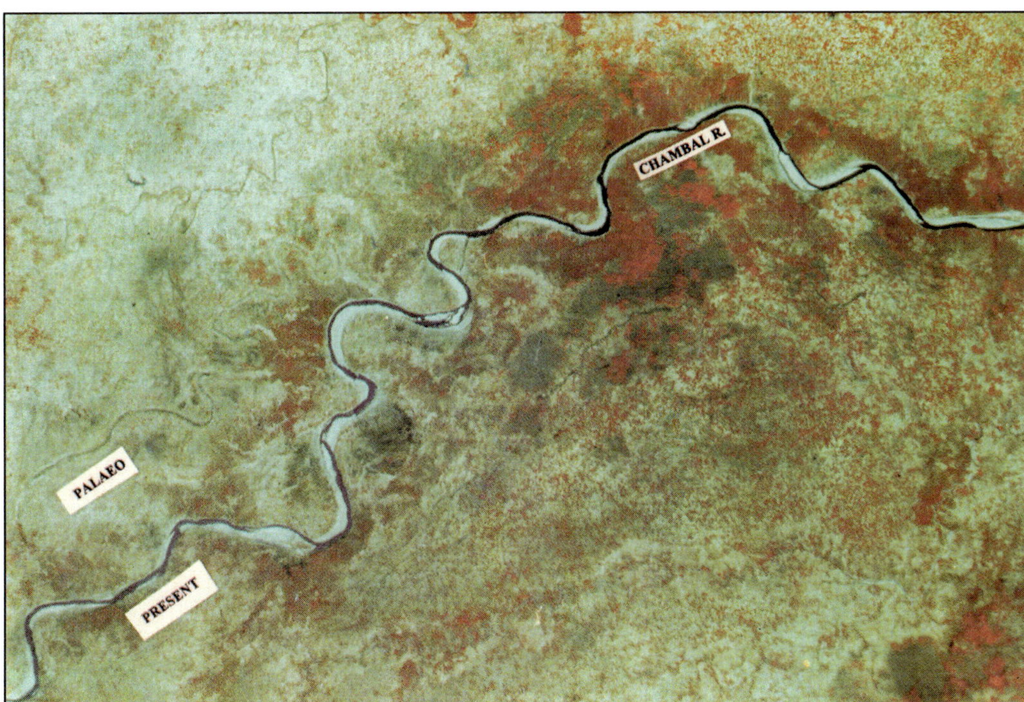
CHAMBAL R.
PALAEO
PRESENT

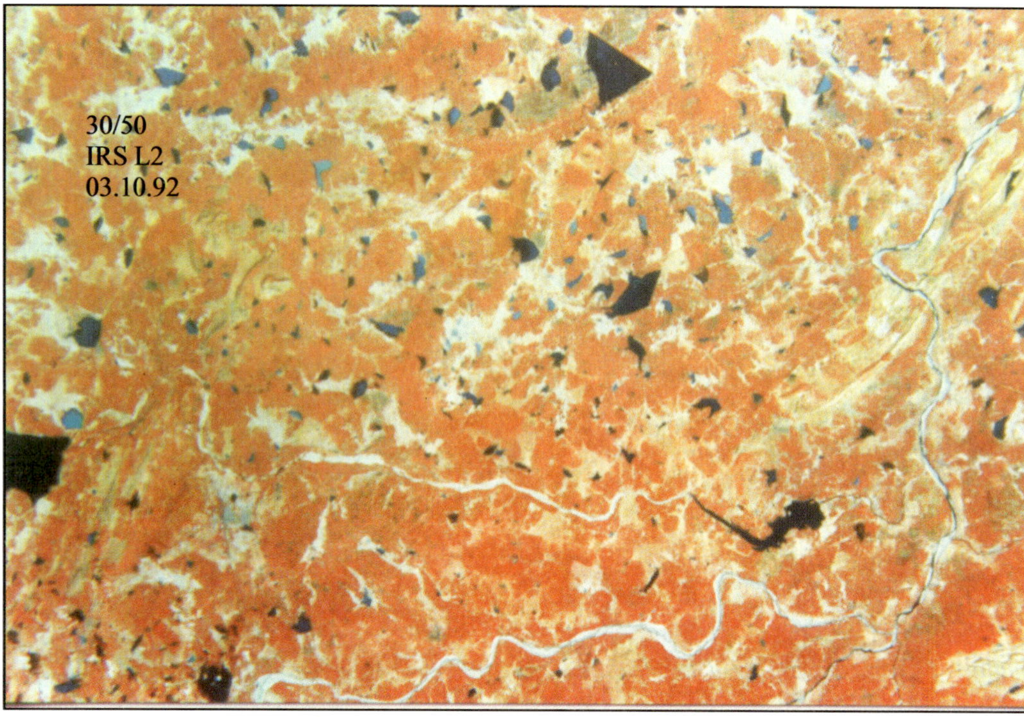
30/50
IRS L2
03.10.92

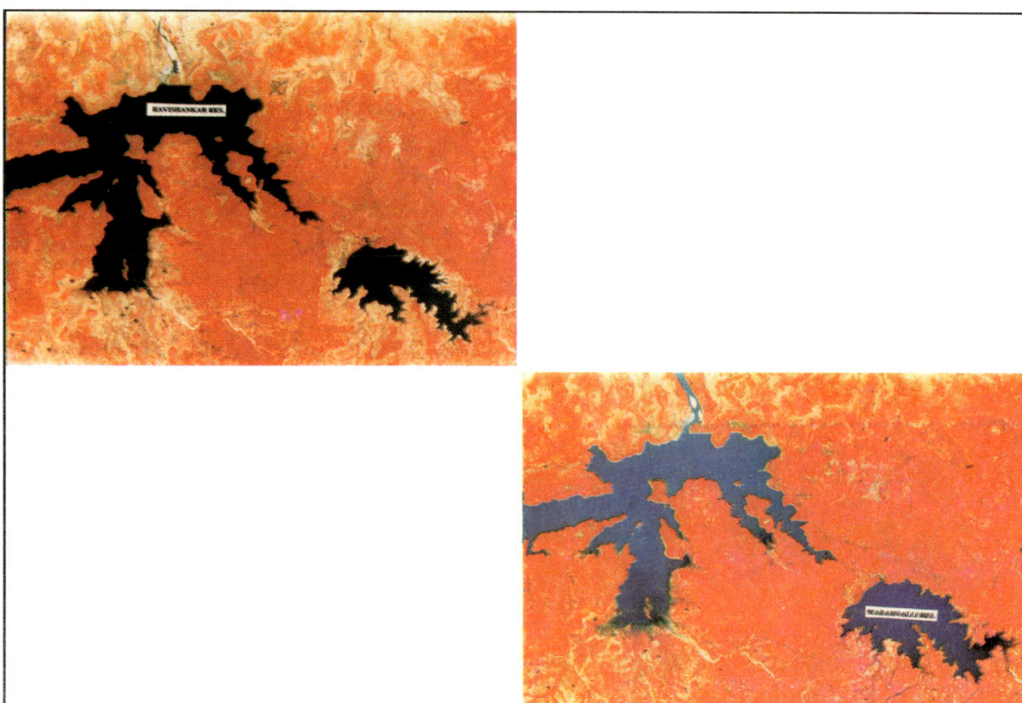

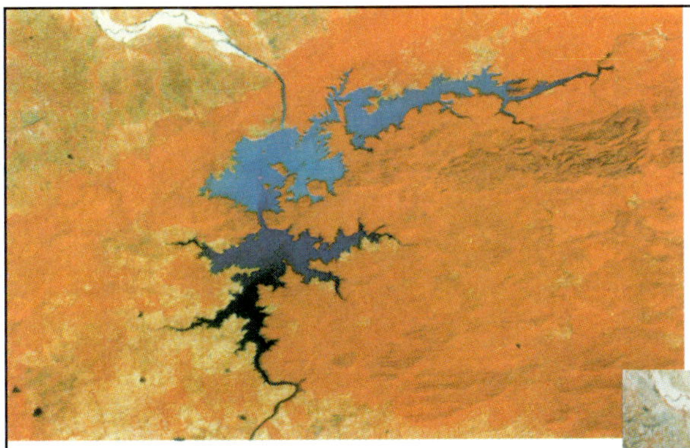

NUMBER OF WETLANDS WITH AREAL EXTENT > 56.25 HA		
TYPE	NUMBER	AREA IN HA.
NATURAL LAKES	22	5409
RESERVOIR	227	266864
TANKS	157	21655
ASH/COOLING POND	1	190
TOTAL	407	294118

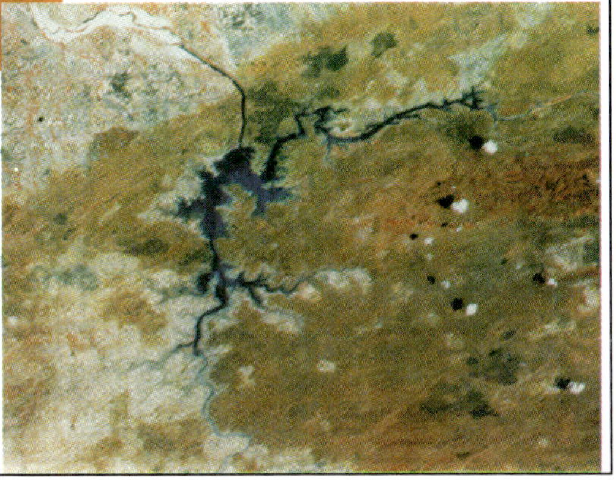

NUMBER OF WETLANDS HAVING TURBIDITY	
INTENSITY	NUMBER
HIGH	74
MODERATE	124
LOW	218

NUMBER OF WETLANDS HAVING AQUATIC VEGETATION	
TYPE	NUMBER
NATURAL LAKE	3
RESERVOIR	23
TANKS	9

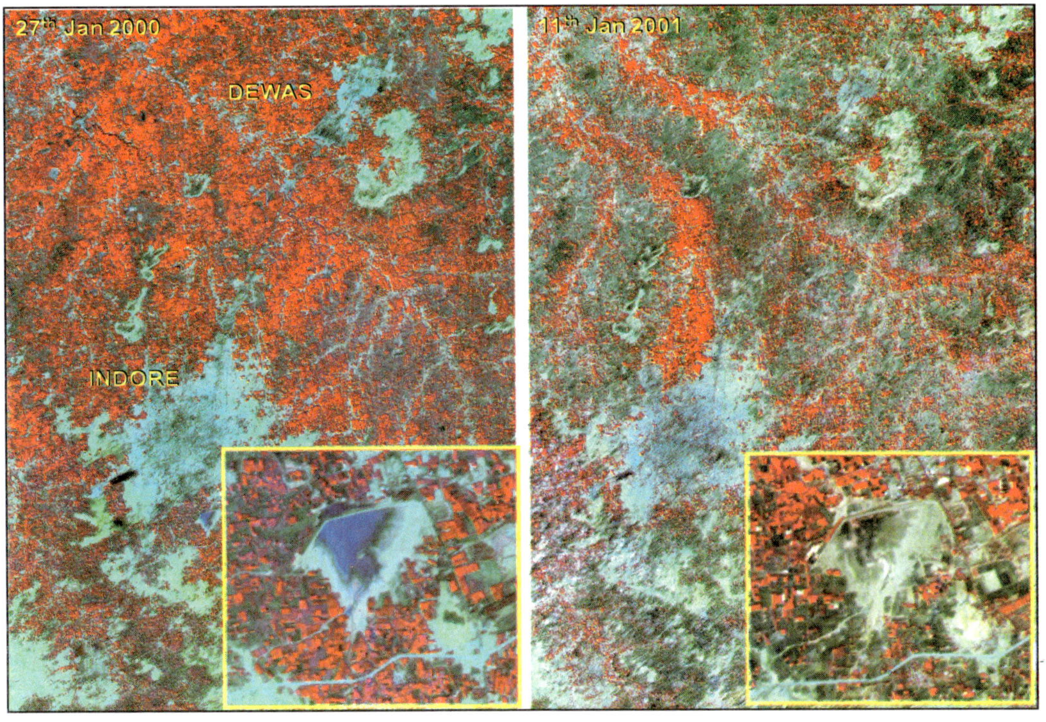

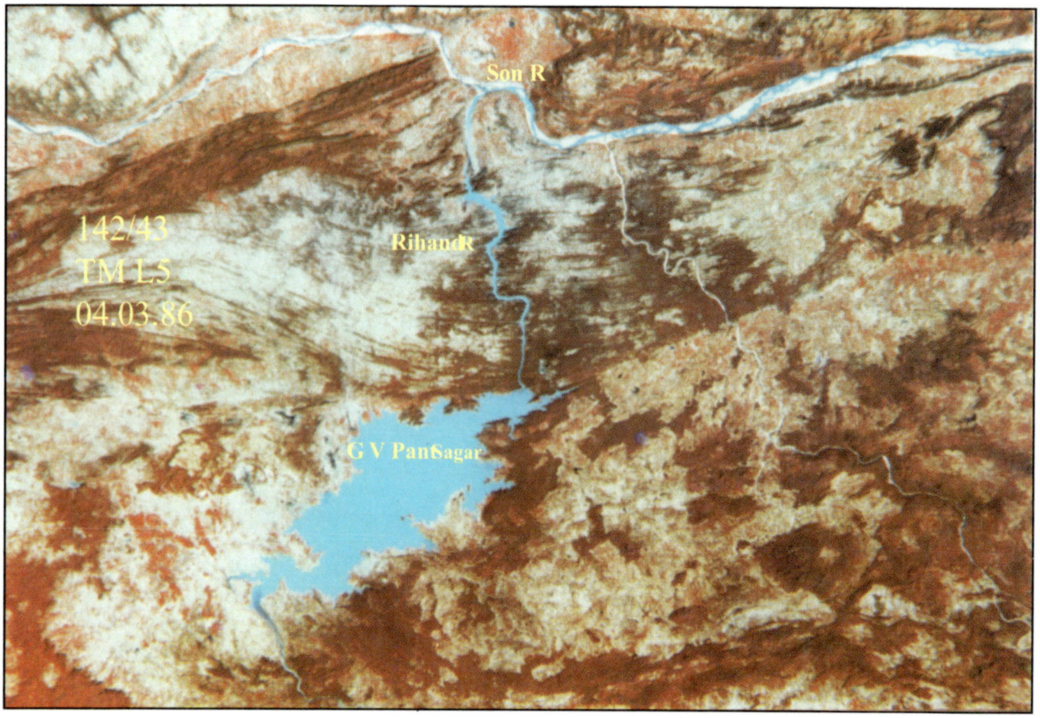

IRS LISS III IMAGE OF BHOPAL AND SURROUNDING AREA

IRS-1D LISS-III DATA (23.5 M RESOLUTION)

BHOPAL AS SEEN ON 25th NOV 1998

IRS-1D (LISSIII+PAN) MERGED SATELLITE IMAGE

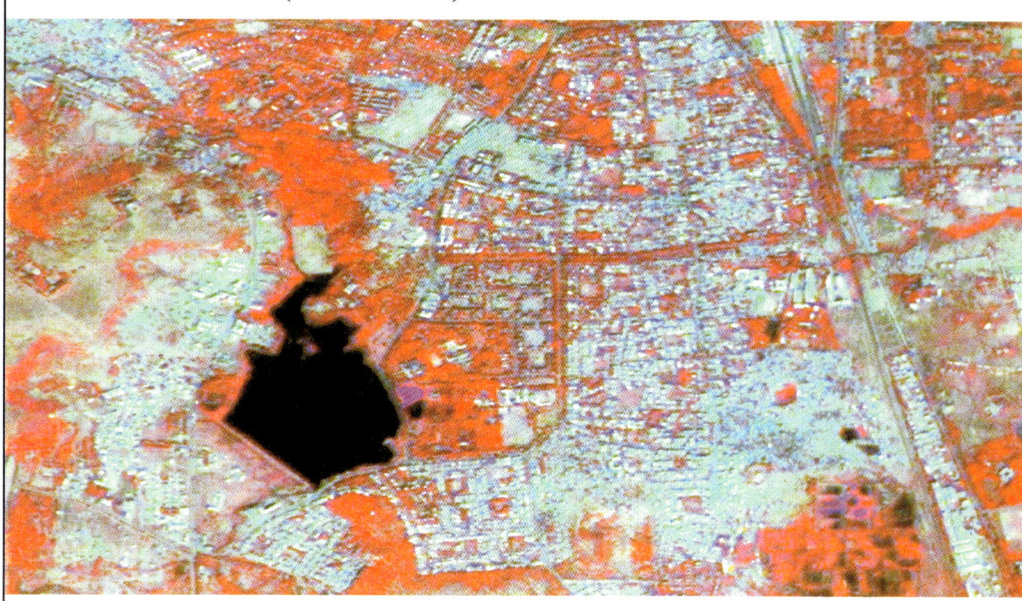

ACQUIRED ON 14TH NOV 2002, RESOLUTION 6M

IRS-1C PAN DATA (5.8 M RESOLUTION)

BHOPAL AS SEEN ON 11th NOV 1998

STATUS OF BHOPAL LAKES

NOV 25, 1998

OCT 16, 1999

MAR 24, 2001

FEB 12, 2002

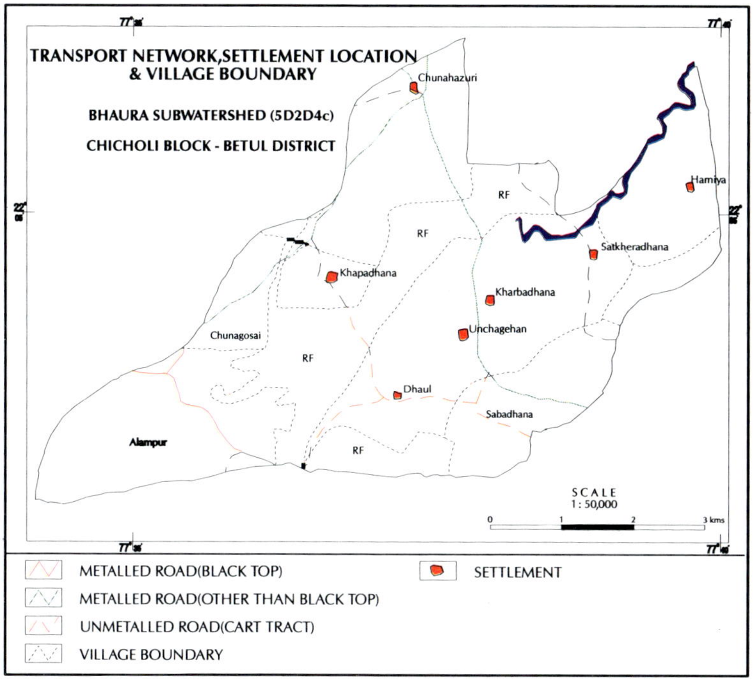
TRANSPORT NETWORK,SETTLEMENT LOCATION
& VILLAGE BOUNDARY
BHAURA SUBWATERSHED (5D2D4c)
CHICHOLI BLOCK - BETUL DISTRICT
Chunahazuri
Hamiya
RF
Satkheradhana
Khapadhana
Kharbadhana
Unchagehan
Chunagosai
Dhaul
Sabadhana
Alampur
SCALE
1 : 50,000
0
1
2
3 kms
METALLED ROAD(BLACK TOP)
SETTLEMENT
METALLED ROAD(OTHER THAN BLACK TOP)
UNMETALLED ROAD(CART TRACT)
VILLAGE BOUNDARY

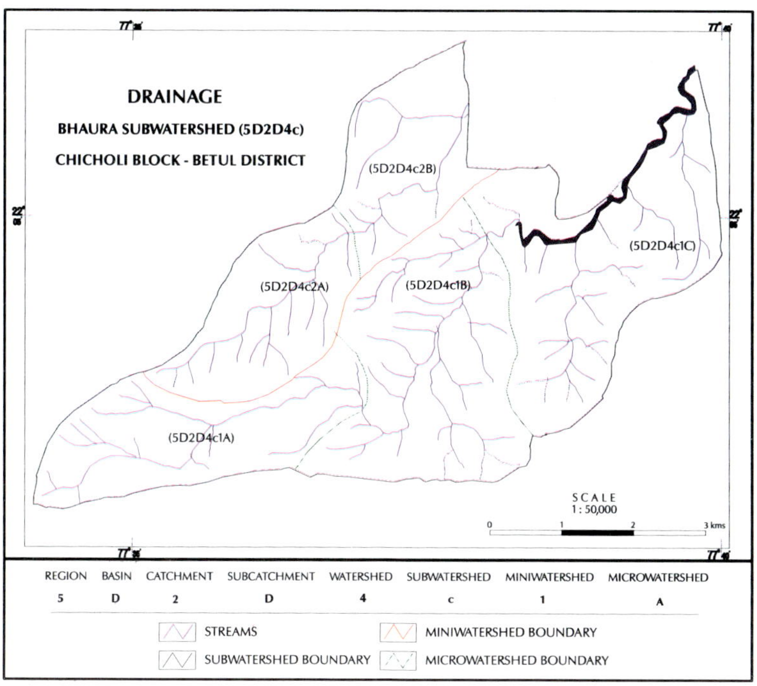
DRAINAGE
BHAURA SUBWATERSHED (5D2D4c)
CHICHOLI BLOCK - BETUL DISTRICT
(5D2D4c2B)
(5D2D4c1C)
(5D2D4c2A)
(5D2D4c1B)
(5D2D4c1A)
SCALE
1 : 50,000
0
1
2
3 kms
REGION BASIN CATCHMENT SUBCATCHMENT WATERSHED SUBWATERSHED MINIWATERSHED MICROWATERSHED
5 D 2 D 4 c 1 A
STREAMS
MINIWATERSHED BOUNDARY
SUBWATERSHED BOUNDARY
MICROWATERSHED BOUNDARY

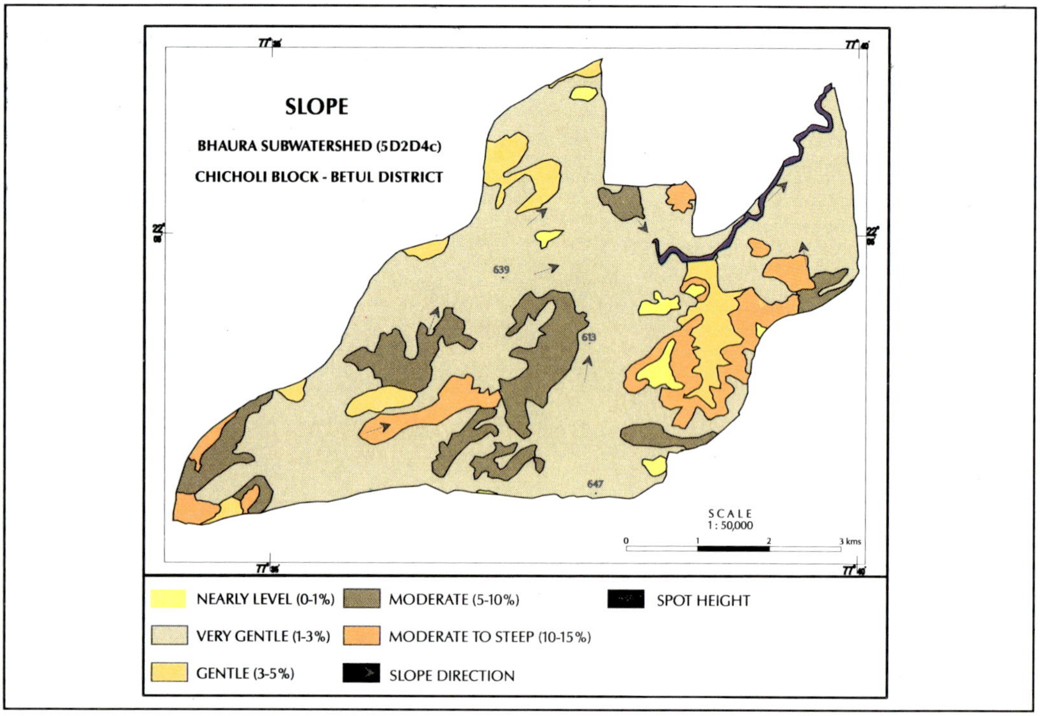
SLOPE
BHAURA SUBWATERSHED (5D2D4c)
CHICHOLI BLOCK - BETUL DISTRICT
639
613
647
SCALE
1 : 50,000
0
1
2
3 kms
NEARLY LEVEL (0-1%)
MODERATE (5-10%)
SPOT HEIGHT
VERY GENTLE (1-3%)
MODERATE TO STEEP (10-15%)
GENTLE (3-5%)
SLOPE DIRECTION

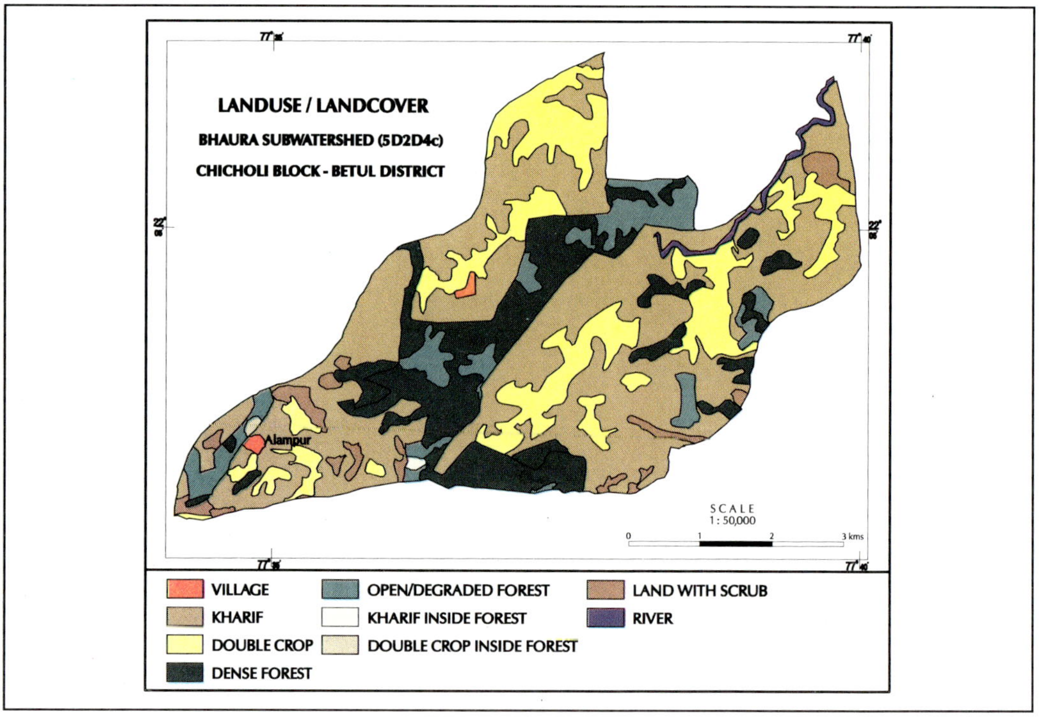
LANDUSE / LANDCOVER
BHAURA SUBWATERSHED (5D2D4c)
CHICHOLI BLOCK - BETUL DISTRICT
Alampur
SCALE
1 : 50,000
0
1
2
3 kms
VILLAGE
OPEN/DEGRADED FOREST
LAND WITH SCRUB
KHARIF
KHARIF INSIDE FOREST
RIVER
DOUBLE CROP
DOUBLE CROP INSIDE FOREST
DENSE FOREST

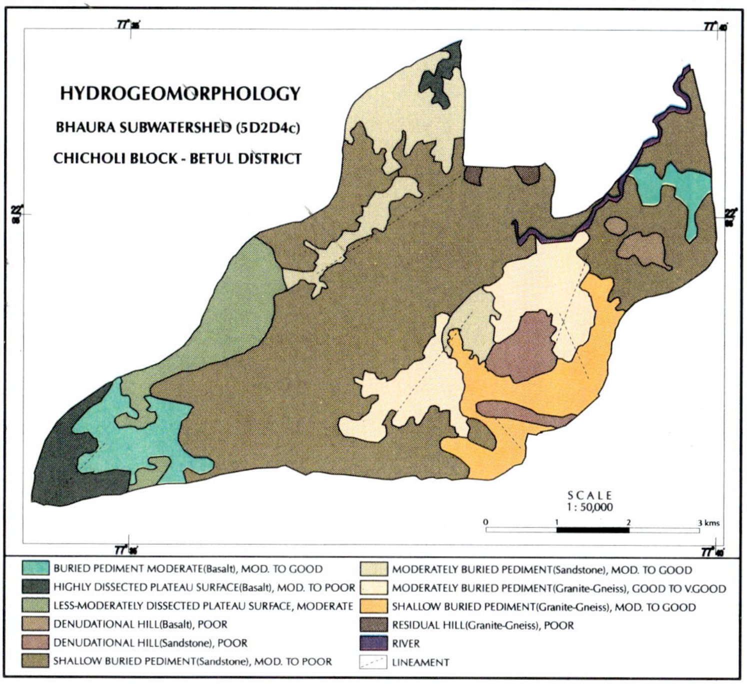
HYDROGEOMORPHOLOGY
BHAURA SUBWATERSHED (5D2D4c)
CHICHOLI BLOCK - BETUL DISTRICT
SCALE
1 : 50,000
0 1 2 3 kms
BURIED PEDIMENT MODERATE(Basalt), MOD. TO GOOD
HIGHLY DISSECTED PLATEAU SURFACE(Basalt), MOD. TO POOR
LESS-MODERATELY DISSECTED PLATEAU SURFACE, MODERATE
DENUDATIONAL HILL(Basalt), POOR
DENUDATIONAL HILL(Sandstone), POOR
SHALLOW BURIED PEDIMENT(Sandstone), MOD. TO POOR
MODERATELY BURIED PEDIMENT(Sandstone), MOD. TO GOOD
MODERATELY BURIED PEDIMENT(Granite-Gneiss), GOOD TO V.GOOD
SHALLOW BURIED PEDIMENT(Granite-Gneiss), MOD. TO GOOD
RESIDUAL HILL(Granite-Gneiss), POOR
RIVER
LINEAMENT

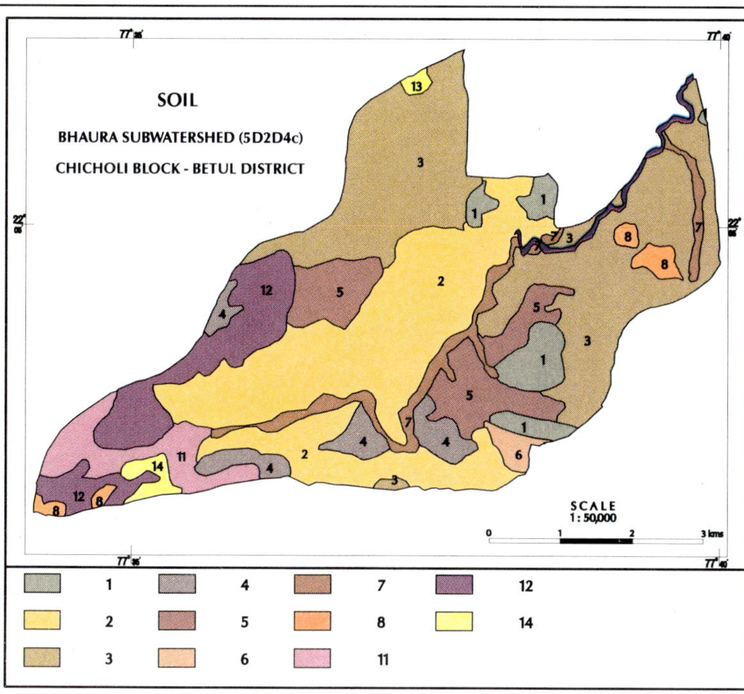
SOIL
BHAURA SUBWATERSHED (5D2D4c)
CHICHOLI BLOCK - BETUL DISTRICT
SCALE
1 : 50,000
0 1 2 3 kms
1 4 7 12
2 5 8 14
3 6 11

DETAILED LEGEND FOR SOIL MAP

LAND CAPABILITY CLASSIFICATION	TAXONOMIC CLASSIFICATION	DESCRIPTION	SOIL SERIES/ ASSOCIATION	MAP SYMBOL
Vles	Loamy skeletal, mixed, Lithic Ustorthents	Shallow, skeletal coarse textured soils very steep hillslopes.	Jantli	1
Ves	Loamy skeletal, mixed, Typic Ustorthents.	Shallow, skeletal coarse textured soils on pediment inselberg complex.	Atari	2
Ves	Loamy skeletal, mixed, Lithic Ustorthents/coarse loamy, mixed, Typic Ustorthents	Shallow, coarse textured soils on pediments.	Atari Harravari	3
IIes	Fine loamy, mixed, Typic, Haplustalfs, Loamy skeletal, mixed Typic Haplustalfs.	Deep, fine textured soils on upper pediplain.	Chicholi-Dhaul	4
IIes	Fine mixed, vertic Ustochrepts, Fine loamy, mixed, Typic Haplustalfs.	Deep, fine textured soils on Lower pediplain.	Kondhar-Chicholi	5
IIw	Fine montmorillonitic Typic Haplusterts Fine montmorillonitic Leptic Haplusterts.	Deep to very deep, fine textured soils on buried pediplain.	Jogli-Dhanora	6
IIe	Fine loamy, mixed, Typic, Fine loamy Fluventic Ustochrepts.	Very deep medium to fine textured soils in valley fills.	Sitadongri/Jir	7
VIes	Loamy skeletal, mixed Lithic Ustorthents.	Shallow, skeletal, medium textured soils on hillslopes.	Patakhera	8
IIIes	Fine loamy, mixed, Typic, Loamy skeletal, Ustochrepts, mixed Typic Ustochrepts.	Medium deep, fine, textured soils moderately eroded undulating plains.	Khapa-Chunhuzuri	11
IIIes	Loamy skeletal mixed Typic Ustorthents/Loamy skeletal mixed Typic Ustochrepts.	Shallow to medium deep, fine textured soils on severely eroded undulating plains.	Chunagosai-Chuna Huzuri	12
IIw	Fine montmorillonitic Typic Haplusterts.	Very deep, fine textured soils on gently sloping plains.	Khokhrakhera	14

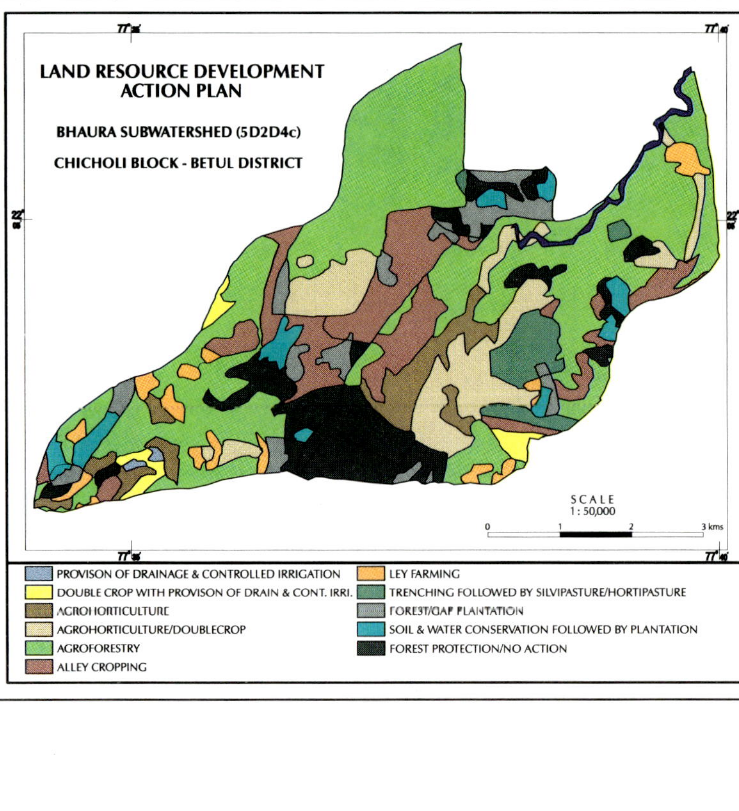

Surface Water Management

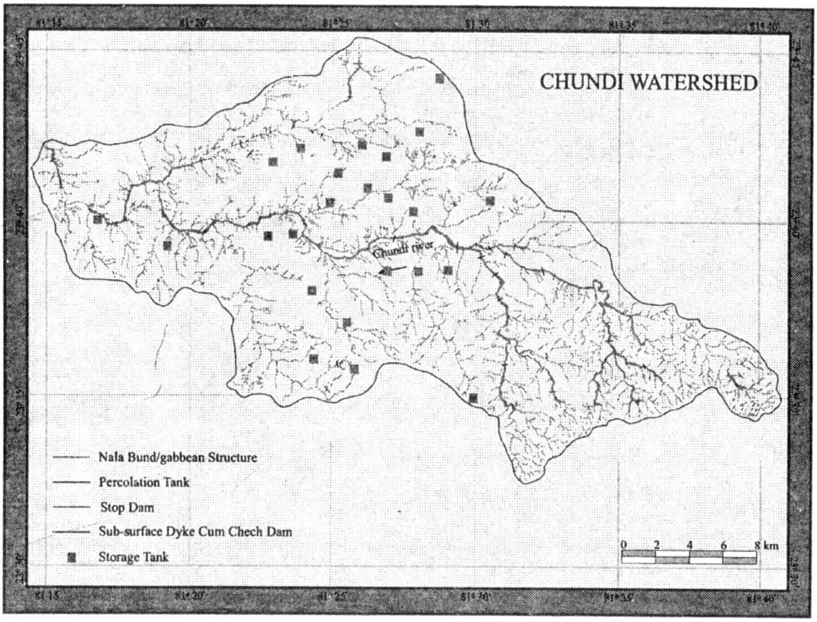

National (Natural) Resources Information System

National (Natural) Resource Information System

Objectives

- To create digital data base for all thematic maps of 1:50,000 scale as per standards for the district node
- To integrate and relate non spatial and spatial data
- To integrate thematic information in gis environment to drop a locale - specific action plans for development of land and water resources
- To generate query shells for data retrieval and generation of action plans interactively for decision making

NRIS Implementation

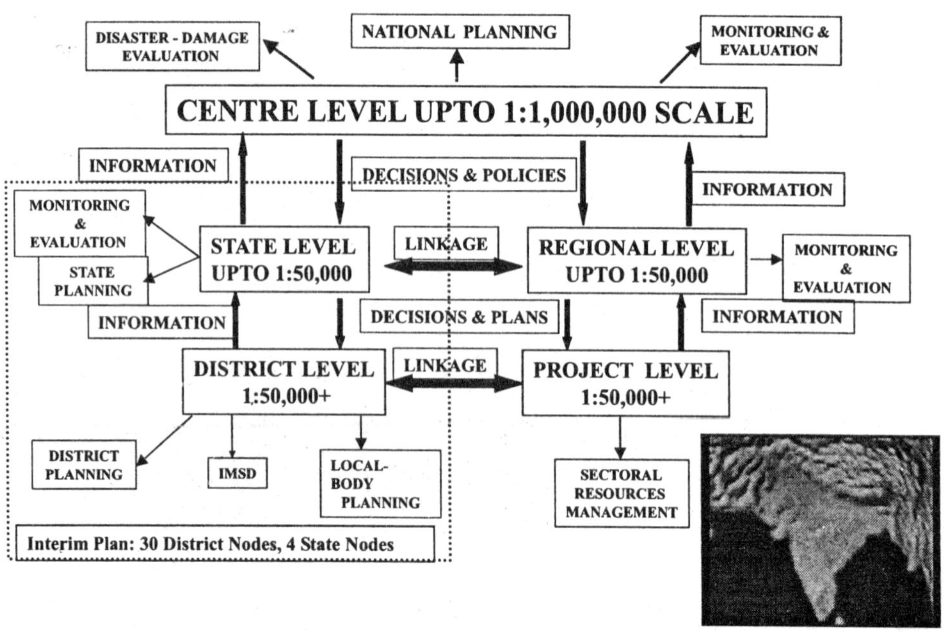

NRIS Database - Primary Elements

SPATIAL
- Landuse/cover map
- Geomorphic units and Landforms
- Lithological units
- Geological structures
- Soil type, depth, texture etc
- Drainage
- Canals
- Elevation contours/spot height points
- Watershed hierarchy polygons
- State/District boundaries
- District/Taluka boundaries
- Village boundaries
- Forest management boundaries
- Location of Wells
- Settlement locations
- Roads
- Rail network
- Rail network along with Railway Stations
- Rainfall/temp observation
- Minerals
- SOI toposheet references

ASPATIAL
- Demography
- Occupation
- Education Facilities
- Medical Facilities
- Communications
- General
- Land Use
- Power

NRIS Database - Derived Elements

SPATIAL
- Slope
- Groundwater Prospects
- Composite Land Development Units
- Land Resources Development Plan
- Water Resources Development Plan
- Land Capability
- Watershed Prioritisation
- Run-off Potential
- Polygonised Meteorological Data
- Soil Irrigability
- Land Irrigability
- Water Quality
  - Irrigation
  - Drinking
- Service Centre Hierarchy

ASPATIAL
- Composite Functionality Index
- Agricultural Development Index
- Village Development Index

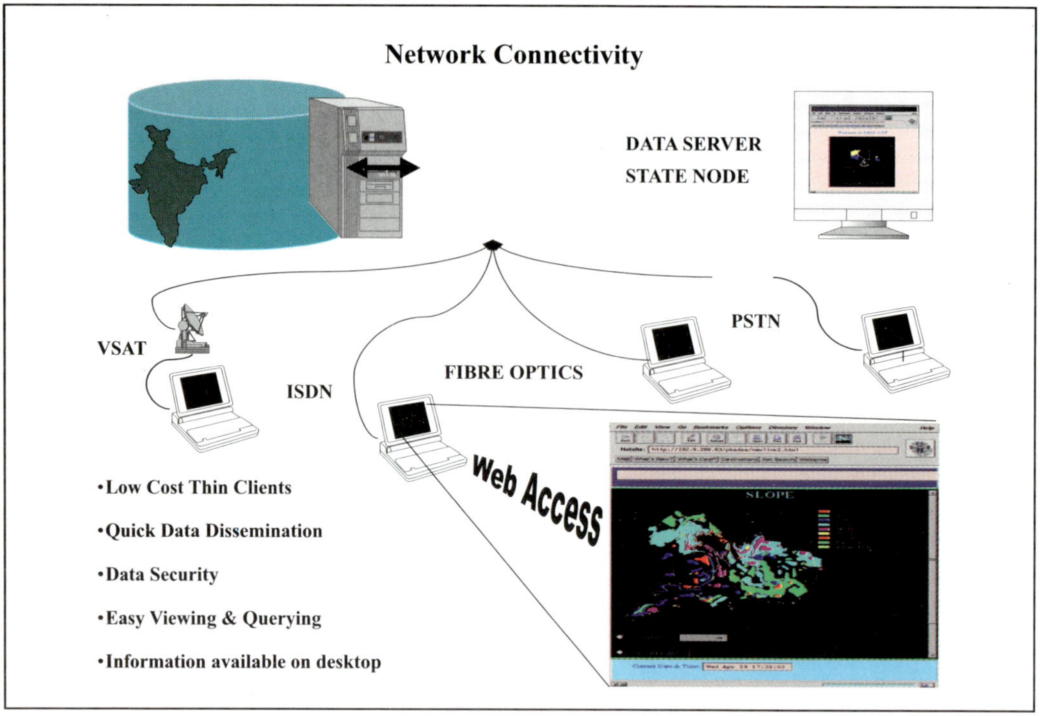
Network Connectivity
DATA SERVER
STATE NODE
VSAT
ISDN
FIBRE OPTICS
PSTN
Web Access
SLOPE
•Low Cost Thin Clients
•Quick Data Dissemination
•Data Security
•Easy Viewing & Querying
•Information available on desktop

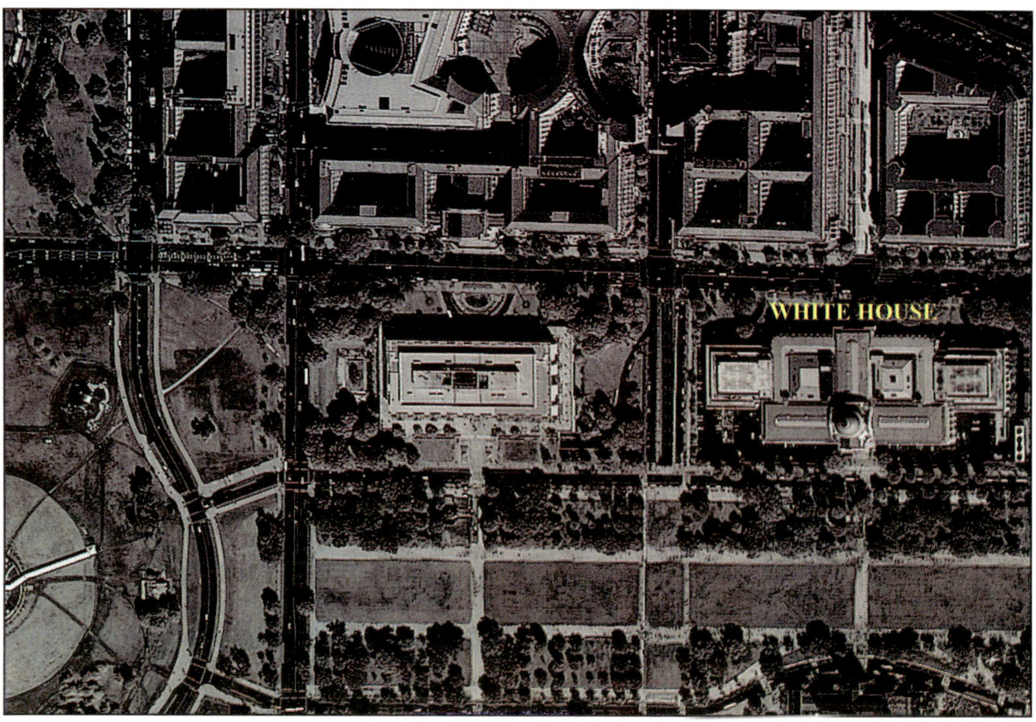
WHITE HOUSE

Ikonos Satellite Data (1m Resolution)

Gandhinagar thermal power plant
chimneys with and without smoke chimney rims are clearly seen

Ikonos Satellite Data (1m Resolution)

Rashtrapati Bhavan and its Environs
New Delhi

Pentagon As Viewed By Resourcesat - 1

Southern India Viewed by aWifs

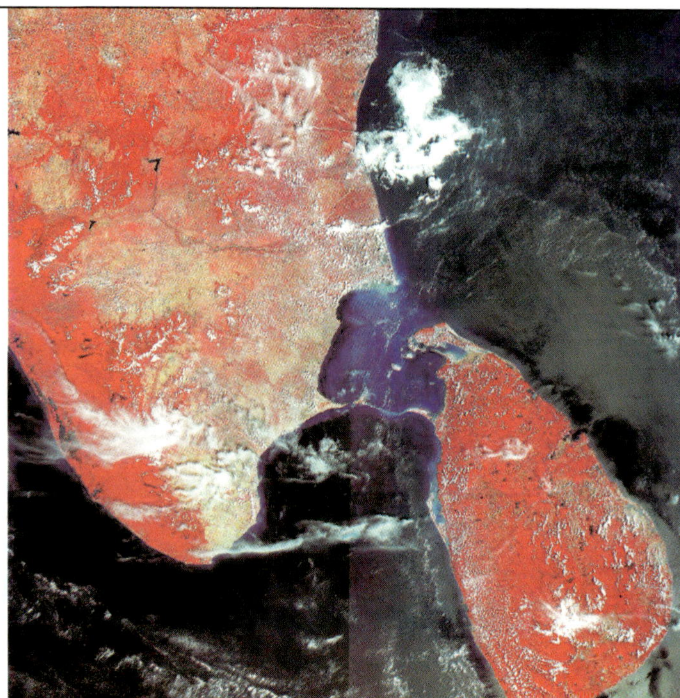

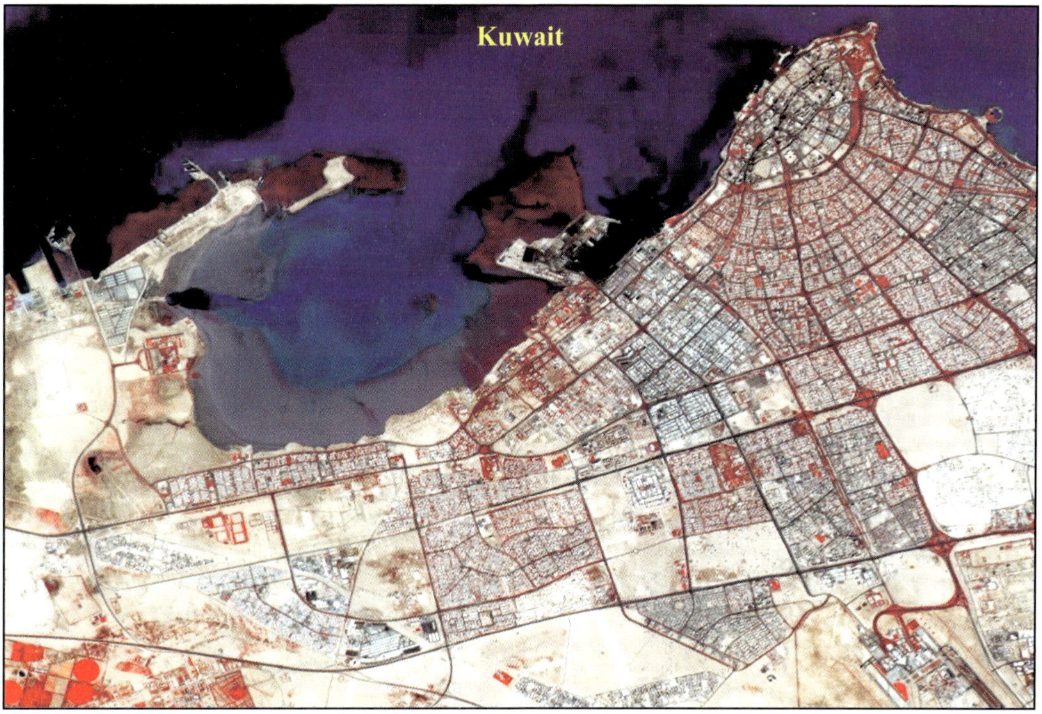
Kuwait

Washington

Mansarover

Sahara Airport

Dock Yard

Railway Station

Bus Stand

Planned Residential Areas

Unplanned Residential Area

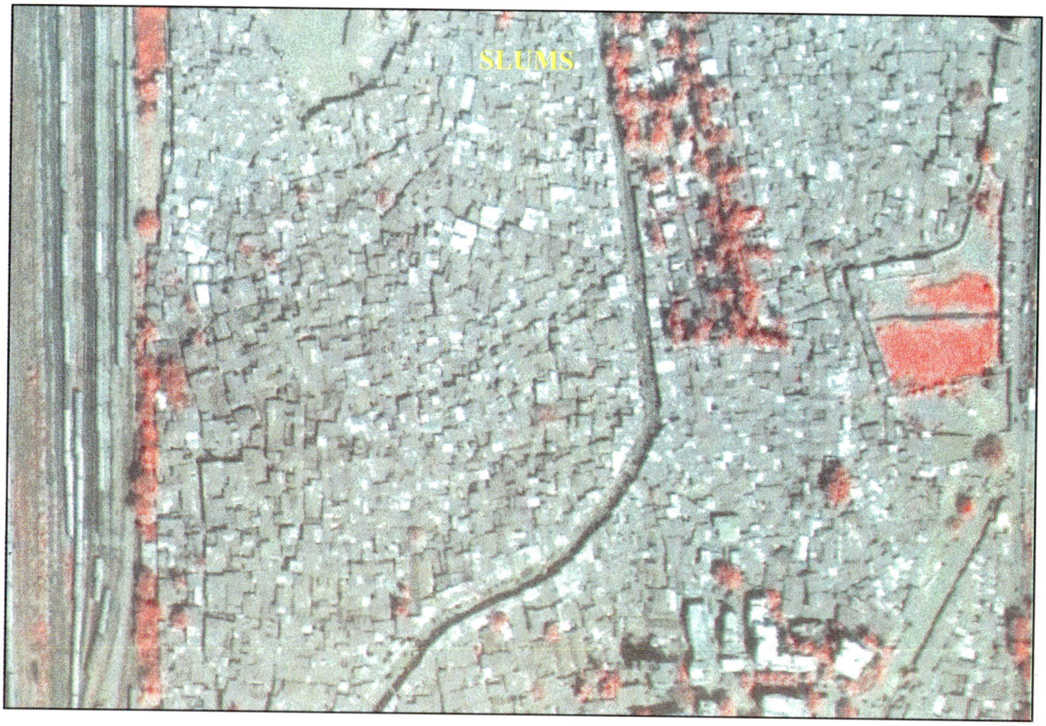
SLUMS

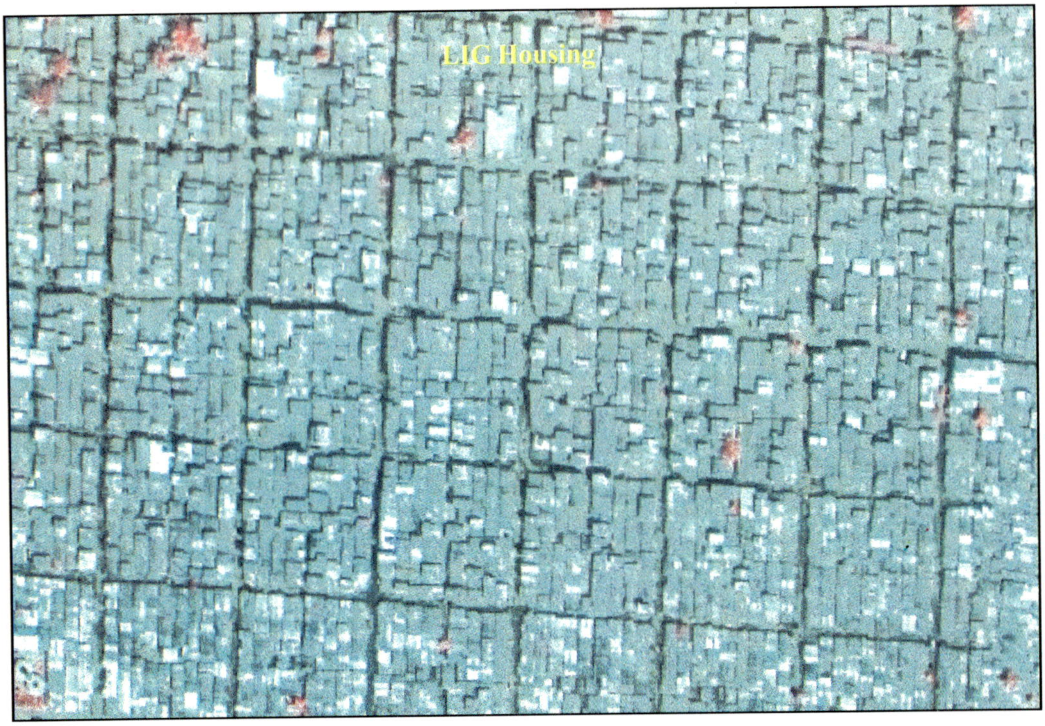
LIG Housing

MIG Housing Scheme

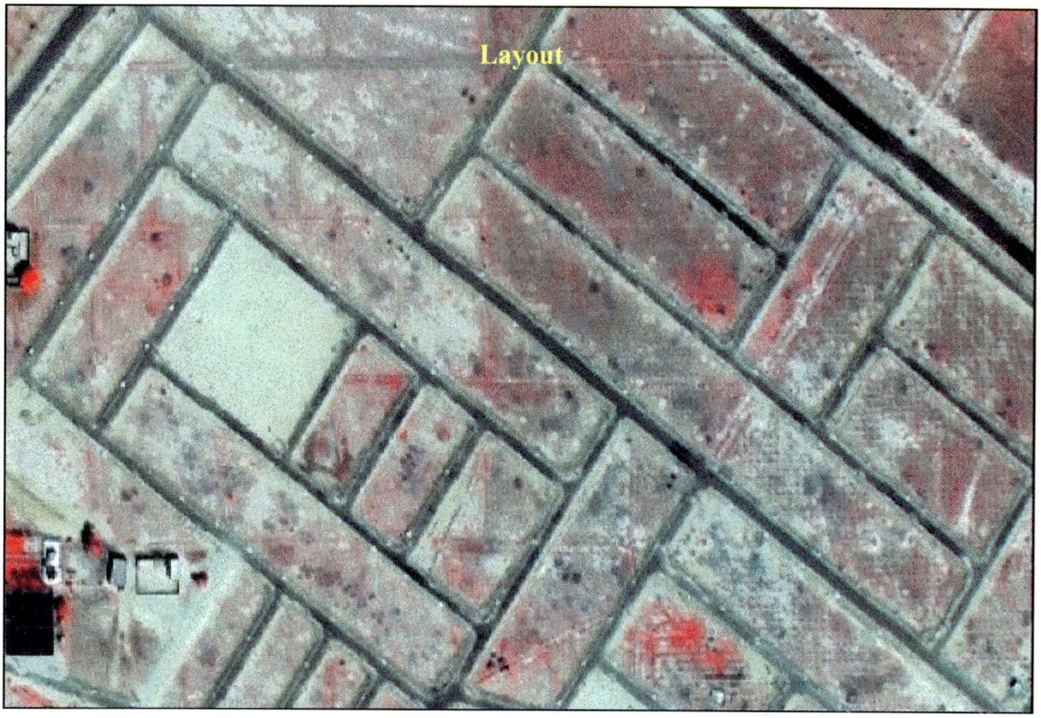
Layout

Express Highway

Flyover

Flyover Under Construction

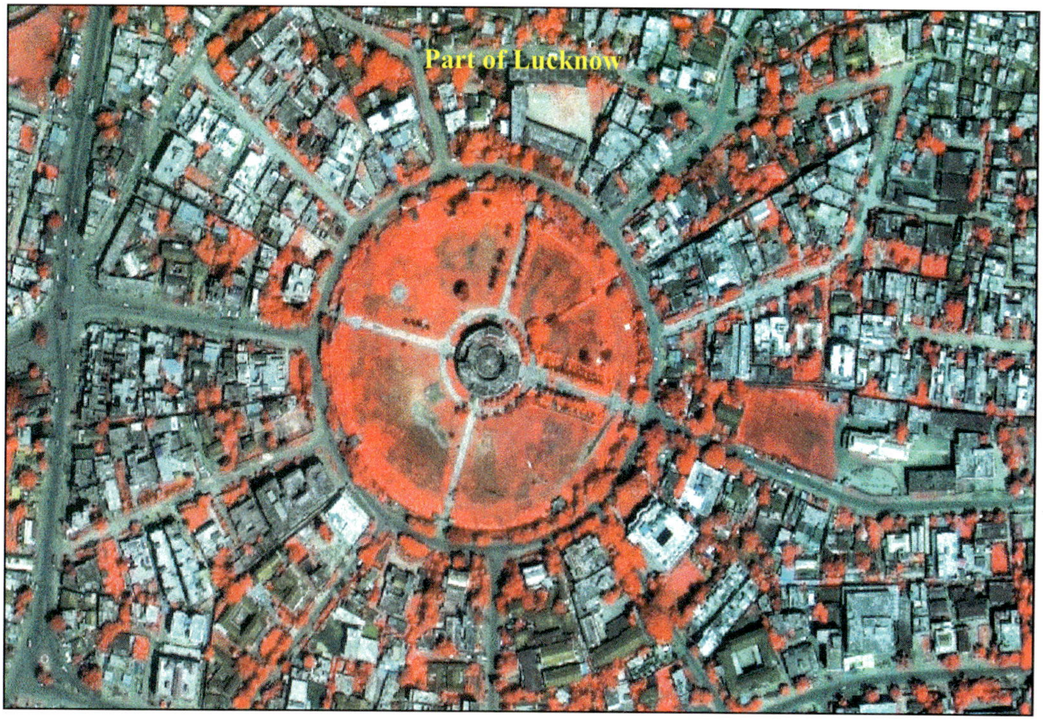
Part of Lucknow

Part of Panjim (Goa)

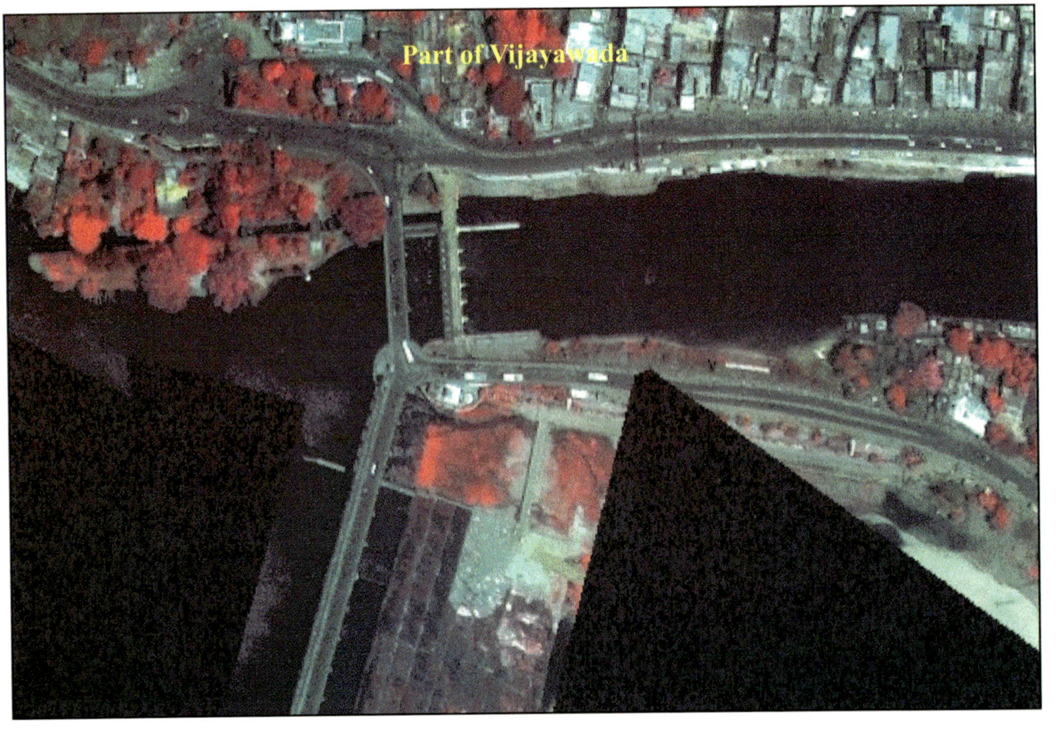
Part of Vijayawada

Salt Pans

Water Polution

Farms

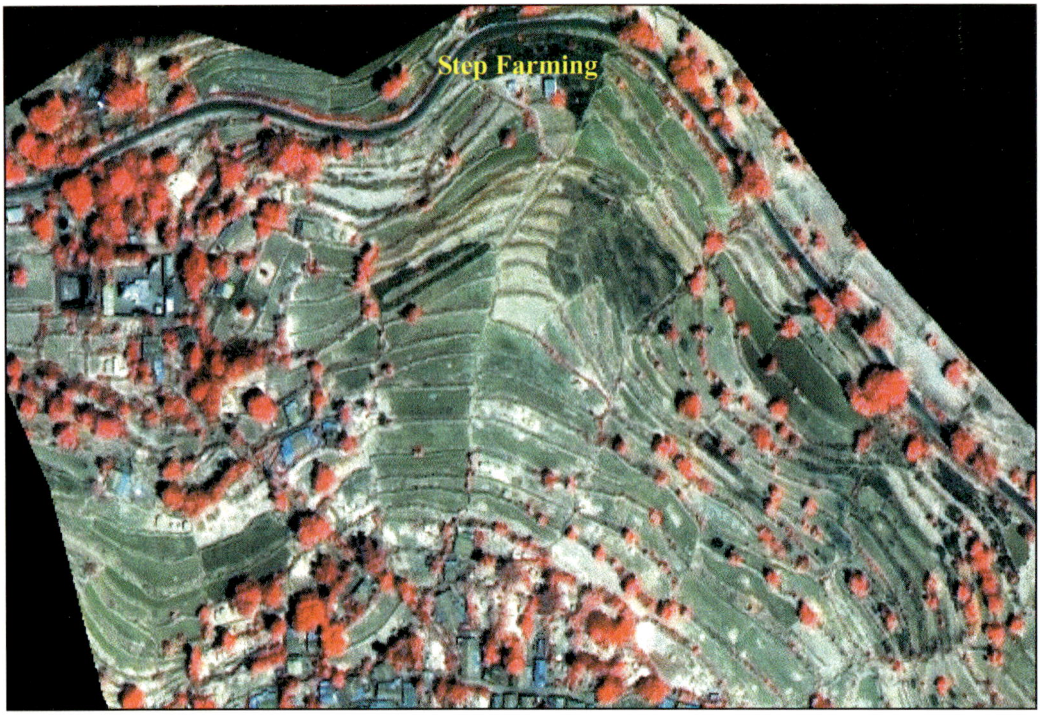
Step Farming

Hill

34

ACQUISITION AND EVALUATION OF MAP AND CARTOGRAPHIC DATA FROM DATA SHEET

SANDEEP GOYAL

WHAT IS GIS?

Powerful tool for solving spatial problems

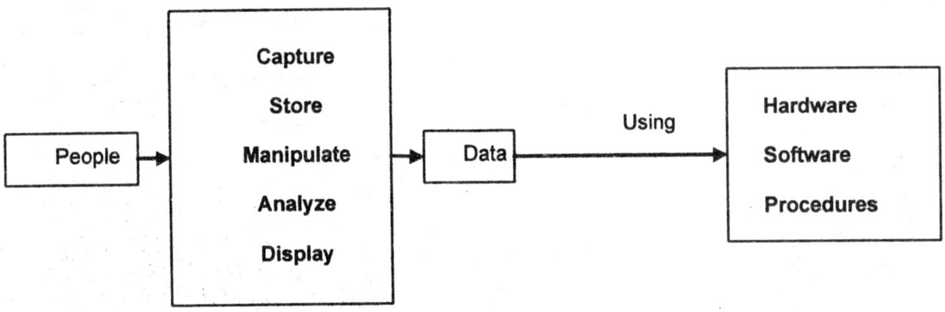

GIS

"A Geographical Information System is an Integrated Database Management System in which large volumes of "Geo-referenced Spatial Data" and corresponding "Attribute Data" can be effectively stored, updated, manipulated, analyzed and displayed as per user defined specifications".

Geographic Information Systems (GIS)

It is a computer based tool—an organised collection of computer hardware, software, geographic data and personnel designed to efficiently capture, store, update, manipulate, analyse and display all forms of geographically referenced information.

Not simply a computer system for making maps. It is an analysis tool.
Helps identify the spatial relationships between map features.
Links spatial data with geographic information about a particular feature on a map.

GIS used for different levels of planning exercises, transport modelling, pollution modelling, land revenue exercises, epidemiological studies, ground water contouring and other geographically related studies.

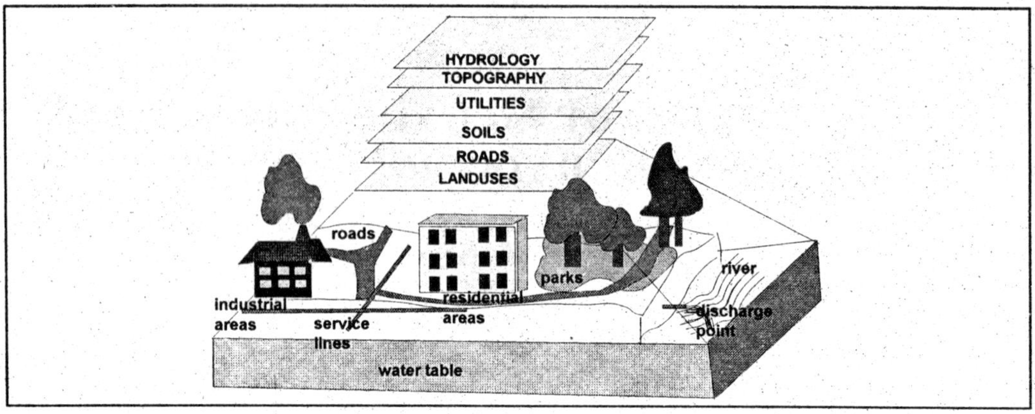

What is a Geographic Information System (GIS)

Computer-based tool for mapping and analyzing features that exist and events that happen on earth.

Integration of common database operations with visualization and geographic analysis
"Spatially enable" otherwise tabular data or "aspatial data"
Create maps and visualize aspatial and spatial data together
- Visualize scenarios
- Develop planning strategies
- Predict outcomes
- Solve complicated problems
- Present powerful ideas
- Develop effective solutions

GIS is a tool to Organise, Analyse, and Visualize Spatially Referenced Information

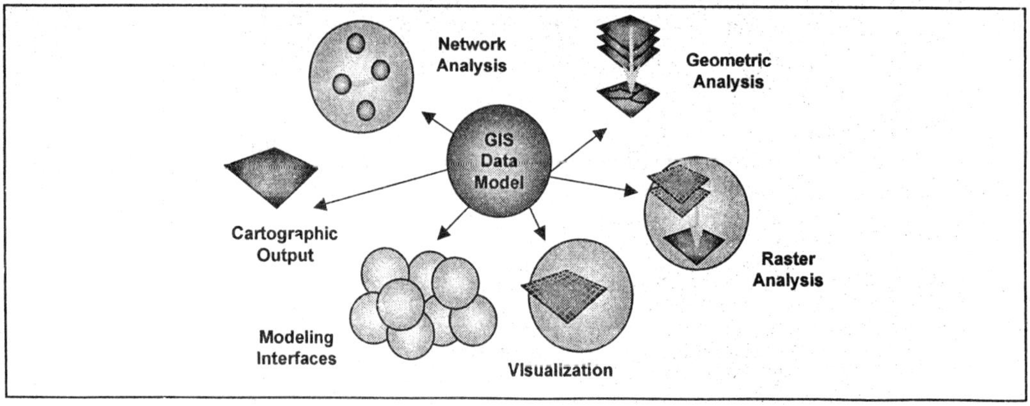

GIS integrates all the parts... to see the whole!

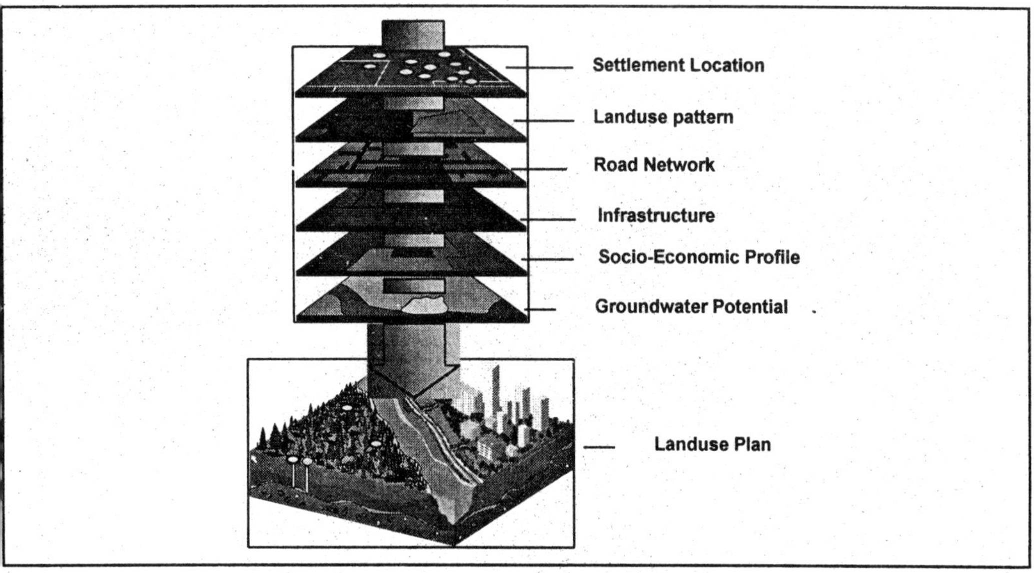

WHAT ARE THE COMPONENTS OF A GIS?

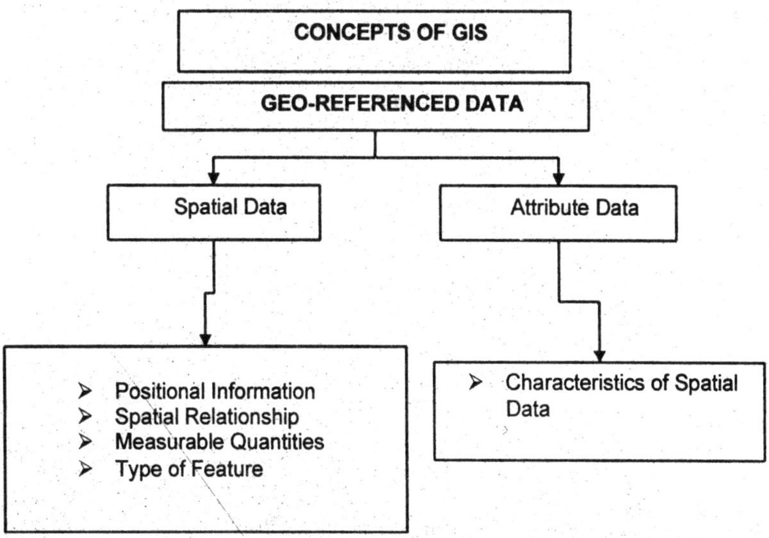

GIS COMPONENTS

1. Hardware
2. Software
3. Data

4. Personnel
5. Methodology.

WHAT ARE THE COMPONENTS OF A GIS?

Hardware

Centralized computing servers
Desktop computers networked or stand-alone

Software

Tools for the input and manipulation of geographic data
Database management system (DBMS)
Tools that support geographic query, analysis, and visualization
A graphical user interface (GUI)

Data

Collected in-house or purchased from commercial data provider
GIS will integrate spatial data with other data resources
Established DBMS can be used for aspatial data

People

System management and enhancement
Develop plans and applications used in real-world solutions
Everyday users to technical specialists

Methods

Well designed usage plans
Well defined business rules
Operational practices unique to each organization

HOW DOES A GIS WORK?

GIS stores information as a collection of thematic layers

Thematic layers are linked together by geography

Explicit geographic reference (latitude and longitude)

Implicit reference (address, postal code, FIPS code, census tract, road name)

GIS can create explicit geographic features (e.g., customer) from implicit references like customer address

Geographic features (points, lines, polygons) used for visualization and analysis.

THE GIS PROCESS

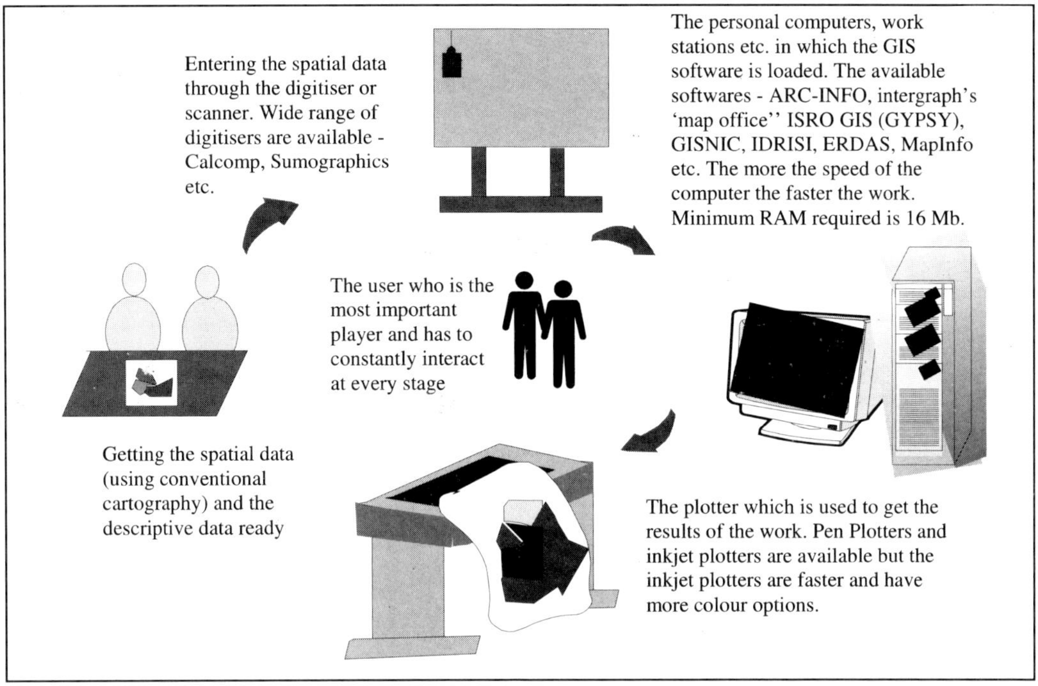

GIS CAPABILITY

1. Capture
2. Store
3. Display and Query
4. Analyze
5. Manipulate

WHAT CAN A GIS DO FOR YOU?

Perform Geographic Queries and Analysis
- Which store is most profitable, how close are competitors?
- Where are all suitable sites to build a new store?
- What is the total number of customers within 10 km of a store?

Improve Organizational Integration
- Improved management of resources
- Interdepartmental information sharing and communication

Maps, Modelling, and Analysis
- Visualization of corporate assets geographically
- Digital continuous and scale free cartographic GIS database

Make Better Decisions and Create Better Solutions
- Tools to query, analyze, and map data in support of the decision making process

DATA FOR GIS

Base Maps

Political boundaries, postal areas
Municipal boundaries
Highways, streets, rivers
Lakes, parks, landmarks

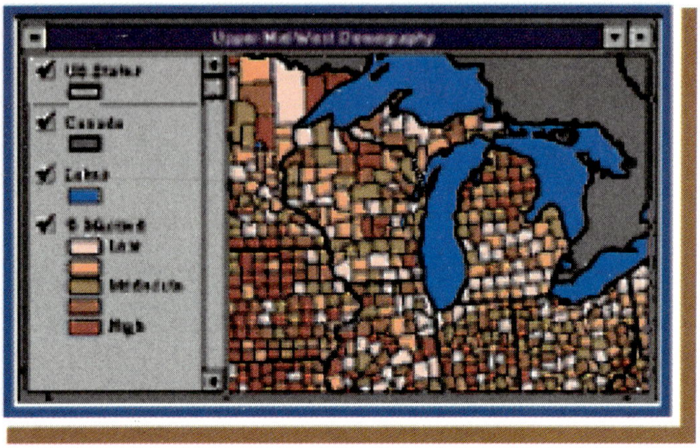

Business Maps and Data

Business/Customer locations
Census/Demography
Consumer products, financial services
Health care, real estate
Your SAP R/3 data!

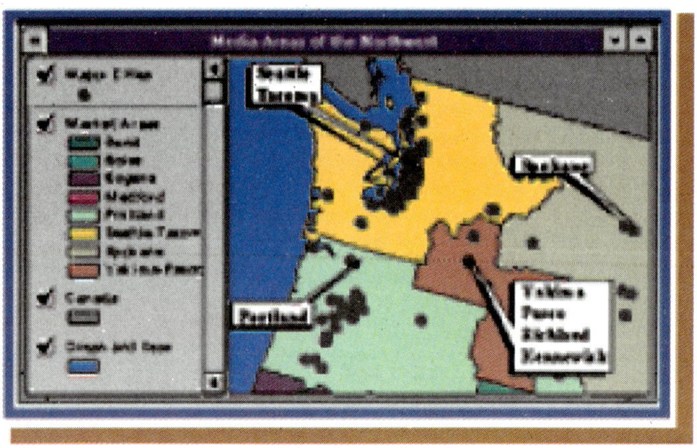

Environmental Maps and Data

Environmental risk
Satellite imagery, weather
Topography, natural resource

General Reference Maps

World boundaries
Country boundaries
City locations, time zones

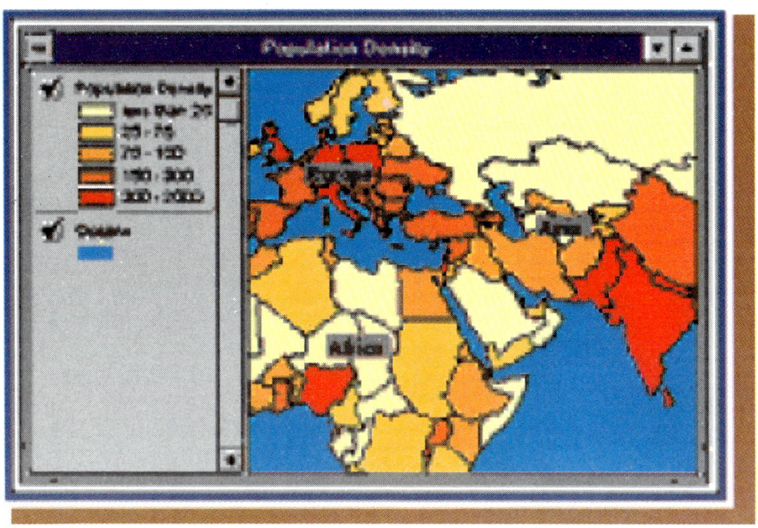

SOURCE OF GEOGRAPHIC DATA

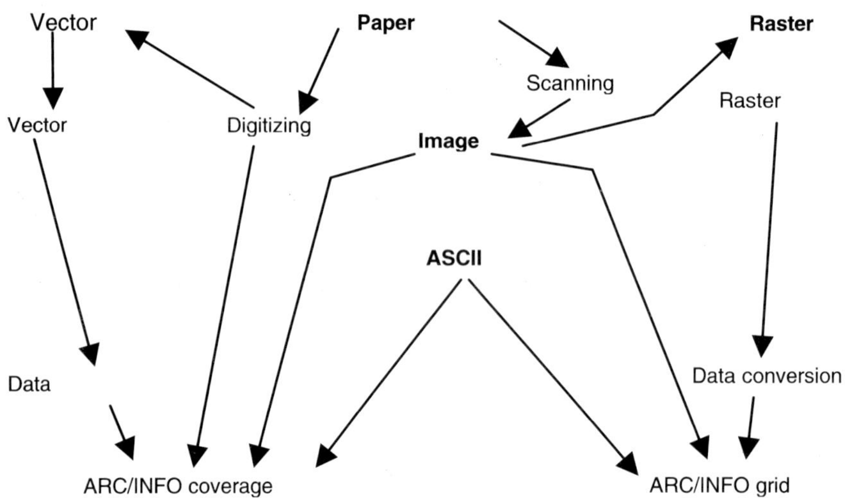

DATA CAPTURING

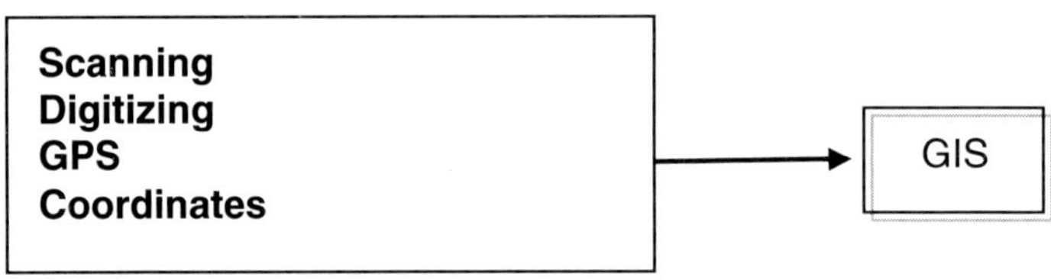

DATA STRUCTURES IN GIS

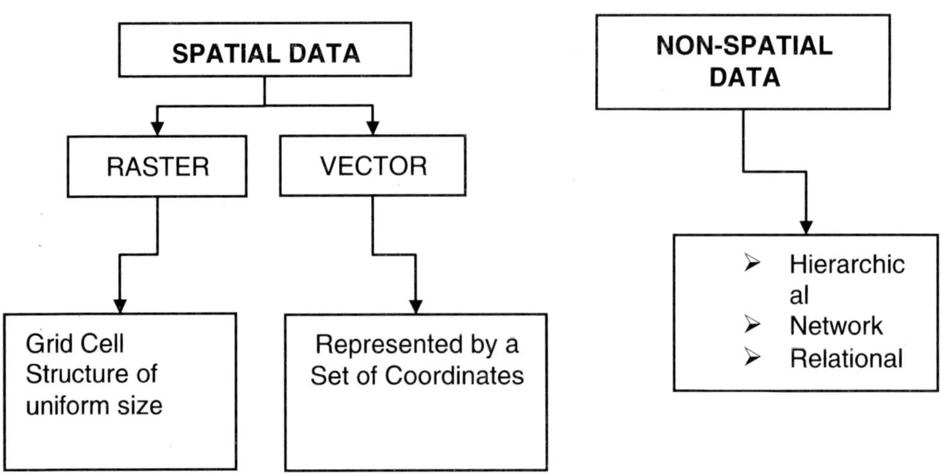

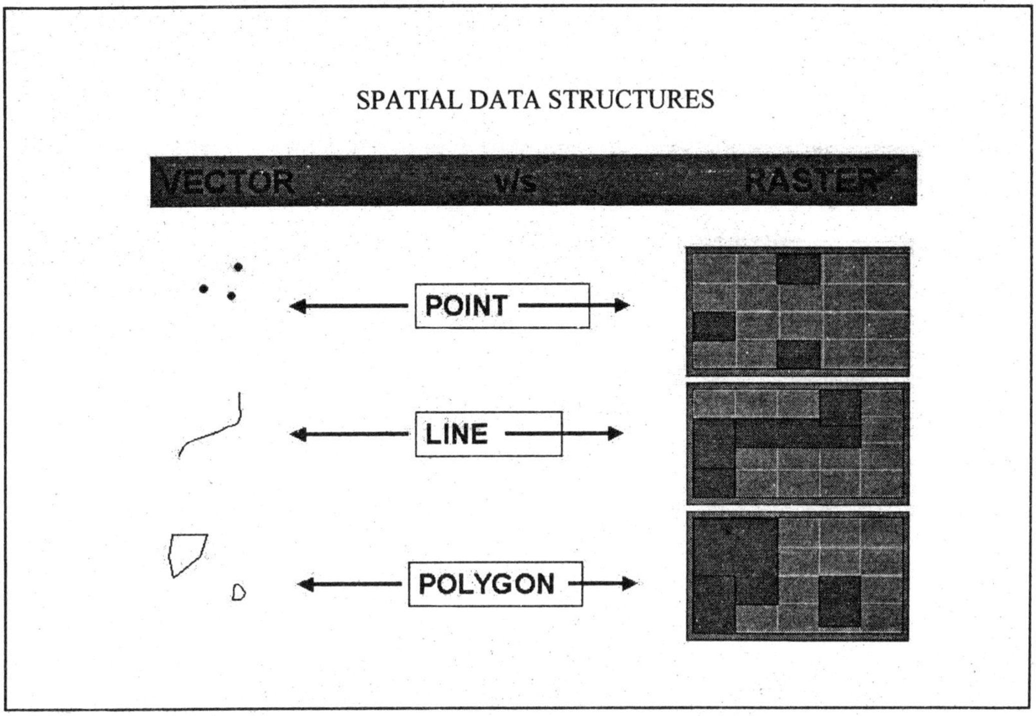
SPATIAL DATA STRUCTURES
VECTOR
v/s
RASTER
POINT
LINE
POLYGON

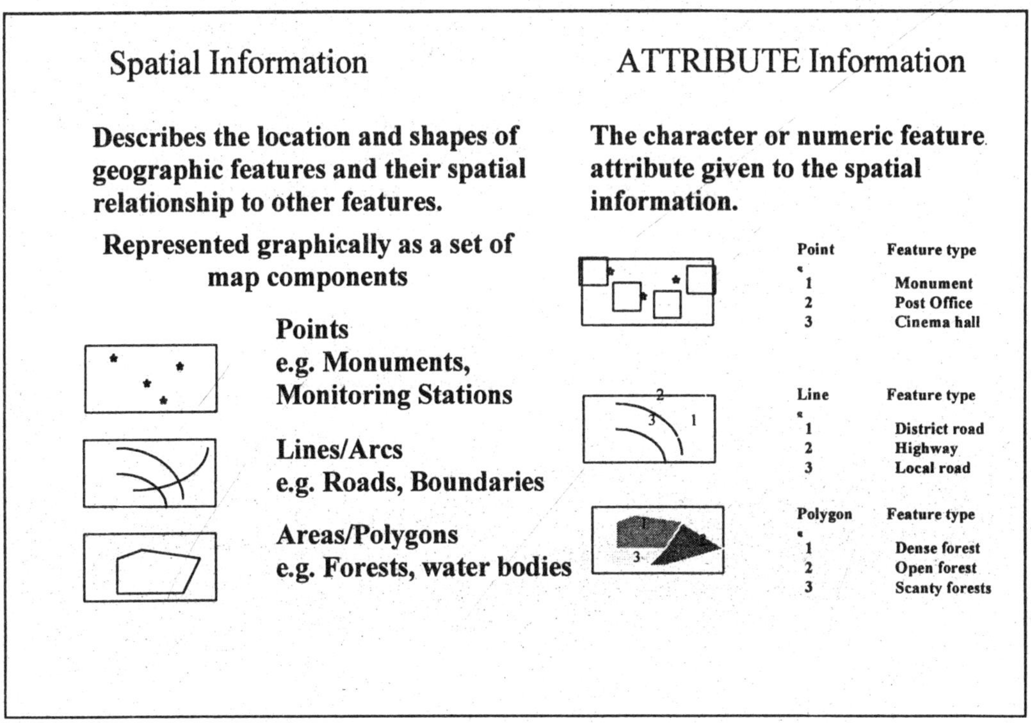
Spatial Information
ATTRIBUTE Information
Describes the location and shapes of geographic features and their spatial relationship to other features.
Represented graphically as a set of map components
Points
e.g. Monuments, Monitoring Stations
Lines/Arcs
e.g. Roads, Boundaries
Areas/Polygons
e.g. Forests, water bodies
The character or numeric feature attribute given to the spatial information.
Point
Feature type
1
Monument
2
Post Office
3
Cinema hall
Line
Feature type
1
District road
2
Highway
3
Local road
Polygon
Feature type
1
Dense forest
2
Open forest
3
Scanty forests

DATA LINKING

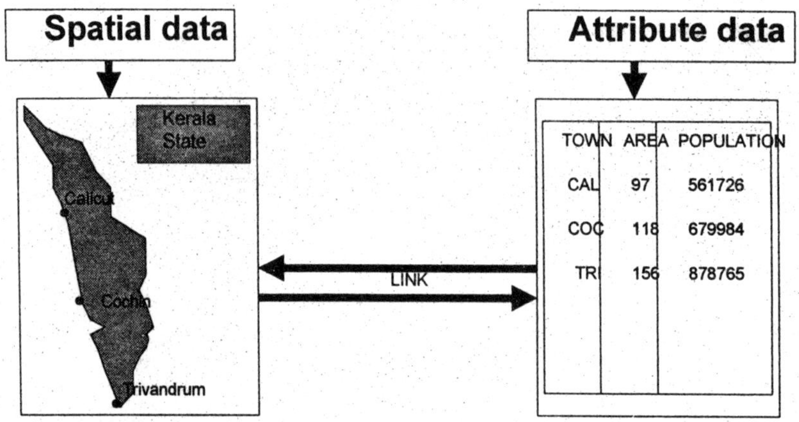

ORGANISING MAP INFORMATION

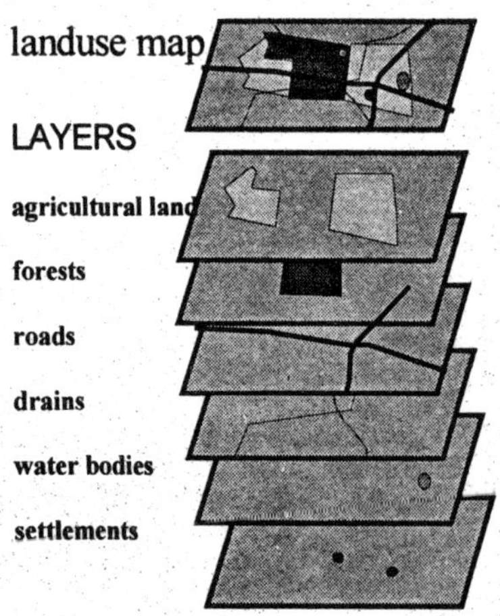

Map features are logically organised into sets of layers or themes of information.

Each layer or theme usually has only one type of map feature—polygon, line or point

Thus any map for e.g. a landuse map is a composite of different layers i.e. forests, agricultural lands, roads, villages, settlements, etc.

TOPOLOGY CREATION

Helps the computer identify relationships between 2 or more map features. Also only after the topology is created the spatial and the descriptive data become usable for spatial analysis.

Arc - Node Topology (Connectivity) - **Arcs intersect at nodes i.e. whenever the topology of a cover consisting of arcs is created a node is created at the intersection points. For e.g. Road no. 1 and 4 meet at A.**

Polygon - Arcs Topology (Area definition) - **A polygon is a set of arcs. The co-ordinates of the initial node of the starting arc and the final node of the closing arc should coincide. If not the shape does not act as an area feature but a line or arcs feature.**

Left - Right Topology (Contiguity) - **Every arc has contiguity or direction. The direction is specified by defining the left and right polygons adjacent to the arc. Thus adjacent polygons share a common arc.**

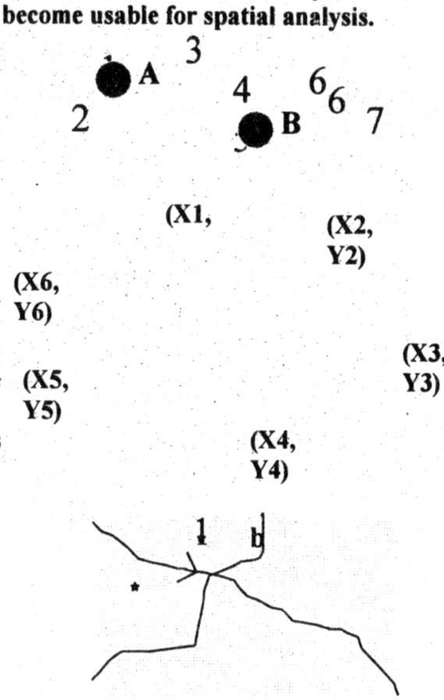

DATA QUERY

Spatial

Which hospitals lie within 5 miles from an accident spot?
What is the shortest path between the airport and the hotel?

Aspatial

What is the average number of people working with GIS in each location?
How many male voters fall in the age group 30-35 in a constituency?
Workstation Based

HARDWARE REQUIREMENT

Workstation Based

RAM-256 MB
Processor-Silicon graphics and its equivalent
HDD-20 GB
Monitor-SVGA color monitor-21″
Drives-Floppy, DAT, Cartridge, CD-ROM
Operating system – UNIX

PC BASED

RAM-128 MB
Processor-INTEL PII and its equivalent
HDD-2.0 GB
Monitor-SVGA color monitor-15″
Drives-Floppy, DAT, Cartridge, CD-ROM
Op.system-DOS & WINDOWS

PERIPHERALS

Disk drive: The CPU is linked to disk storage unit which gives space for storing data and programs.

Digitizer: It is used to convert data from maps and documents into digital form and send them to the computer.

Plotter: It is used to display the results of the data processing.

Tape drive: It is used to communicate with other systems.

The visual: It is used to control the computer and peripherals. Otherwise known as 'terminals'.

GIS SOFTWARE

The GIS software consists of 4 basic technical modules:

A. Data input and verification (Data capture)
B. Data storage and management
C. Data analysis and modelling
D. Data output and presentation

GIS SOFTWARES

ESRI—ARC/INFO, ARC VIEW
Autodesk, Inc.,—AUTOCAD MAP
Bentley Systems—MICROSTATION GEOGRAPHICS
Mapinfo Corporation—MAPINFO
Pamap—PAMAP
Integraph Corporation—GEOMEDIA, MGE
PCI Geomatics Group—SPANS

GIS APPLICATIONS

Questions GIS can answer

Location.....What is at...?
Answer what exist at a particular location.

Condition.....Where is it...?
Identify locations by specifying conditions.
Trends.....What has changed since...?
Highlight changes over a period of time.
Patterns.....What spatial patterns exist...?
Distribution and reasons behind a phenomenon.
Modeling.....What if...?
Evaluate different scenarios.

INDUSTRIES USING GIS SOLUTIONS

Mapping	Federal Government
Land Use Planning	National Agencies
Real Estate and Cadastral	Local Government
Site Location	Public Sector
Agriculture	Retail
Forestry	CAD
Land Use Planning	Electric/Gas Utilities
Risk Management	Business/Marketing
Environmental Mgmt.	Telecommunications
Military/Intelligence	Transportation Logistics
Petroleum and Mining	Pipeline
Water and Wastewater	Health Care

GIS AND MAPPING

A. Detailed Regional Road maps
B. City Street maps
C. Shaded relief maps (Differing elevations can be instantly appreciated by using the GIS database to create color-shaded relief maps)
D. 3 D maps (Displays the data 3-dimensionally enhances the visualization by the reader and gives a more realistic interpretation)
E. Soil map
F. Electricity network map...etc.

GIS AND SITE PLANNING

To locate sites for new facilities or locate alternate sites for existing facilities:

A. Site for recycling facilities
B. Site for Nuclear waste disposal
C. Regulate anti social activities

GIS AND RISK MANAGEMENT

GIS can help with risk management and analysis by showing you which areas will be prone to natural or man-made disasters:

A. Estimate flood damage
B. Estimate landslide susceptibility
C. Determine seismic hazard zones

GIS AND TRANSPORTATION

GIS can be used to help you manage transportation infrastructure or help you manage your logistical problems:

A. Find the best route
B. Compare the costs of different routes
C. Balance workloads and services
D. Plan facility investments
E. Monitor traffic volume
F. Show complete transportation networks
G. Define alternate transportation routes
H. Plan new rail lines
I. Plan air routes etc.

GIS AND WATER/WASTE WATER INDUSTRY

For the planning, engineering, operations, maintenance, finance, and administration functions of the water/waste water networks:

A. Assess sanitary sewer systems
B. Inventory and monitor water/waste water systems
C. Assess the water quality
D. Determine the effects disasters have on water and sewer lines

GIS AND FORESTRY

Managing forests is becoming a more complex and demanding challenge. With GIS foresters can easily see the forest as an ecosystem and manage it responsibly.

A. Inventory and current conditions
B. Assist strategic management planning
C. Assess old growth forest
D. Manage forest harvesting

GIS AND LANDUSE PLANNING

Use GIS to help visualize and plan the landuse needs of cities, regions, or even national governments:

A. Develop master plans for the city
B. Assess redevelopment plans
C. Develop land use plans
D. Depict visual impact development

E. Plan international development projects
F. Analyze building trends

GIS AND GEOLOGY

To study geologic features, analyse soils and strata, assess seismic information, or create 3D displays of geographic features:

A. Assess the morphology for an area
B. Analyze geologic structures

GIS AND AGRICULTURE

GIS software is the most critical component of a precision agriculture system:

A. Manage crop production
B. Manage irrigation systems
C. Manage crop rotation techniques
D. Projecting soil loss for individual farms or entire agricultural regions
A. Information about sales
B. Information about customers
C. Inventory information
D. Plan media
E. Planning trade areas
F. Identifying delivery route
G. Design sales territory
H. Product promotions
I. Market analysis and planning, etc.

GIS AND ASSET MANAGEMENT

Commercial real estate companies that recommend:

A. Acquisition and manage portfolios
B. Delivers latest market data to customers

GIS AND DEFENCE

A. Creating standard base maps
B. Aid in tactical solutions
C. Clarifying specific feature needed for navigation
D. Realistic rendition of the area using Aerial photo and map

GIS AND ELECTRIC/GAS UTILITIES

A. Inventory and manage entire system
B. Track maintenance

C. Monitor regulatory compliance
D. Model distribution analysis
E. Transformer analysis
F. Load analysis
H. Analyze trouble calls

GIS AND AM/FM

Facilities Management can answer the following questions:

Where are the facilities located?
How well they are being used?
Where is the money invested in assets in relationship to revenue?
How can you provide better services to customers for less money?

GIS FOR PLANNING AND ENGINEERING

A. Watershed modelling
B. Water distribution system master planning
C. Population and demand projections
D. Groundwater management/modelling
E. Water quality monitoring
F. Well log and data management
G. Site analysis
H. Water flow analysis
I. Hazardous material tracking
J. Underground tank management etc.

OTHER APPLICATIONS

A. **GIS for engineering plan and design** to plan, design and choose cost effective solutions.
B. **GIS for fire management** to understand the direction, rate of speed, intensity etc.
C. **GIS for ecosystem management** to describe, analyze and model ecosystem processes.
D. **GIS for strategic planning** to forecast the future prospects relative to alternative managment activities,
E. **GIS for demographic analysis** is the basis for many other functions like customer service, site analysis etc.
F. **GIS for Regulatory compliance** to assist in identifying possible discriminatory lending patterns.
G. **GIS for infrastructure management** to support planning, design, construction, operations and maintenance.
H. **GIS for fleet and logistics management** knowing where a vehicle, or delivery is at any given time leverages assets for optimum deployment and cost customer satisfaction, competitive position, timely response etc.

INDEX